MW01148959

Probability and Its Applications

Published in association with the Applied Probability Trust

Editors: J. Gani, C.C. Heyde, P. Jagers, T.G. Kurtz

Probability and Its Applications

Azencott et al.: Series of Irregular Observations. Forecasting and Model Building. 1986
Bass: Diffusions and Elliptic Operators. 1997
Bass: Probabilistic Techniques in Analysis. 1995
Berglund/Gentz: Noise-Induced Phenomena in Slow-Fast Dynamical Systems: A Sample-Paths Approach. 2006
Biagini/Hu/Øksendal/Zhang: Stochastic Calculus for Fractional Brownian Motion and Applications. 2008
Chen: Eigenvalues, Inequalities and Ergodic Theory. 2005
Costa/Fragoso/Marques: Discrete-Time Markov Jump Linear Systems. 2005
Daley/Vere-Jones: An Introduction to the Theory of Point Processes I: Elementary Theory and Methods. 2nd ed. 2003, corr. 2nd printing 2005
Daley/Vere-Jones: An Introduction to the Theory of Point Processes II: General Theory and Structure. 2nd ed. 2008
de la Peña/Gine: Decoupling: From Dependence to Independence, Randomly Stopped Processes U-Statistics and Processes Martingales and Beyond. 1999
de la Peña/Lai/Shao: Self-Normalized Processes. 2009
Del Moral: Feynman-Kac Formulae. Genealogical and Interacting Particle Systems with Applications. 2004
Durrett: Probability Models for DNA Sequence Evolution. 2002, 2nd ed. 2008
Galambos/Simonelli: Bonferroni-Type Inequalities with Equations. 1996
Gani (ed.): The Craft of Probabilistic Modelling. A Collection of Personal Accounts. 1986
Gut: Stopped Random Walks. Limit Theorems and Applications. 1987
Guyon: Random Fields on a Network. Modeling, Statistics and Applications. 1995
Kallenberg: Foundations of Modern Probability. 1997, 2nd ed. 2002
Kallenberg: Probabilistic Symmetries and Invariance Principles. 2005
Last/Brandt: Marked Point Processes on the Real Line. 1995
Molchanov: Theory of Random Sets. 2005
Nualart: The Malliavin Calculus and Related Topics, 1995, 2nd ed. 2006
Schmidli: Stochastic Control in Insurance. 2008
Schneider/Weil: Stochastic and Integral Geometry. 2008
Shedler: Regeneration and Networks of Queues. 1986
Silvestrov: Limit Theorems for Randomly Stopped Stochastic Processes. 2004
Rachev/Rueschendorf: Mass Transportation Problems. Volume I: Theory and Volume II: Applications. 1998
Resnick: Extreme Values, Regular Variation and Point Processes. 1987
Thorisson: Coupling, Stationarity and Regeneration. 2000

Victor H. de la Peña · Tze Leung Lai · Qi-Man Shao

Self-Normalized Processes

Limit Theory and Statistical Applications

Victor H. de la Peña
Department of Statistics
Columbia University
Mail Code 4403
New York, NY 10027
USA
vp@stat.columbia.edu

Tze Leung Lai
Department of Statistics
Sequoia Hall, 390 Serra Mall
Stanford University
Stanford, CA 94305-4065
USA
lait@stat.stanford.edu

Qi-Man Shao
Department of Mathematics
Hong Kong University of Science and Technology
Clear Water Bay
Kowloon, Hong Kong
People's Republic of China
maqmshao@ust.hk

Series Editors:

Joe Gani
Chris Heyde
Centre for Mathematics and its Applications
Mathematical Sciences Institute
Australian National University
Canberra, ACT 0200
Australia
gani@maths.anu.edu.au

Thomas G. Kurtz
Department of Mathematics
University of Wisconsin - Madison
480 Lincoln Drive
Madison, WI 53706-1388
USA
kurtz@math.wisc.edu

Peter Jagers
Mathematical Statistics
Chalmers University of Technology
and Göteborg (Gothenburg) University
412 96 Göteborg
Sweden
jagers@chalmers.se

ISBN: 978-3-642-09926-7 e-ISBN: 978-3-540-85636-8

Probability and Its Applications ISSN print edition: 1431-7028

Mathematics Subject Classification (2000): Primary: 60F10, 60F15, 60G50, 62E20;
Secondary: 60E15, 60G42, 60G44, 60G40, 62L10

© 2009 Springer-Verlag Berlin Heidelberg
Softcover reprint of the hardcover 1st edition 2009
This work is subject to copyright. All rights are reserved, whether the whole or part of the material is concerned, specifically the rights of translation, reprinting, reuse of illustrations, recitation, broadcasting, reproduction on microfilm or in any other way, and storage in data banks. Duplication of this publication or parts thereof is permitted only under the provisions of the German Copyright Law of September 9, 1965, in its current version, and permission for use must always be obtained from Springer. Violations are liable to prosecution under the German Copyright Law.

The use of general descriptive names, registered names, trademarks, etc. in this publication does not imply, even in the absence of a specific statement, that such names are exempt from the relevant protective laws and regulations and therefore free for general use.

Cover design: WMXDesign GmbH, Heidelberg

9 8 7 6 5 4 3 2 1

springer.com

To our families

for V.H.P., Colleen, Victor, Mary-Margaret and Patrick

for T.L.L., Letitia, Peter and David

for Q.-M.S., Jiena and Wenqi

Preface

This year marks the centennial of Student's seminal 1908 paper, "On the probable error of a mean," in which the t-statistic and the t-distribution were introduced. During the past century, the t-statistic has evolved into much more general Studentized statistics and self-normalized processes, and the t-distribution generalized to the multivariate case, leading to multivariate processes with matrix self-normalization and bootstrap-t methods for tests and confidence intervals. The past two decades have also witnessed the active development of a rich probability theory of self-normalized processes, beginning with laws of the iterated logarithm, weak convergence, large and moderate deviations for self-normalized sums of independent random variables, and culminating in exponential and moment bounds and a universal law of the iterated logarithm for self-normalized processes in the case of dependent random variables. An important goal of this book is to present the main techniques and results of these developments in probability and to relate them to the asymptotic theory of Studentized statistics and to other statistical applications.

Another objective of writing this book is to use it as course material for a Ph.D. level course on selected topics in probability theory and its applications. Lai and Shao co-taught such a course for Ph.D. students in the Department of Statistics at Stanford University in the summer of 2007. These students had taken the Ph.D. core courses in probability (at the level of Durrett's *Probability: Theory and Examples*) and in theoretical statistics (at the level of Lehmann's *Testing Statistical Hypotheses* and *Theory of Point Estimation*). They found the theory of self-normalized processes an attractive topic, supplementing and integrating what they had learned from their core courses in probability and theoretical statistics and also exposing them to new techniques and research topics in both areas. The success of the experimental course STATS 300 (Advanced Topics in Statistics and Probability) prompted Lai and Shao to continue offering it periodically at Stanford and Hong Kong University of Science and Technology. A similar course is being planned at Columbia University by de la Peña. With these courses in mind, we have included exercises and supplements for the reader to explore related concepts and methods not covered in introductory Ph.D.-level courses, besides providing basic references related to these topics. We also plan to update these periodically at the Web site for the book: `http://www.math.ust.hk/~maqmshao/book-self/SNP.html`.

We acknowledge grant support for our research projects related to this book from the National Science Foundation (DMS-0505949 and 0305749) and the Hong Kong Research Grants Council (CERG-602206 and 602608). We thank three anonymous reviewers for their valuable suggestions, and all the students who took STATS 300 for their interest in the subject and comments on an earlier draft of certain chapters of the book that were used as lecture notes. We also thank our collaborators Hock Peng Chan, Bing-Yi Jing, Michael Klass, David Siegmund, Qiying Wang and Wang Zhou for working with us on related projects and for their helpful comments. We are particularly grateful to Cindy Kirby who helped us to coordinate our writing efforts and put together the separate chapters in an efficient and timely fashion. Without her help, this book would not have been completed in 2008 to commemorate Student's centennial.

Department of Statistics, Columbia University — *Victor H. de la Peña*
Department of Statistics, Stanford University — *Tze Leung Lai*
Department of Mathematics, Hong Kong University of Science & Technology — *Qi-Man Shao*

Contents

Part II Martingales and Dependent Random Vectors

Chapter 1
Introduction

A prototypical example of a self-normalized process is Student's t-statistic based on a sample of normal i.i.d. observations $X_1,\dots,X_n$, dating back to 1908 when William Gosset ("Student") considered the problem of statistical inference on the mean μ when the standard deviation σ of the underlying distribution is unknown. Let $\bar{X}_n = n^{-1}\sum_{i=1}^n X_i$ be the sample mean and $s_n^2 = (n-1)^{-1}\sum_{i=1}^n (X_i - \bar{X}_n)^2$ be the sample variance. Gosset (1908) derived the distribution of the t-statistic $T_n = \sqrt{n}(\bar{X}_n - \mu)/s_n$ for normal X_i; this is the t-distribution with $n-1$ degrees of freedom. The t-distribution converges to the standard normal distribution, and in fact T_n has a limiting standard normal distribution as $n \to \infty$ even when the X_i are nonnormal. When nonparametric methods were subsequently introduced, the t-test was compared with the nonparametric tests (e.g., the sign test and rank tests), in particular for "fat-tailed" distributions with infinite second or even first absolute moments. It has been found that the t-test of $\mu = \mu_0$ is robust against non-normality in terms of the Type I error probability but not the Type II error probability. Without loss of generality, consider the case $\mu_0 = 0$ so that

$$T_n = \frac{\sqrt{n}\bar{X}_n}{s_n} = \frac{S_n}{V_n}\left\{\frac{n-1}{n-(S_n/V_n)^2}\right\}^{1/2}, \tag{1.1}$$

where $S_n = \sum_{i=1}^n X_i, V_n^2 = \sum_{i=1}^n X_i^2$. Efron (1969) and Logan et al. (1973) have derived limiting distributions of self-normalized sums S_n/V_n. In view of (1.1), if T_n or S_n/V_n has a limiting distribution, then so does the other, and it is well known that they coincide; see, e.g., Proposition 1 of Griffin (2002).

Active development of the probability theory of self-normalized processes began in the 1990s with the seminal work of Griffin and Kuelbs (1989, 1991) on laws of the iterated logarithm for self-normalized sums of i.i.d. variables belonging to the domain of attraction of a normal or stable law. Subsequently, Bentkus and Götze (1996) derived a Berry–Esseen bound for Student's t-statistic, and Giné et al. (1997) proved that the t-statistic has a limiting standard normal distribution if and only if X_i is in the domain of attraction of a normal law. Moreover, Csörgő et al. (2003a)

V.H. de la Peña et al., *Self-Normalized Processes: Limit Theory and Statistical Applications*, Probability and its Applications,

© Springer-Verlag Berlin Heidelberg 2009

proved a self-normalized version of the weak invariance principle under the same necessary and sufficient condition. Shao (1997) proved large deviation results for S_n/V_n without moment conditions and moderate deviation results when X_i is the domain of attraction of a normal or stable law. Subsequently Shao (1999) obtained Cramér-type large deviation results when $E|X_1|^3 < \infty$. Jing et al. (2004) derived saddlepoint approximations for Student's t-statistic with no moment assumptions. Bercu et al. (2002) obtained large and moderate deviation results for self-normalized empirical processes. Self-normalized sums of independent but non-identically distributed X_i have been considered by Bentkus et al. (1996), Wang and Jing (1999), Jing et al. (2003) and Csörgő et al. (2003a).

Part I of the book presents in Chaps. 3–7 the basic ideas and results in the probability theory of self-normalized sums of independent random variables described above. It also extends in Chap. 8 the theory to self-normalized U-statistics based on independent random variables. Part II considers self-normalized processes in the case of dependent variables. Like Part I that begins by introducing some basic probability theory for sums of independent random variables in Chap. 2, Part II begins by giving in Chap. 9 an overview of martingale inequalities and related results which will be used in the subsequent chapters. Chapter 10 provides a general framework for self-normalization, which links the approach of de la Peña et al. (2000, 2004) for general self-normalized processes to that of Shao (1997) for large deviations of self-normalized sums of i.i.d. random variables. This general framework is also applicable to dependent random vectors that involve matrix normalization, as in Hotelling's T^2-statistic which generalizes Student's t-statistic to the multivariate case. In particular, it is noted in Chap. 10 that a basic ingredient in Shao's (1997) self-normalized large deviations theory is $e^{\psi(\theta,\rho)} := E\exp\{\theta X_1 - \rho\theta^2 X_1^2\}$, which is always finite for $\rho > 0$. This can be readily extended to the multivariate case by replacing θX_1 with $\theta' X_1$, where θ and X_1 are d-dimensional vectors. Under the assumptions $EX_1 = 0$ and $E\|X_1\|^2 < \infty$, Taylor's theorem yields

$$\psi(\theta,\rho) = \log\left(E\exp\{\theta' X_1 - \rho(\theta' X_1)^2\}\right) = \left\{\left(\frac{1}{2} - \rho + o(1)\right)\theta' E(X_1 X_1')\theta\right\}$$

as $\theta \to 0$. Let $\gamma > 0, C_n = (1+\gamma)\Sigma_{i=1}^n X_i X_i', A_n = \Sigma_{i=1}^n X_i$. It then follows that ρ and ε can be chosen sufficiently small so that

$$\begin{gathered}\{\exp(\theta' A_n - \theta' C_n \theta/2),\ \mathscr{F}_n, n \geq 1\} \\ \text{is a supermartingale with mean } \leq 1, \text{ for } \|\theta\| < \varepsilon.\end{gathered} \tag{1.2}$$

Note that (1.2) implies that $\{\int_{\|\theta\|<\varepsilon} e^{\theta' A_n - \theta' C_n \theta/2} f(\theta) d\theta,\ \mathscr{F}_n, n \geq 1\}$ is also a supermartingale, for any probability density f on the ball $\{\theta : \|\theta\| < \varepsilon\}$.

In Chap. 11 and its multivariate extension given in Chap. 14, we show that the supermartingale property (1.2), its weaker version $E\{\exp(\theta' A_n - \theta' C_n \theta/2)\} \leq 1$ for $\|\theta\| < \varepsilon$, and other variants given in Chap. 10 provide a general set of conditions from which we can derive exponential bounds and moment inequalities for self-normalized processes in dependent settings. A key tool is the *pseudo-maximization* method which involves Laplace's method for evaluating integrals of the form $\int_{\|\theta\|<\varepsilon} e^{\theta' A_n - \theta' C_n \theta/2} f(\theta) d\theta$. If the random function $\exp\{\theta' A_n - \theta' C_n \theta/2\}$ in (1.2) could be maximized over θ inside the expectation $E\{\exp(\theta' A_n - \theta' C_n \theta/2)\}$, taking the maximizing value $\theta = C_n^{-1} A_n$ would yield the expectation of the self-normalized variable $\exp\{A_n C_n^{-1} A_n/2\}$. Although this argument is not valid, integrating $\exp\{\theta' A_n - \theta' C_n \theta/2\}$ with respect to $f(\theta)d\theta$ and applying Laplace's method to evaluate the integral basically achieves the same effect as in the heuristic argument. This method is used to derive exponential and L_p-bounds for self-normalized processes in Chap. 12. The exponential bounds are used to derive laws of the iterated logarithm for self-normalized processes in Chap. 13.

Student's t-statistic $\sqrt{n}(\bar{X}_n - \mu)/s_n$ has also undergone far-reaching generalizations in the statistics literature during the past century. Its generalization is the *Studentized statistic* $(\hat{\theta}_n - \theta)/\widehat{\mathrm{se}}_n$, where θ is a functional $g(F)$ of the underlying distribution function F, $\hat{\theta}_n$ is usually chosen to be the corresponding functional $g(\hat{F}_n)$ of the empirical distribution, and $\widehat{\mathrm{se}}_n$ is a consistent estimator of the standard error of $\hat{\theta}_n$. Its multivariate generalization, which replaces $1/\widehat{\mathrm{se}}_n$ by $\hat{\Sigma}_n^{-1/2}$, where $\hat{\Sigma}_n$ is a consistent estimator of the covariance matrix of the vector $\hat{\theta}_n$ or its variant, is ubiquitous in statistical applications. Part III of the book, which is on statistical applications of self-normalized processes, begins with an overview in Chap. 15 of the distribution theory of the t-statistic and its multivariate extensions, for samples first from normal distributions and then from general distributions that may have infinite second moments. Chapter 15 also considers the asymptotic theory of general Studentized statistics in time series and control systems and relates this theory to that of self-normalized martingales. An alternative to inference based on asymptotic distributions of Studentized statistics is to make use of bootstrapping. Chapter 16 describes the role of self-normalization in deriving approximate pivots for the construction of bootstrap confidence intervals, whose accuracy and correctness are analyzed by Edgeworth and Cornish–Fisher expansions. Chapter 17 introduces generalized likelihood ratio statistics as another class of self-normalized statistics. It also relates the pseudo-maximization approach and the method of mixtures in Part II to the close connections between likelihood and Bayesian inference. Whereas the framework of Part I covers the classical setting of independent observations sampled from a population, that of Part II is applicable to time series models and stochastic dynamic systems, and examples are given in Chaps. 15, 17 and 18. Moreover, the probability theory in Parts I and II is related not only to samples of fixed size, but also to sequentially generated samples that are associated with asymptotically optimal stopping rules. Part III concludes with Chap. 18 which considers self-normalized processes in sequential analysis and the associated boundary crossing problems.

Part I
Independent Random Variables

Chapter 2
Classical Limit Theorems, Inequalities and Other Tools

This chapter summarizes some classical limit theorems, basic probability inequalities and other tools that are used in subsequent chapters. Throughout this book, all random variables are assumed to be defined on the same probability space $(\Omega, \mathscr{F}, P)$ unless otherwise specified.

2.1 Classical Limit Theorems

The law of large numbers, the central limit theorem and the law of the iterated logarithm form a trilogy of the asymptotic behavior of sums of independent random variables. They are closely related to moment conditions and deal with three modes of convergence of a sequence of random variables Y_n to a random variable Y. We say that Y_n converges to Y *in probability*, denoted by $Y_n \xrightarrow{P} Y$, if, for any $\varepsilon > 0$, $P(|Y_n - Y| > \varepsilon) \to 0$ as $n \to \infty$. We say that Y_n converges *almost surely* to Y (or Y_n converges to Y with probability 1), denoted by $Y_n \xrightarrow{a.s.} Y$, if $P(\lim_{n\to\infty} Y_n = Y) = 1$. Note that almost sure convergence is equivalent to $P(\max_{k\ge n} |Y_k - Y| > \varepsilon) \to 0$ as $n \to \infty$ for any given $\varepsilon > 0$. We say that Y_n converges *in distribution* (or *weakly*) to Y, and write $Y_n \xrightarrow{D} Y$ or $Y_n \Rightarrow Y$, if $P(Y_n \le x) \to P(Y \le x)$, at every continuity point of the cumulative distribution function of Y. If the cumulative distribution $P(Y \le x)$ is continuous, then $Y_n \xrightarrow{D} Y$ not only means $P(Y_n \le x) \to P(Y \le x)$ for every x, but also implies that the convergence is uniform in x, i.e.,

$$\sup_x |P(Y_n \le x) - P(Y \le x)| \to 0 \qquad \text{as } n \to \infty.$$

The three modes of convergence are related by

$$Y_n \xrightarrow{a.s.} Y \Longrightarrow Y_n \xrightarrow{P} Y \Longrightarrow Y_n \xrightarrow{D} Y.$$

V.H. de la Peña et al., *Self-Normalized Processes: Limit Theory and Statistical Applications*, Probability and its Applications,

© Springer-Verlag Berlin Heidelberg 2009

The reverse relations are not true in general. However, $Y_n \xrightarrow{D} c$ is equivalent to $Y_n \xrightarrow{P} c$ when c is a constant. Another relationship is provided by Slutsky's theorem: If $Y_n \xrightarrow{D} Y$ and $\xi_n \xrightarrow{P} c$, then $Y_n + \xi_n \xrightarrow{D} Y + c$ and $\xi_n Y_n \xrightarrow{D} cY$.

2.1.1 The Weak Law, Strong Law and Law of the Iterated Logarithm

Let $X_1, X_2, \ldots$ be *independent and identically distributed* (i.i.d.) random variables and let $S_n = \sum_{i=1}^n X_i$. Then we have Kolmogorov's strong law of large numbers and Feller's weak law of large numbers.

Theorem 2.1. $n^{-1}S_n \xrightarrow{a.s.} c < \infty$ *if and only if* $E(|X_1|) < \infty$, *in which case* $c = E(X_1)$.

Theorem 2.2. *In order that there exist constants* c_n *such that* $n^{-1}S_n - c_n \xrightarrow{P} 0$, *it is necessary and sufficient that* $\lim_{x\to\infty} xP(|X_1| \geq x) = 0$. *In this case,* $c_n = EX_1 I(|X_1| \leq n)$.

The Marcinkiewicz–Zygmund law of large numbers gives the rate of convergence in Theorem 2.1.

Theorem 2.3. *Let* $1 < p < 2$. *If* $E(|X_1|) < \infty$, *then*

$$n^{1-1/p}\left(n^{-1}S_n - E(X_1)\right) \xrightarrow{a.s.} 0 \tag{2.1}$$

if and only if $E(|X_1|^p) < \infty$.

When $p = 2$, (2.1) is no longer valid. Instead, we have the Hartman–Wintner *law of the iterated logarithm* (LIL), the converse of which is established by Strassen (1966).

Theorem 2.4. *If* $EX_1^2 < \infty$ *and* $EX_1 = \mu$, $Var(X_1) = \sigma^2$, *then*

$$\limsup_{n\to\infty} \frac{S_n - n\mu}{\sqrt{2n\log\log n}} = \sigma \ a.s.,$$

$$\liminf_{n\to\infty} \frac{S_n - n\mu}{\sqrt{2n\log\log n}} = -\sigma \ a.s.,$$

$$\limsup_{n\to\infty} \frac{\max_{1\leq k\leq n}|S_k - k\mu|}{\sqrt{2n\log\log n}} = \sigma \ a.s.$$

Conversely, if there exist finite constants a *and* τ *such that*

$$\limsup_{n\to\infty} \frac{S_n - na}{\sqrt{2n\log\log n}} = \tau \ a.s.,$$

then $a = E(X_1)$ *and* $\tau^2 = Var(X_1)$.

The following is an important tool for proving Theorems 2.1, 2.3 and 2.4.

Lemma 2.5 (Borel–Cantelli Lemma).

(1) Let $A_1, A_2, \ldots$ be an arbitrary sequence of events on $(\Omega, \mathscr{F}, P)$. Then $\sum_{i=1}^{\infty} P(A_i) < \infty$ implies $P(A_n \text{ i.o.}) = 0$, where $\{A_n \text{ i.o.}\}$ denotes the event $\bigcap_{k \ge 1} \bigcup_{n \ge k} A_n$, i.e., A_n occurs infinitely often.
(2) Let $A_1, A_2, \ldots,$ be a sequence of independent events on $(\Omega, \mathscr{F}, P)$. Then $\sum_{i=1}^{\infty} P(A_i) = \infty$ implies $P(A_n \text{ i.o.}) = 1$.

The strong law of large numbers and LIL have also been shown to hold for independent but not necessarily identically distributed random variables $X_1, X_2, \ldots$.

Theorem 2.6.

(1) If $b_n \uparrow \infty$ and $\sum_{i=1}^{\infty} \operatorname{Var}(X_i)/b_i^2 < \infty$, then $(S_n - ES_n)/b_n \xrightarrow{a.s.} 0$.
(2) If $b_n \uparrow \infty$, $\sum_{i=1}^{\infty} P(|X_i| \ge b_i) < \infty$ and $\sum_{i=1}^{\infty} b_i^{-2} EX_i^2 I(|X_i| \le b_i) < \infty$, then $(S_n - a_n)/b_n \xrightarrow{a.s.} 0$, where $a_n = \sum_{i=1}^{n} EX_i I(|X_i| \le b_i)$.

The "if" part in Theorems 2.1 and 2.3 can be derived from Theorem 2.6, which can be proved by making use Kolmogorov's three-series theorem and the Kronecker lemma in the following.

Theorem 2.7 (Three-series Theorem). *The series $\sum_{i=1}^{\infty} X_i$ converges a.s. if and only if the three series*

$$\sum_{i=1}^{\infty} P(|X_i| \ge c), \quad \sum_{i=1}^{\infty} EX_i I(|X_i| \le c), \quad \sum_{i=1}^{\infty} \operatorname{Var}\{X_i I(|X_i| \le c)\}$$

converge for some $c > 0$.

Lemma 2.8 (Kronecker's Lemma). *If $\sum_{i=1}^{\infty} x_i$ converges and $b_n \uparrow \infty$, then $b_n^{-1} \sum_{i=1}^{n} b_i x_i \to 0$.*

We end this subsection with Kolmogorov's LIL for independent but not necessarily identically distributed random variables; see Chow and Teicher (1988, Sect. 10.2). *Assume that $EX_i = 0$ and $EX_i^2 < \infty$ and put $B_n^2 = \sum_{i=1}^{n} EX_i^2$. If $B_n \to \infty$ and $X_n = o(B_n (\log\log B_n)^{-1/2})$ a.s., then*

$$\limsup_{n \to \infty} \frac{S_n}{B_n \sqrt{2 \log\log B_n}} = 1 \quad a.s. \tag{2.2}$$

2.1.2 The Central Limit Theorem

For any sequence of random variables X_i with finite means, the sequence $X_i - E(X_i)$ has zero means and therefore we can assume, without loss of generality, that the mean of X_i is 0. For i.i.d. X_i, we have the classical central limit theorem (CLT).

Theorem 2.9. *If $X_1,\dots,X_n$ are i.i.d. with $E(X_1)=0$ and $\mathrm{Var}(X_1)=\sigma^2<\infty$, then*

$$\frac{S_n}{\sqrt{n}\sigma} \xrightarrow{D} N(0,1).$$

The Berry–Esseen inequality provides the convergence rate in the CLT.

Theorem 2.10. *Let Φ denote the standard normal distribution function and $W_n = S_n/(\sqrt{n}\sigma)$. Then*

$$\begin{aligned} &\sup_x |P(W_n \le x) - \Phi(x)| \\ &\le 4.1\left\{\sigma^{-2}EX_1^2 I\left(|X_1| > \sqrt{n}\sigma\right) + n^{-1/2}\sigma^{-3}E|X_1|^3 I\left(|X_1| \le \sqrt{n}\sigma\right)\right\}. \end{aligned} \tag{2.3}$$

In particular, if $E|X_1|^3<\infty$, then

$$\sup_x |P(W_n \le x) - \Phi(x)| \le \frac{0.79E|X_1|^3}{\sqrt{n}\sigma^3}. \tag{2.4}$$

For general independent not necessarily identically distributed random variables, the CLT holds under the Lindeberg condition, under which a non-uniform Berry–Esseen inequality of the type in (2.3) still holds.

Theorem 2.11 (Lindberg–Feller CLT). *Let X_n be independent random variables with $E(X_i)=0$ and $E(X_i^2)<\infty$. Let $W_n = S_n/B_n$, where $B_n^2=\sum_{i=1}^n E(X_i^2)$. If the Lindberg condition*

$$B_n^{-2}\sum_{i=1}^n EX_i^2 I(|X_i| \ge \varepsilon B_n) \longrightarrow 0 \qquad \textit{for all } \varepsilon > 0 \tag{2.5}$$

holds, then $W_n \xrightarrow{D} N(0,1)$. Conversely, if $\max_{1\le i\le n} EX_i^2 = o(B_n^2)$ and $W_n \xrightarrow{D} N(0,1)$, then the Lindberg condition (2.5) is satisfied.

Theorem 2.12. *With the same notations as in Theorem 2.11,*

$$\begin{aligned} &\sup_x |P(W_n \le x) - \Phi(x)| \\ &\le 4.1\left(B_n^{-2}\sum_{i=1}^n EX_i^2 I\{|X_i| > B_n\} + B_n^{-3}\sum_{i=1}^n E|X_i|^3 I\{|X_i| \le B_n\}\right) \end{aligned} \tag{2.6}$$

and

$$\begin{aligned} &|P(W_n \le x) - \Phi(x)| \\ &\le C\left(\sum_{i=1}^n \frac{EX_i^2 I\{|X_i| > (1+|x|)B_n\}}{(1+|x|)^2 B_n^2} + \sum_{i=1}^n \frac{E|X_i|^3 I\{|X_i| \le (1+|x|)B_n\}}{(1+|x|)^3 B_n^3}\right), \end{aligned} \tag{2.7}$$

where C is an absolute constant.

2.1.3 Cramér's Moderate Deviation Theorem

The Berry–Esseen inequality gives a bound on the absolute error in approximating the distribution of W_n by the standard normal distribution. The usefulness of the bound may be limited when $\Phi(x)$ is close to 0 or 1. Cramér's theory of moderate deviations provides the relative errors. Petrov (1975, pp. 219–228) gives a comprehensive treatment of the theory and introduces the *Cramér series*, which is a power series whose coefficients can be expressed in terms of the cumulants of the underlying distribution and which is used in part (a) of the following theorem.

Theorem 2.13.

(a) *Let* $X_1, X_2, \ldots$ *be i.i.d. random variables with* $E(X_1) = 0$ *and* $Ee^{t_0|X_1|} < \infty$ *for some* $t_0 > 0$. *Then for* $x \geq 0$ *and* $x = o(n^{1/2})$,

$$\frac{P(W_n \geq x)}{1 - \Phi(x)} = \exp\left\{x^2 \lambda\left(\frac{x}{\sqrt{n}}\right)\right\}\left(1 + O\left(\frac{1+x}{\sqrt{n}}\right)\right), \tag{2.8}$$

where $\lambda(t)$ *is the Cramér series.*

(b) *If* $Ee^{t_0\sqrt{|X_1|}} < \infty$ *for some* $t_0 > 0$, *then*

$$\frac{P(W_n \geq x)}{1 - \Phi(x)} \to 1 \quad \text{as } n \to \infty \text{ uniformly in } x \in \left[0, o(n^{1/6})\right). \tag{2.9}$$

(c) *The converse of (b) is also true; that is, if* (2.9) *holds, then* $Ee^{t_0\sqrt{|X_1|}} < \infty$ *for some* $t_0 > 0$.

In parts (a) and (b) of Theorem 2.13, $P(W_n \geq x)/(1 - \Phi(x))$ can clearly be replaced by $P(W_n \leq -x)/\Phi(-x)$. Moreover, similar results are also available for standardized sums S_n/B_n of independent but not necessarily identically distributed random variables with bounded moment generating functions in some neighborhood of the origin; see Petrov (1975). In Chap. 7, we establish Cramér-type moderate deviation results for *self-normalized* (rather than standardized) sums of independent random variables under much weaker conditions.

2.2 Exponential Inequalities for Sample Sums

2.2.1 Self-Normalized Sums

We begin by considering independent Rademacher random variables.

Theorem 2.14. *Assume that* ε_i *are independent and* $P(\varepsilon_i = 1) = P(\varepsilon_i = -1) = 1/2$. *Then*

$$P\left(\frac{\sum_{i=1}^n a_i \varepsilon_i}{\left(\sum_{i=1}^n a_i^2\right)^{1/2}} \geq x\right) \leq e^{-x^2/2} \tag{2.10}$$

for $x > 0$ *and real numbers* $\{a_i\}$.

Proof. Without loss of generality, assume $\sum_{i=1}^n a_i^2 = 1$. Observe that

$$\frac{1}{2}(e^{-t}+e^{t}) \le e^{t^2/2}$$

for $t \in \mathbb{R}$. We have

$$\begin{aligned} P\left(\sum_{i=1}^n a_i\varepsilon_i \ge x\right) &\le e^{-x^2} E e^{x\sum_{i=1}^n a_i\varepsilon_i} \\ &= e^{-x^2}\prod_{i=1}^n \frac{1}{2}(e^{-a_ix}+e^{a_ix}) \\ &\le e^{-x^2}\prod_{i=1}^n e^{a_i^2x^2/2} = e^{-x^2/2}. \end{aligned}$$

□

Let X_n be independent random variables and let $V_n^2 = \sum_{i=1}^n X_i^2$. If we further assume that X_i is symmetric, then X_i and $\varepsilon_i X_i$ have the same distribution, where $\{\varepsilon_i\}$ are i.i.d. Rademacher random variables independent of $\{X_i\}$. Hence the self-normalized sum S_n/V_n has the same distribution as $(\sum_{i=1}^n X_i\varepsilon_i)/V_n$. Given $\{X_i, 1 \le i \le n\}$, applying (2.10) to $a_i = X_i$ yields the following.

Theorem 2.15. *If X_i is symmetric, then for $x > 0$,*

$$P(S_n \ge xV_n) \le e^{-x^2/2}. \tag{2.11}$$

The next result extends the above "sub-Gaussian" property of the self-normalized sum S_n/V_n to general (not necessarily symmetric) independent random variables.

Theorem 2.16. *Assume that there exist $b > 0$ and a such that*

$$P(S_n \ge a) \le 1/4 \quad \text{and} \quad P(V_n^2 \ge b^2) \le 1/4. \tag{2.12}$$

Then for $x > 0$,

$$P\{S_n \ge x(a+b+V_n)\} \le 2e^{-x^2/2}. \tag{2.13}$$

In particular, if $E(X_i) = 0$ and $E(X_i^2) < \infty$, then

$$P\{|S_n| \ge x(4B_n+V_n)\} \le 4e^{-x^2/2} \qquad \text{for } x > 0, \tag{2.14}$$

where $B_n = (\sum_{i=1}^n EX_i^2)^{1/2}$.

Proof. When $x \le 1$, (2.13) is trivial. When $x > 1$, let $\{Y_i, 1 \le i \le n\}$ be an independent copy of $\{X_i, 1 \le i \le n\}$. Then

$$\begin{aligned} P\left(\sum_{i=1}^n Y_i \le a, \sum_{i=1}^n Y_i^2 \le b^2\right) &\ge 1 - P\left(\sum_{i=1}^n Y_i > a\right) - P\left(\sum_{i=1}^n Y_i^2 > b^2\right) \\ &\ge 1 - 1/4 - 1/4 = 1/2. \end{aligned}$$

Noting that

$$\left\{S_n \geq x(a+b+V_n),\ \sum_{i=1}^n Y_i \leq a, \sum_{i=1}^n Y_i^2 \leq b^2\right\}$$
$$\subset \left\{\sum_{i=1}^n (X_i - Y_i) \geq x\left(a+b+\left(\sum_{i=1}^n (X_i-Y_i)^2\right)^{1/2} - \left(\sum_{i=1}^n Y_i^2\right)^{1/2}\right) - a, \sum_{i=1}^n Y_i^2 \leq b^2\right\}$$
$$\subset \left\{\sum_{i=1}^n (X_i - Y_i) \geq x\left(\sum_{i=1}^n (X_i-Y_i)^2\right)^{1/2}\right\}$$

and that $\{X_i - Y_i, 1 \leq i \leq n\}$ is a sequence of independent symmetric random variables, we have

$$\begin{aligned} P(S_n \geq x(a+b+V_n)) &= \frac{P\left(S_n \geq x(a+b+V_n), \sum_{i=1}^n Y_i \leq a, \sum_{i=1}^n Y_i^2 \leq b^2\right)}{P\left(\sum_{i=1}^n Y_i \leq a, \sum_{i=1}^n Y_i^2 \leq b^2\right)} \\ &\leq 2P\left(\sum_{i=1}^n (X_i - Y_i) \geq x\left(\sum_{i=1}^n (X_i-Y_i)^2\right)^{1/2}\right) \\ &\leq 2e^{-x^2/2} \end{aligned}$$

by (2.11). This proves (2.13), and (2.14) follows from (2.13) with $a = b = 2B_n$. □

2.2.2 Tail Probabilities for Partial Sums

Let X_n be independent random variables and let $S_n = \sum_{i=1}^n X_i$. The following theorem gives the *Bennett–Hoeffding inequalities*.

Theorem 2.17. *Assume that* $EX_i \leq 0$, $X_i \leq a$ $(a > 0)$ *for each* $1 \leq i \leq n$, *and* $\sum_{i=1}^n EX_i^2 \leq B_n^2$. *Then*

$$Ee^{tS_n} \leq \exp\left(a^{-2}(e^{ta} - 1 - ta)B_n^2\right) \qquad \text{for } t > 0, \tag{2.15}$$

$$P(S_n \geq x) \leq \exp\left(-\frac{B_n^2}{a^2}\left\{\left(1+\frac{ax}{B_n^2}\right)\log\left(1+\frac{ax}{B_n^2}\right) - \frac{ax}{B_n^2}\right\}\right) \tag{2.16}$$

and

$$P(S_n \geq x) \leq \exp\left(-\frac{x^2}{2(B_n^2 + ax)}\right) \qquad \text{for } x > 0. \tag{2.17}$$

Proof. It is easy to see that $(e^s - 1 - s)/s^2$ is an increasing function of s. Therefore

$$e^{ts} \leq 1 + ts + (ts)^2(e^{ta} - 1 - ta)/(ta)^2 \tag{2.18}$$

for $s \le a$, and hence

$$\begin{aligned} Ee^{tS_n} &= \prod_{i=1}^n Ee^{tX_i} \le \prod_{i=1}^n \left(1 + tEX_i + a^{-2}(e^{ta} - 1 - ta)EX_i^2\right) \\ &\le \prod_{i=1}^n \left(1 + a^{-2}(e^{ta} - 1 - ta)EX_i^2\right) \le \exp\left(a^{-2}(e^{ta} - 1 - ta)B_n^2\right). \end{aligned}$$

This proves (2.15). To prove (2.16), let $t = a^{-1}\log(1 + ax/B_n^2)$. Then, by (2.15),

$$\begin{aligned} P(S_n \ge x) &\le e^{-tx} Ee^{tS_n} \\ &\le \exp\left(-tx + a^{-2}(e^{ta} - 1 - ta)B_n^2\right) \\ &= \exp\left(-\frac{B_n^2}{a^2}\left\{\left(1 + \frac{ax}{B_n^2}\right)\log\left(1 + \frac{ax}{B_n^2}\right) - \frac{ax}{B_n^2}\right\}\right), \end{aligned}$$

proving (2.16). To prove (2.17), use (2.16) and

$$(1+s)\log(1+s) - s \ge \frac{s^2}{2(1+s)} \qquad \text{for } s > 0.$$

□

The inequality (2.17) is often called *Bernstein's inequality*. From the Taylor expansion of e^x, it follows that

$$e^x \le 1 + x + x^2/2 + |x|^3 e^x/6. \tag{2.19}$$

Let $\beta_n = \sum_{i=1}^n E|X_i|^3$. Using (2.19) instead of (2.18) in the above proof, we have

$$Ee^{tS_n} \le \exp\left(\frac{1}{2}t^2 B_n^2 + \frac{1}{6}t^3 \beta_n e^{ta}\right), \tag{2.20}$$

$$P(S_n \ge x) \le \exp\left(-tx + \frac{1}{2}t^2 B_n^2 + \frac{1}{6}t^3 \beta_n e^{ta}\right) \tag{2.21}$$

for all $t > 0$, and in particular

$$P(S_n \ge x) \le \exp\left(-\frac{x^2}{2B_n^2} + \frac{x^3}{6B_n^6}\beta_n e^{ax/B_n^2}\right). \tag{2.22}$$

When X_i is not bounded above, we can first truncate it and then apply Theorem 2.17 to prove the following inequality.

Theorem 2.18. *Assume that $EX_i \le 0$ for $1 \le i \le n$ and that $\sum_{i=1}^n EX_i^2 \le B_n^2$. Then*

$$\begin{aligned} P(S_n \ge x) \le P\left(\max_{1\le i\le n} X_i \ge b\right) + \exp\left(-\frac{B_n^2}{a^2}\left\{\left(1 + \frac{ax}{B_n^2}\right)\log\left(1 + \frac{ax}{B_n^2}\right) - \frac{ax}{B_n^2}\right\}\right) \\ + \sum_{i=1}^n P(a < X_i < b) P(S_n - X_i > x - b) \end{aligned} \tag{2.23}$$

for $x > 0$ and $b \geq a > 0$. In particular,

$$P(S_n \geq x) \leq P\left(\max_{1\leq i\leq n} X_i > \delta x\right) + \left(\frac{3B_n^2}{B_n^2 + \delta x^2}\right)^{1/\delta} \tag{2.24}$$

for $x > 0$ and $\delta > 0$.

Proof. Let $\bar{X}_i = X_i I(X_i \leq a)$ and $\bar{S}_n = \sum_{i=1}^n \bar{X}_i$. Then

$$\begin{aligned}
P(S_n \geq x) &\leq P\left(\max_{1\leq i\leq n} X_i \geq b\right) + P\left(S_n \geq x, \max_{1\leq i\leq n} X_i \leq a\right) \\
&\quad + P\left(S_n \geq x, \max_{1\leq i\leq n} X_i > a, \max_{1\leq i\leq n} X_i < b\right) \\
&\leq P\left(\max_{1\leq i\leq n} X_i \geq b\right) + P(\bar{S}_n \geq x) \\
&\quad + \textstyle\sum_{i=1}^n P(S_n \geq x, a < X_i < b) \qquad (2.25)\\
&\leq P\left(\max_{1\leq i\leq n} X_i \geq b\right) + P(\bar{S}_n \geq x) \\
&\quad + \textstyle\sum_{i=1}^n P(S_n - X_i \geq x - b, a < X_i < b) \\
&= P\left(\max_{1\leq i\leq n} X_i \geq b\right) + P(\bar{S}_n \geq x) \\
&\quad + \textstyle\sum_{i=1}^n P(a < X_i < b) P(S_n - X_i \geq x - b).
\end{aligned}$$

Applying (2.16) to $\bar{S}_n$ gives

$$P(\bar{S}_n \geq x) \leq \exp\left(-\frac{B_n^2}{a^2}\left[\left(1 + \frac{ax}{B_n^2}\right)\log\left(1 + \frac{ax}{B_n^2}\right) - \frac{ax}{B_n^2}\right]\right),$$

which together with (2.26) yields (2.23). From (2.23) with $a = b = \delta x$, (2.24) follows. □

The following two results are about nonnegative random variables.

Theorem 2.19. *Assume that $X_i \geq 0$ with $E(X_i^2) < \infty$. Let $\mu_n = \sum_{i=1}^n EX_i$ and $B_n^2 = \sum_{i=1}^n EX_i^2$. Then for $0 < x < \mu_n$,*

$$P(S_n \leq x) \leq \exp\left(-\frac{(\mu_n - x)^2}{2B_n^2}\right). \tag{2.26}$$

Proof. Note that $e^{-a} \leq 1 - a + a^2/2$ for $a \geq 0$. For any $t \geq 0$ and $x \leq \mu_n$, we have

$$\begin{aligned}
P(S_n \leq x) &\leq e^{tx} E e^{-tS_n} = e^{tx} \prod_{i=1}^n E e^{-tX_i} \\
&\leq e^{tx} \prod_{i=1}^n E(1 - tX_i + t^2 X_i^2/2) \\
&\leq \exp\left(-t(\mu_n - x) + t^2 B_n^2/2\right).
\end{aligned}$$

Letting $t = (\mu_n - x)/B_n^2$ yields (2.26). □

Theorem 2.20. *Assume that $P(X_i = 1) = p_i$ and $P(X_i = 0) = 1 - p_i$. Then for $x > 0$,*

$$P(S_n \geq x) \leq \left(\frac{\mu e}{x}\right)^x, \tag{2.27}$$

where $\mu = \sum_{i=1}^n p_i$.

Proof. Let $t > 0$. Then

$$\begin{aligned} P(S_n \geq x) &\leq e^{-tx} \prod_{i=1}^n E e^{tX_i} = e^{-tx} \prod_{i=1}^n \left(1 + p_i(e^t - 1)\right) \\ &\leq \exp\left(-tx + (e^t - 1) \textstyle\sum_{i=1}^n p_i\right) = \exp\left(-tx + (e^t - 1)\mu\right). \end{aligned}$$

Since the case $x \leq \mu$ is trivial, we assume that $x > \mu$. Then letting $t = \log(x/\mu)$ yields

$$\exp\left(-tx + (e^t - 1)\mu\right) = \exp\left(-x \log(x/\mu) + x - \mu\right) \leq (\mu e/x)^x.$$

□

We end this section with the Ottaviani maximal inequality.

Theorem 2.21. *Assume that there exists a such that* $\max_{1 \leq k \leq n} P(S_k - S_n \geq a) \leq 1/2$. *Then*

$$P\left(\max_{1 \leq k \leq n} S_k \geq x\right) \leq 2P(S_n \geq x - a). \tag{2.28}$$

In particular, if $E(X_i) = 0$ and $E(X_i^2) < \infty$, then

$$P\left(\max_{1 \leq k \leq n} S_k \geq x\right) \leq 2P(S_n \geq x - \sqrt{2}B_n), \tag{2.29}$$

where $B_n = \sqrt{\sum_{i=1}^n E(X_i^2)}$.

Proof. Let $A_1 = \{S_1 \geq x\}$ and $A_k = \{S_k \geq x, \max_{1 \leq i \leq k-1} S_i < x\}$. Then $\{\max_{1 \leq k \leq n} S_k \geq x\} = \cup_{k=1}^n A_k$ and

$$\begin{aligned} P\left(\max_{1 \leq k \leq n} S_k \geq x\right) &\leq P(S_n \geq x - a) + \sum_{k=1}^n P(A_k, S_n < x - a) \\ &\leq P(S_n \geq x - a) + \sum_{k=1}^n P(A_k, S_n - S_k < -a) \\ &= P(S_n \geq x - a) + \sum_{k=1}^n P(A_k) P(S_n - S_k < -a) \\ &\leq P(S_n \geq x - a) + (1/2) \sum_{k=1}^n P(A_k) \\ &= P(S_n \geq x - a) + (1/2) P\left(\max_{1 \leq k \leq n} S_k \geq x\right), \end{aligned}$$

which gives (2.28). (2.29) follows from (2.28) with $a = \sqrt{2}B_n$. □

The proof of Kolmogorov's LIL (2.2) involves upper exponential bounds like those in Theorem 2.17 and the following lower exponential bound, whose proof is given in Chow and Teicher (1988, pp. 352–354) and uses the "conjugate method" that will be described in Sect. 3.1.

Theorem 2.22. *Assume that* $EX_i = 0$ *and* $|X_i| \leq a_i$ *a.s. for* $1 \leq i \leq n$ *and that* $\sum_{i=1}^{n} EX_i^2 = B_n^2$. *Let* $c_n \geq c_0 > 0$ *be such that* $\lim_{n\to\infty} a_n c_n / B_n = 0$. *Then for every* $0 < \gamma < 1$, *there exists* $0 < \delta_\gamma < 1/2$ *such that for all large n,*

$$P\{S_n \geq (1-\gamma)^2 c_n B_n\} \geq \delta_\gamma \exp\{-(1-\gamma)(1-\gamma^2)c_n^2/2\}.$$

2.3 Characteristic Functions and Expansions Related to the CLT

Let Y be a random variable with distribution function F. The *characteristic function* of Y is defined by $\varphi(t) = Ee^{itY} = \int_{-\infty}^{\infty} e^{ity} dF(y)$ for $t \in \mathbb{R}$. In view of *Lévy's inversion formula*

$$\lim_{T\to\infty} \frac{1}{2\pi} \int_{-T}^{T} \frac{e^{-ita} - e^{-itb}}{it} \varphi(t)dt = P(a < Y < b) + \frac{1}{2}\{P(Y = a) + P(Y = b)\} \tag{2.30}$$

for $a < b$ (see Durrett, 2005, pp. 93–94), the characteristic function uniquely determines the distribution function. The characteristic function φ is continuous, with $\varphi(0) = 1$, $|\varphi(t)| \leq 1$ for all $t \in \mathbb{R}$. There are three possibilities concerning solutions to the equation $|\varphi(t)| = 1$ (see Durrett, 2005, p. 129):

(a) $|\varphi(t)| < 1$ for all $t \neq 0$.
(b) $|\varphi(t)| = 1$ for all $t \in \mathbb{R}$. In this case, $\varphi(t) = e^{ita}$ and Y puts all its mass at a.
(c) $|\varphi(\tau)| = 1$ and $|\varphi(t)| < 1$ for $0 < t < \tau$. In this case $|\varphi|$ has period τ and there exists $b \in \mathbb{R}$ such that the support of Y is the lattice $\{b + 2\pi j/\tau : j = 0, \pm 1, \pm 2, \ldots\}$, i.e., Y is *lattice with span* $2\pi/\tau$.

A random variable Y is called *non-lattice* if its support is not a lattice, which corresponds to case (a) above. It is said to be *strongly non-lattice* if it satisfies *Cramér's condition*

$$\limsup_{|t|\to\infty} |\varphi(t)| < 1. \tag{2.31}$$

Note that (2.31), which is only concerned with the asymptotic behavior of $|\varphi(t)|$ as $|t| \to \infty$, is stronger than (a) because it rules out (b) and (c).

If the characteristic function φ of Y is integrable, i.e., $\int_{-\infty}^{\infty} |\varphi(t)| dt < \infty$, then Y has a bounded continuous density function f with respect to Lebesgue measure and

$$f(y) = \frac{1}{2\pi} \int_{-\infty}^{\infty} e^{-ity} \varphi(t) dt. \tag{2.32}$$

This is the *Fourier inversion formula*; see Durrett (2005, p. 95). In this case, since $\varphi(t) = \int_{-\infty}^{\infty} e^{ity} f(y)dy$ and f is integrable,

$$\lim_{|t|\to\infty} \varphi(t) = 0 \tag{2.33}$$

by the Riemann–Lebesgue lemma; see Durrett (2005, p. 459). Hence, if Y has an integrable characteristic function, then Y satisfies Cramér's condition (2.31).

In the case of lattice distributions with support $\{b + hk : k = 0, \pm 1, \pm 2, \dots\}$, let $p_k = P(Y = b + hk)$. Then the characteristic function is a Fourier series $\varphi(t) = \sum_{k=-\infty}^{\infty} p_k e^{it(b+hk)}$, with

$$p_k = \frac{h}{2\pi} \int_{-\pi/h}^{\pi/h} e^{-it(b+hk)} \varphi(t) dt, \tag{2.34}$$

noting that the span h corresponds to $2\pi/\tau$ (or $\tau = 2\pi/h$) in (b).

2.3.1 Continuity Theorem and Weak Convergence

Theorem 2.23. *Let φ_n be the characteristic function of Y_n.*

(a) If $\varphi_n(t)$ converges, as $n \to \infty$, to a limit $\varphi(t)$ for every t and if φ is continuous at 0, then φ is the characteristic function of a random variable Y and $Y_n \Rightarrow Y$.
(b) If $Y_n \Rightarrow Y$ and φ is the characteristic function of Y, then $\lim_{n\to\infty} \varphi_n(t) = \varphi(t)$ for all $t \in \mathbb{R}$.

For independent random variables $X_1, \dots, X_n$, the characteristic function of the sum $S_n = \sum_{k=1}^{n} X_k$ is the product of their characteristic functions $\varphi_1, \dots, \varphi_n$. If X_i has mean 0 and variance σ_i^2, quadratic approximation of $\varphi_i(t)$ in a neighborhood of the origin by Taylor's theorem leads to the central limit theorem under the Lindeberg condition (2.5). When the X_k have infinite second moments, the limiting distribution of $(S_n - b_n)/a_n$, if it exists for suitably chosen centering and scaling constants, is an *infinitely divisible* distribution, which is characterized by the property that its characteristic function is the nth power of a characteristic function for every integer $n \geq 1$. Equivalently, Y is infinitely divisible if for every $n \geq 1$, $Y \stackrel{D}{=} X_{n1} + \cdots + X_{nn}$, where X_{ni} are i.i.d. random variables and $\stackrel{D}{=}$ denotes equality in distribution (i.e., both sides having the same distribution). Another equivalent characterization of infinite divisibility is the Lévy–Khintchine representation of the characteristic function φ of Y:

$$\varphi(t) = \exp\left\{ i\gamma t + \int_{-\infty}^{\infty} \left(e^{itu} - 1 - \frac{itu}{1+u^2} \right) \left(\frac{1+u^2}{u^2} \right) dG(u) \right\}, \tag{2.35}$$

where $\gamma \in \mathbb{R}$ and G is nondecreasing, left continuous with $G(-\infty) = 0$ and $G(\infty) < \infty$. Examples of infinitely divisible distributions include the normal, gamma, Poisson, and stable distributions; see Durrett (2005, Sect. 2.8) and Chow and Teicher (1988, Chap. 12).

A random variable Y is said to have a *stable distribution* if for every integer $n \geq 1$, $Y \stackrel{D}{=} (X_{n1} + \cdots + X_{nn} - \beta_n)/\alpha_n$, where X_{ni} are i.i.d. and $\alpha_n > 0$ and β_n are constants. In this case, α_n must be of the form $n^{1/\alpha}$ for some $0 < \alpha \leq 2$; α is called the *index* of the stable distribution. For the sum S_n of i.i.d. random variables $X_1, \ldots, X_n$, if $(S_n - b_n)/a_n$ converges in distribution for some constants $a_n \neq 0$ and b_n, then the limiting distribution must be stable.

The following theorem (see Durrett, 2005, p. 151) gives necessary and sufficient conditions for the common distribution of X_i to belong to the *domain of attraction* of a stable distribution with exponent $0 < \alpha < 2$ (i.e., for $(S_n - b_n)/a_n$ to converge weakly to the stable distribution). A function $L : (0, \infty) \to \mathbb{R}$ is said to be *slowly varying* (at ∞) if

$$\lim_{x \to \infty} \frac{L(cx)}{L(x)} = 1 \qquad \text{for all } c > 0. \tag{2.36}$$

Theorem 2.24. *X belongs to the domain of attraction of a stable distribution with exponent $0 < \alpha < 2$ if and only if there exist $0 \leq \theta \leq 1$ and a slowly varying function L such that*

(a) $P(|X| \geq x) = x^{-\alpha} L(x)$,
(b) $\lim_{x \to \infty} P(X \geq x)/P(|X| \geq x) = \theta$.

In this case, $(S_n - b_n)/a_n$ converges weakly to the stable distribution, with

$$a_n = \inf\{x : \ P(|X| \geq x) \leq n^{-1}\}, \qquad b_n = nEXI(|X| \leq a_n).$$

There are analogous results for the domain of attraction of a normal distribution (with $\alpha = 2$). Further details on slowly varying functions and domain of attraction are given in Chap. 4 where we consider weak convergence of self-normalized sums of i.i.d. random variables.

2.3.2 Smoothing, Local Limit Theorems and Expansions

As noted in the previous subsection, the usual proof of the central limit theorem for sums of independent random variables involves quadratic approximations of the characteristic functions. Higher-order Taylor expansions of the characteristic functions will lead to more redefined approximations of the characteristic function of $W_n = S_n/B_n$ in Sect. 16.2.1. The Fourier inversion formula (2.32) can be used to derive asymptotic expansions of $f_n - \phi$, where f_n is the density function of W_n and ϕ is the standard normal density, if the characteristic functions of Ee^{itX_k} are integrable. Without such integrability assumptions, we can still perform a modified Fourier inversion to estimate the difference between the distribution functions of W_n and $N(0,1)$, by using the following bound that was first introduced to prove the Berry–Esseen inequality (Theorem 2.10).

Theorem 2.25. *Let F be a distribution function and G a function such that $G(-\infty)=0$, $G(\infty)=1$ and $|G'(x)|\le C<\infty$. Then*

$$\sup_x |F(x)-G(x)| \le \frac{1}{\pi}\int_{-T}^{T}\left|\frac{\varphi(t)-\gamma(t)}{t}\right| dt + \frac{24C}{T} \tag{2.37}$$

for $T>0$, where $\varphi(t)=\int_{-\infty}^{\infty} e^{itx}dF(x)$ and $\gamma(t)=\int_{-\infty}^{\infty} e^{itx}dG(x)$ are the Fourier transforms of F and G, respectively.

For details of the proof of Theorem 2.25, see Chow and Teicher (1988, pp. 301–302). Here we summarize the main "smoothing" idea behind the bound. First note that if U is independent of Y and has an integrable characteristic function, then the characteristic function of $Y+U$ (which is a product of the two individual characteristic functions) is integrable and therefore the Fourier inversion formula (2.32) can be used to evaluate the density function (and hence also the distribution function) of $Y+U$. Choosing U such that its characteristic function vanishes outside $[-T,\ T]$ and its density function is concentrated around 0 (so that U is small) is the basic idea behind (2.37). Specifically, we choose U with density function

$$u_T(x)=\frac{1}{\pi}\frac{1-\cos(xT)}{x^2T}, \qquad -\infty<x<\infty, \tag{2.38}$$

whose characteristic function is $\omega_T(t)=(1-|t|/T)I(|t|\le T)$. Instead of the distribution function of $Y+U$ in the preceding discussion, we let $\Delta=F-G$ and consider the more general convolution $\int_{-\infty}^{\infty}\Delta(t-x)u_T(x)dx$, noting that the Fourier transform of $F-G$ is $\varphi-\gamma$. In Chap. 16, we apply Theorem 2.25 to derive Edgeworth expansions related to central limit theorems for a wide variety of statistics. Here we illustrate its application and that of the smoothing density (2.38) in the following.

Theorem 2.26. *Let $X, X_1, X_2, \ldots X_n$ be i.i.d. random variables with $EX=0$ and $EX^2=\sigma^2>0$, and assume that X is non-lattice. Let $S_n=X_1+\cdots+X_n$.*

(a) If $E|X|^3<\infty$, then

$$P\left(\frac{S_n}{\sigma\sqrt{n}}\le x\right)=\Phi(x)-\frac{EX^3}{6\sigma^3\sqrt{n}}(x^2-1)\phi(x)+o\left(\frac{1}{\sqrt{n}}\right) \tag{2.39}$$

as $n\to\infty$, uniformly in $x\in\mathbb{R}$.

(b) For any $x\in\mathbb{R}$ and $h>0$,

$$P\left(\sqrt{n}x\le S_n\le\sqrt{n}x+h\right)=\frac{h+o(1)}{\sqrt{n}}\frac{1}{\sigma}\phi\left(\frac{x}{\sigma}\right) \qquad \text{as } n\to\infty. \tag{2.40}$$

Proof. To prove (2.39), we apply Theorem 2.25 with F equal to the distribution function of $S_n/(\sigma\sqrt{n})$, $G(x)$ equal to the right hand side of (2.39) with the $o(1/\sqrt{n})$ term removed, and $T=a\sqrt{n}$, where for given $\varepsilon>0$, a is chosen so that $24|G'(x)|<\varepsilon a$ for all x. The Fourier transform of G is

$$\gamma(t) = e^{-t^2/2}\left\{1 + \frac{EX^3}{6\sigma^3\sqrt{n}}(it)^3\right\}.$$

Hence by Theorem 2.25,

$$\sup_x |F(x) - G(x)| \le \int_{-a\sqrt{n}}^{a\sqrt{n}} \left| \frac{\phi^n(t/\sigma\sqrt{n}) - \gamma(t)}{t} \right| dt + \frac{\varepsilon}{\sqrt{n}}. \tag{2.41}$$

Split the integral into $\int_{\delta\sigma \le |t|/\sqrt{n} \le a} + \int_{|t| \le \delta\sigma\sqrt{n}}$, which we denote by $(I) + (II)$. Since X is non-lattice, $\max_{\delta\sigma \le |t|/\sqrt{n} \le a} |\varphi(t/\sigma\sqrt{n})| < 1$, from which it follows that $(I) = O(\rho^n)$ for some $0 < \rho < 1$. For $|t| \le \delta\sigma\sqrt{n}$ with $\delta > 0$ chosen sufficiently small, we can apply the Taylor expansion to $\varphi(t/\sigma\sqrt{n})$ and choose n sufficiently large so that $(II) < \varepsilon/\sqrt{n}$. Since ε is arbitrary, (2.39) follows from (2.41). The proof of (2.40) does not use Theorem 2.25; it uses the smoothing density (2.38) more directly in the Fourier inversion formula (2.32); see Durrett (2005, pp. 132–134) for details. □

Theorem 2.26(b) is called a *local limit theorem*. It says that the probability that S_n belongs to an interval of width $h/\sqrt{n}$ is asymptotically equal to the $N(0,1)$ density function at any value of interval multiplied by the width $h/\sqrt{n}$. In the case of lattice X with support $\{b + hk : k = 0, \pm 1, \pm 2, \dots\}$, we can apply the Fourier inversion formula (2.32) to obtain a similar result for the probability density of S_n. The preceding ideas and results can be readily extended to the multivariate case. The characteristic function φ of a $d \times 1$ random vector Y is given by $\varphi(t) = Ee^{it'Y}$, $t \in \mathbb{R}^d$. There are corresponding Fourier inversion formulas, smoothing bounds and local limit theorems. In particular, Stone (1965) has established a local limit theorem for the sum S_n of i.i.d. random vectors Xn that belong to the domain of attraction of a multivariate stable distribution.

2.4 Supplementary Results and Problems

1. Prove the following relation among the three models of convergence:

$$Y_n \xrightarrow{a.s.} Y \Longrightarrow Y_n \xrightarrow{P} Y \Longrightarrow Y_n \xrightarrow{D} Y.$$

   Also give counterexamples to show that the reverse relations are not true in general.
2. Prove that if $W_n \xrightarrow{D} N(0,1)$, then $\sup_x |P(W_n \le x) - \Phi(x)| \to 0$ as $n \to \infty$.
3. Let $\{X_i,\ 1 \le i \le n\}$ be independent random variables. Prove that for $x > 0$,

$$P(\max_{1 \le k \le n} |S_k| \ge x) \le 3 \max_{1 \le k \le n} P(|S_k| \ge x/3).$$

4. Montgomery-Smith (1993): Let $X_1, \dots, X_n$ be i.i.d. random variables. Prove that for $x > 0$,

$$\max_{1 \le k \le n} P(|S_k| \ge x) \le 3P(|S_n| \ge x/10).$$

5. Let $\{X_i,\ 1 \le i \le n\}$ be independent random variables with $E(X_i) = 0$ and $E|X_i|^p < \infty$, where $1 < p \le 3$. Prove that

$$E|S_n|^p \le 2^{2-p} \sum_{i=1}^{n} E|X_i|^p \qquad \text{for } 1 < p \le 2,$$

and

$$E|S_n|^p \le (p-1)B_n^p + \sum_{i=1}^{n} E|X_i|^p \qquad \text{for } 2 < p \le 3,$$

where $B_n^2 = \sum_{i=1}^n E|X_i|^2$.
Hint: $|1+x|^p \le 1 + px + 2^{2-p}|x|^p$ for $1 < p \le 2$ and $x \in \mathbb{R}$.

6. Prove Rosenthal's inequality: *Let $p \ge 2$ and let $X_1, \ldots, X_n$ be independent random variables with $EX_i = 0$ and $E|X_i|^p < \infty$ for $1 \le i \le n$. Then there exists a constant A_p depending only on p such that*

$$E|S_n|^p \le A_p \left((ES_n^2)^{p/2} + \sum_{i=1}^{n} E|X_i|^p \right). \tag{2.42}$$

Hint: Write $E|S_n|^p = \int_0^\infty px^{p-1}P(S_n > x)dx + \int_0^\infty px^{p-1}P(-S_n > x)dx$ and apply (2.24).

7. Let $X_1, \ldots, X_n$ be i.i.d. random variables with $E(X_i) = 0$ and $\mathrm{Var}(X_i) = 1$. The *functional form* of the central limit theorem (Theorem 2.9) says that if we define for $0 \le t \le 1$,

$$W_n(t) = \begin{cases} S_i/\sqrt{n} & \text{for } t = i/n \quad (S_0 = 0), \\ \text{linear} & \text{for } i/n \le t \le (i+1)/n, \end{cases} \tag{2.43}$$

then W_n converges weakly to Brownian motion W in $C[0,1]$ (the space of real-valued continuous functions on $[0,1]$ with metric $\rho(f,g) = \max_{0 \le x \le 1} |f(x) - g(x)|$), i.e., $Ef(W_n) \to Ef(W)$ for all bounded continuous functions $f : C[0,1] \to \mathbb{R}$; see Durrett (2005, pp. 401–407). In particular, use this result to show that

$$n^{-3/2} \sum_{i=1}^{n} (n-i)X_i \Longrightarrow N(0, 1/3). \tag{2.44}$$

Also prove (2.44) by applying the Lindeberg–Feller theorem.

8. There is also a *functional LIL* due to Strassen; see Durrett (2005, p. 435): With the same notation and assumptions as in the preceding problem, let $Z_n(\cdot) = W_n(\cdot)/\sqrt{2 \log\log n}$ for $n \ge 3$. Then with probability 1, $\{Z_n,\ n \ge 3\}$ is relatively compact in $C[0,1]$ and its set of limit points in $C[0,1]$ is

$$\left\{ f \in C[0,1] : f(0) = 0,\ f \text{ is absolutely continuous and } \int_0^1 (f'(t))^2 dt \le 1 \right\}. \tag{2.45}$$

The set (2.45) is related to Brownian motion (see the preceding problem) as the unit ball of the reproducing kernel Hilbert space of the covariance function $\mathrm{Cov}(W(t),W(s)) = t \wedge s$ of Brownian motion. Making use of the functional LIL, prove the *compact LIL* that the set of limit points of $\{Z_n(1),\ n \geq 3\}$ is $[-1,1]$ and find the set of limit points of $\{n^{-3/2}\sum_{i=1}^{n}(n-i)X_i/\sqrt{\log\log n},\ n \geq 3\}$.

9. Another refinement of Theorem 2.4 is the *upper-lower class test*. A sequence b_n of positive numbers such that $b_n/\sqrt{n}$ is nondecreasing for all large n is said to belong to the *upper class* with respect to the random walk $\{S_n\}$ if $P\{S_n - n\mu \geq b_n\ i.o.\} = 0$ and to the *lower class* otherwise. Show that

$$P\{S_n - n\mu \geq b_n\ i.o.\} = \begin{cases} 0 & \text{if } \sum_{n=1}^{\infty} n^{-3/2} b_n \exp\left(-b_n^2/(2\sigma^2 n)\right) < \infty, \\ \infty & \text{otherwise.} \end{cases}$$

Chapter 3
Self-Normalized Large Deviations

In this chapter we first review the classical large deviation theorem (LDT) for sums of i.i.d. random variables $X_1, X_2, \ldots, X_n$. As shown in Sect. 3.1, a key ingredient of LDT is a finite moment generating function of X_1 in a right neighborhood of the origin. Surprisingly, Shao (1997) shows that for self-normalized sums of the X_i, the LDT holds without any moment assumption on X_i. The main results and proofs of this self-normalized large deviation theory are given in Sect. 3.2.

3.1 A Classical Large Deviation Theorem for Sample Sums

Let $X, X_1, \ldots, X_n$ be i.i.d. random variables with $P(X \neq 0) > 0$ and let $S_n = \sum_{i=1}^n X_i$. The Cramér–Chernoff large deviation theorem states that if

$$Ee^{\theta_0 X} < \infty \qquad \text{for some } \theta_0 > 0 \tag{3.1}$$

then for every $x > EX$,

$$\lim_{n\to\infty} n^{-1} \log P\left(\frac{S_n}{n} \geq x\right) = \log \rho(x),$$

or equivalently,

$$\lim_{n\to\infty} P\left(\frac{S_n}{n} \geq x\right)^{1/n} = \rho(x), \tag{3.2}$$

where $\rho(x) = \inf_{t\geq 0} e^{-tx} Ee^{tX}$; see Bahadur and Ranga Rao (1960) who give references to previous literature and also develop a more precise estimate of $P(S_n/n \geq x)$ as described below. See also Dembo and Zeitouni (1998) for subsequent extensions and related large deviation results.

Let $\bar{X}_n = S_n/n$. Bahadur and Ranga Rao (1960) make use of a change of measures to prove the Cramér–Chernoff large deviation theorem and derive a more precise asymptotic approximation to $P(\bar{X}_n \geq x)$, which is often called *exact asymptotics* for

© Springer-Verlag Berlin Heidelberg 2009

large deviations; see Dembo and Zeitouni (1998). Letting $e^{\psi(\theta)} = Ee^{\theta X}$, the basic idea is to embed P in a family of measures P_θ under which $X_1, X_2, \dots$ are i.i.d. with density function $f_\theta(x) = e^{\theta x - \psi(\theta)}$ with respect to P. Then for any event A,

$$P(A) = \int_A \frac{dP}{dP_\theta} dP_\theta = E_\theta \left\{ e^{-(\theta S_n - n\psi(\theta))} I(A) \right\}, \tag{3.3}$$

since the Radon–Nikodym derivative (or likelihood ratio) dP_θ/dP is equal to $\prod_{i=1}^n f_\theta(X_i) = e^{\theta S_n - n\psi(\theta)}$.

The family of density functions f_θ is an exponential family with the following properties:

$$E_\theta X = \psi'(\theta), \qquad \mathrm{Var}_\theta(X) = \psi''(\theta). \tag{3.4}$$

The change-of-measure formula (3.3) is a special case of likelihood ratio identities which will be discussed further in Sect. 18.1.1. In particular, for $A = \{\bar{X}_n \geq x\}$ with $x > EX$, we choose $\theta = \theta_x$ such that $E_\theta X = x$ (and therefore $x = \psi'(\theta)$ by (3.4)). For this choice of θ in (3.3), which is often called the "conjugate method" (see Petrov, 1965),

$$\begin{aligned} &E_{\theta_x} \left\{ e^{-n(\theta_x \bar{X}_n - \psi(\theta_x))} I(\bar{X}_n \geq x) \right\} \\ &\quad = e^{-n(\theta_x x - \psi(\theta_x))} E_{\theta_x} \left\{ e^{-n\theta_x(\bar{X}_n - x)} I(\bar{X}_n \geq x) \right\} \\ &\quad = e^{-n\mathscr{I}(x)} E_{\theta_x} \left\{ e^{-\sqrt{n}\,\theta_x(\sqrt{n}(\bar{X}_n - x))} I(\bar{X}_n \geq x) \right\}, \end{aligned} \tag{3.5}$$

where

$$\mathscr{I}(x) := \theta_x x - \psi(\theta_x) = \sup_\theta (\theta x - \psi(\theta)); \tag{3.6}$$

see Problem 3.1. Since $x = E_{\theta_x}(X)$, $Z_n := \sqrt{n}(\bar{X}_n - x)$ converges in distribution to $N(0, \mathrm{Var}_{\theta_x}(X))$ under P_{θ_x} by the central limit theorem. However, because (3.5) involves $E_{\theta_x} e^{-\sqrt{n}\theta_x Z_n} I(Z_n \geq 0)$, which contains $-\theta_x \sqrt{n} Z_n$ in the exponent, we need a local limit theorem for Z_n when it is in a $O(1/\sqrt{n})$-neighborhood of 0. If X is non-lattice, we can use Theorem 2.26(b) to conclude from (3.5) that

$$\begin{aligned} P(\bar{X}_n \geq x) &= e^{-n\mathscr{I}(x)} E_{\theta_x} \left\{ e^{-\theta_x \sqrt{n} Z_n} I(Z_n \geq 0) \right\} \\ &\sim \frac{e^{-n\mathscr{I}(x)}}{\sigma_x \sqrt{2\pi n}} \int_0^\infty e^{-\theta_x h} dh = \frac{e^{-n\mathscr{I}(x)}}{\theta_x \sigma_x \sqrt{2\pi n}}, \end{aligned} \tag{3.7}$$

where $\sigma_x^2 = \mathrm{Var}_{\theta_x}(X) = \psi''(\theta_x)$ by (3.4); see Bahadur and Ranga Rao (1960) who have also obtained higher-order expansions for $P(\bar{X}_n \geq x)$ when X is strongly non-lattice and when X is lattice.

3.2 A Large Deviation Theorem for Self-Normalized Sums

In this section we prove Shao's (1997) large deviation theorem for the self-normalized sum S_n/V_n without any moment assumption.

Theorem 3.1. *Assume that either $EX \geq 0$ or $EX^2 = \infty$. Let $V_n^2 = \sum_{i=1}^n X_i^2$. Then*

$$\lim_{n\to\infty} P(S_n \geq x\sqrt{n}V_n)^{1/n} = \sup_{b\geq 0}\inf_{t\geq 0} Ee^{t(bX - x(X^2+b^2)/2)} \tag{3.8}$$

for $x > EX/(EX^2)^{1/2}$, where $EX/(EX^2)^{1/2} = 0$ if $EX^2 = \infty$.

Corollary 3.2. *Assume that either $EX = 0$ or $EX^2 = \infty$. Then*

$$\lim_{n\to\infty} P\left(S_n \geq x\sqrt{n}V_n\right)^{1/n} = \sup_{b\geq 0}\inf_{t\geq 0} Ee^{t(bX - x(X^2+b^2)/2)} \tag{3.9}$$

for $x > 0$.

Note that for any random variable X, either $EX^2 < \infty$ or $EX^2 = \infty$. If $EX^2 < \infty$, which obviously implies $E|X| < \infty$, the assumption $EX \geq 0$ in Theorem 3.1 is clearly needed to describe large deviation probabilities for S_n. Because Theorem 3.1 is also valid when $EX^2 = \infty$, it holds without any moment conditions. Shao (1998) gives a review of further results in this direction. We first outline Shao's main ideas and then provide details of his proof of Theorem 3.1. In Sect. 10.1.1 we describe an alternative approach that is related to the general framework of Part II.

3.2.1 Representation by Supremum over Linear Functions of (S_n, V_n^2)

We start with normalization by $aV_n^2 + nb$ (instead of by $\sqrt{n}V_n$) that reveals the key idea of the proof, where $a > 0$ and $b \geq 0$. Thus we consider $S_n/(aV_n^2 + nb)$ and observe that for $x > 0$,

$$\left\{\frac{S_n}{aV_n^2 + nb} \geq x\right\} = \{S_n - axV_n^2 \geq nbx\} = \left\{\frac{1}{n}\sum_{i=1}^n (X_i - axX_i^2) \geq bx\right\}, \tag{3.10}$$

and that $Ee^{t(X - axX^2)} < \infty$ for $t > 0$. Therefore, by the Cramér–Chernoff large deviation result (3.2),

$$P\left(\frac{S_n}{aV_n^2 + nb} \geq x\right)^{1/n} \longrightarrow \inf_{t\geq 0} e^{-tbx} Ee^{t(X - axX^2)} \tag{3.11}$$

provided $bx > E(X) - axE(X^2)$. This suggests that if one can write $n^{1/2}V_n$ in terms of a functional of $aV_n^2 + bn$, then one should be able to prove (3.8). Since for any positive numbers x and y,

$$xy = \inf_{b>0} \frac{1}{2}\left(\frac{x^2}{b} + y^2 b\right), \tag{3.12}$$

we have the representation

$$\sqrt{n} V_n = \inf_{b>0} \frac{1}{2b}(V_n^2 + nb^2) \qquad \text{if } V_n > 0, \tag{3.13}$$

thereby expressing $\sqrt{n} V_n$ as an extremal functional of $b^{-1}V_n^2 + bn$. From (3.13), it follows that

$$\begin{aligned} &\{S_n \geq x\sqrt{n} V_n\} \qquad (3.14) \\ &= \left\{S_n \geq x \inf_{b>0} \frac{1}{2b}(V_n^2 + nb^2) \text{ or } V_n = 0\right\} \\ &= \left\{\sup_{b>0} \sum_{i=1}^{n} (bX_i - x(X_i^2 + b^2)/2) \geq 0 \text{ or } V_n = 0\right\} \\ &= \left\{\sup_{b\geq 0} \sum_{i=1}^{n} (bX_i - x(X_i^2 + b^2)/2) \geq 0\right\}. \end{aligned}$$

Note the resemblance of (3.14) to (3.5). Both events involve linear functions of (S_n, V_n^2). The nonlinearity in V_n (which is the square root of V_n^2) results in the supremum of a family, parameterized by b, of linear functions of (S_n, V_n^2).

3.2.2 *Proof of Theorem 3.1*

Lemma 3.3. *If $g_1(t), g_2(t), \cdots$ is a non-increasing sequence of functions continuous in the closed interval $[a,b]$, then*

$$\lim_{n\to\infty} \sup_{a\leq t\leq b} g_n(t) = \sup_{a\leq t\leq b} g(t), \tag{3.15}$$

where $g(t) = \lim_{n\to\infty} g_n(t)$.

Proof. It suffices to show that

$$\lim_{n\to\infty} \sup_{a\leq t\leq b} g_n(t) \leq \sup_{a\leq t\leq b} g(t).$$

Since $g_n(t)$ is continuous in the closed interval $[a,b]$, there exists $t_n \in [a,b]$ such that

$$\sup_{a\leq t\leq b} g_n(t) = g_n(t_n). \tag{3.16}$$

By the Bolzano–Weierstrass theorem, there exists a subsequence $\{t_{k_n}\}$ of $\{t_n\}$ that converges to a point $t_0 \in [a,b]$. Since $g_n(t_0) \to g(t_0)$, for any $\varepsilon > 0$, there exists n_0 such that

$$g_{k_{n_0}}(t_0) \le g(t_0) + \varepsilon. \tag{3.17}$$

Furthermore, $g_{k_{n_0}}$ is continuous at t_0, there exists a $\delta > 0$ such that for all $|t - t_0| \le \delta$,

$$|g_{k_{n_0}}(t) - g_{k_{n_0}}(t_0)| \le \varepsilon.$$

Note that $t_{k_n} \to t_0$, there exists $m_0 \ge n_0$ such that for $n \ge m_0$, $|t_{k_n} - t_0| \le \delta$. Hence for $n \ge m_0$

$$g_{k_{n_0}}(t_{k_n}) \le g_{k_{n_0}}(t_0) + \varepsilon. \tag{3.18}$$

Since g_n is non-increasing, we have

$$\lim_{n\to\infty} g_n(t_n) = \lim_{n\to\infty} g_{k_n}(t_{k_n})$$

and for $n \ge m_0$,

$$g_{k_n}(t_{k_n}) \le g_{k_{n_0}}(t_{k_n}) \le g_{k_{n_0}}(t_0) + \varepsilon \le g(t_0) + 2\varepsilon \le \sup_{a\le t\le b} g(t) + 2\varepsilon. \tag{3.19}$$

This proves that $\lim_{n\to\infty} g_{k_n}(t_{k_n}) \le \sup_{a\le t\le b} g(t)$, as desired. □

Note that for $x > EX/(EX^2)^{1/2}$ (≥ 0) and for $b \ge 0$,

$$Ee^{t(bX - x(X^2+b^2)/2)} < \infty \qquad \text{for all } t \ge 0$$

and

$$\begin{aligned} &E\left(bX - x(X^2+b^2)/2\right) \\ &= \begin{cases} -\infty & \text{if } EX^2 = \infty, \\ -\frac{x}{2}\left(b - (EX)/x\right)^2 - \frac{1}{2}\left(xEX^2 - (EX)^2/x\right) < 0 & \text{if } EX^2 < \infty. \end{cases} \end{aligned}$$

Therefore, by (3.14) and (3.2)

$$\begin{aligned} &\liminf_{n\to\infty} P\left(S_n \ge x\sqrt{n}V_n\right)^{1/n} \qquad (3.20)\\ &\ge \liminf_{n\to\infty} \sup_{b\ge 0} P\left(\sum_{i=1}^n \left(bX_i - x(X_i^2+b^2)/2\right) \ge 0\right)^{1/n} \\ &\ge \sup_{b\ge 0}\inf_{t\ge 0} Ee^{t(bX - x(X^2+b^2)/2))}. \end{aligned}$$

To complete the proof of (3.8), it remains to show that

$$\limsup_{n\to\infty} P\left(S_n \ge x\sqrt{n}V_n\right)^{1/n} \le \sup_{b\ge 0}\inf_{t\ge 0} Ee^{t(bX - x(X^2+b^2)/2)}. \tag{3.21}$$

Let $a \geq 1$ and $m = [2a/x] + 1$, where $[y]$ denotes the integer part of y. Then

$$P\left(S_n \geq x\sqrt{n}V_n\right) = I_1 + I_2, \tag{3.22}$$

where

$$I_1 = P\left(S_n \geq x\sqrt{n}V_n, V_n > m\sqrt{n}\right), \qquad I_2 = P\left(S_n \geq x\sqrt{n}V_n, V_n \leq m\sqrt{n}\right).$$

Noting that $na \leq x\sqrt{n}V_n/2$ for $V_n > m\sqrt{n}$, we have

$$\begin{aligned}
I_1 &\leq P\left(\sum_{i=1}^n X_i I(|X_i| > a) + na \geq x\sqrt{n}V_n, V_n > m\sqrt{n}\right) \\
&\leq P\left(\sum_{i=1}^n X_i I(|X_i| > a) \geq (1/2)x\sqrt{n}V_n\right) \\
&\leq P\left(\sum_{i=1}^n I(|X_i| > a) \geq x^2 n/4\right) \\
&\leq \left\{12x^{-2}P(|X| > a)\right\}^{x^2 n/4},
\end{aligned} \tag{3.23}$$

where the last inequality follows from (2.27). Therefore

$$\limsup_{n\to\infty} I_1^{1/n} \leq \left(12x^{-2}P(|X| > a)\right)^{x^2/4}. \tag{3.24}$$

We next estimate I_2. Noting that $\sqrt{n}V_n = \inf_{0 \leq b \leq V_n/\sqrt{n}}(bn + V_n^2/b)/2$, we have

$$\begin{aligned}
I_2 &\leq P\left(\sup_{0 \leq b \leq m}\left\{bS_n - x(V_n^2 + nb^2)/2\right\} \geq 0\right) \\
&\leq \sum_{i=1}^{nm} P\left(\sup_{(i-1)/n \leq b \leq i/n}\left\{bS_n - x(V_n^2 + nb^2)/2\right\} \geq 0\right) \\
&\leq \sum_{i=1}^{mn} P\left((i/n)S_n - x\left(V_n^2 + n((i-1)/n)^2\right)/2 \geq 0\right) \\
&\leq \sum_{i=1}^{mn} P\left((i/n)S_n - x\left(V_n^2 + n(i/n)^2\right)/2 \geq -xm\right) \\
&\leq \sum_{i=1}^{mn} \inf_{0 \leq t \leq k} e^{xmt}\left\{E\exp\left(t(i/n)X - (x/2)\left(X^2 + (i/n)^2\right)\right)\right\}^n \\
&\leq mne^{xmk} \sup_{0 \leq b \leq m}\inf_{0 \leq t \leq k}\left\{E\exp\left(t\left(bX - (x/2)(X^2 + b^2)\right)\right)\right\}^n
\end{aligned} \tag{3.25}$$

for any $k \geq 1$. Hence

$$\limsup_{n\to\infty} I_2^{1/n} \leq \sup_{0 \leq b \leq m}\inf_{0 \leq t \leq k} E\exp\left(t\left(bX - (x/2)(X^2 + b^2)\right)\right). \tag{3.26}$$

By Lemma 3.3,

$$\lim_{k\to\infty} \sup_{0\le b\le m} \inf_{0\le t\le k} E\exp\left(t\left(bX-(x/2)(X^2+b^2)\right)\right)$$
$$= \sup_{0\le b\le m} \inf_{0\le t<\infty} E\exp\left(t\left(bX-(x/2)(X^2+b^2)\right)\right),$$

and therefore

$$\limsup_{n\to\infty} I_2^{1/n} \le \sup_{0\le b\le m} \inf_{t\ge 0} E\exp\left(t\left(bX-(x/2)(X^2+b^2)\right)\right) \tag{3.27}$$
$$\le \sup_{0\le b<\infty} \inf_{t\ge 0} E\exp\left(t\left(bX-(x/2)(X^2+b^2)\right)\right).$$

Letting $a\to\infty$ together with (3.22) and (3.27) yields (3.21). This completes the proof of Theorem 3.1.

3.3 Supplementary Results and Problems

Let $X, X_1, X_2 \ldots$ be i.i.d. random variables and $S_n = \sum_{i=1}^n X_i$:

1. Prove (3.4) and the equality in (3.6).
2. Prove the following generalization of Theorem 3.1: Let μ and ν be two real numbers. Assume that either $EX \ge \mu$ or $EX^2 = \infty$. Then

$$\lim_{n\to\infty} P\left(\frac{\sum_{i=1}^n (X_i-\mu)}{\left(n\sum_{i=1}^n (X_i-\nu)^2\right)^{1/2}} \ge x\right)^{1/n} = \sup_{b\ge 0} \inf_{t\ge 0} Ee^{t(b(X-\mu)-x((X-\nu)^2+b^2)/2)}$$

   for $x > (EX-\mu)/\sqrt{E(X-\nu)^2}$.
3. Let $p>1$. Prove that for any $x>0$ and $y>0$,

$$x^{1/p}y^{1-1/p} = \inf_{b>0}\left(\frac{1}{p}\cdot\frac{x}{b}+\frac{p-1}{p}yb^{1/(p-1)}\right). \tag{3.28}$$

4. Let $p>1$. Assume that either $EX\ge 0$ or $E|X|^p=\infty$. Making use of (3.28), modify the proof of Theorem 3.1 to generalize it to the following result of Shao (1997):

$$\lim_{n\to\infty} P\left(\frac{S_n}{V_{n,p}n^{1-1/p}} \ge x\right)^{1/n} = \sup_{c\ge 0}\inf_{t\ge 0} Ee^{t(cX-x(\frac{1}{p}|X|^p+\frac{p-1}{p}c^{p/(p-1)}))} \tag{3.29}$$

   for $x > EX/(E|X|^p)^{1/p}$, where $V_{n,p} = (\sum_{i=1}^n |X_i|^p)^{1/p}$ and $EX/(E|X|^p)^{1/p}=0$ if $E|X|^p=\infty$.

5. Assume that $E(X) = 0$ or $E(X^2) = \infty$. Let $a_i, 1 \le i \le n$, be a sequence of real numbers. Under what conditions on $\{a_i\}$ does the result

$$\lim_{n\to\infty} n^{-1} \log P\left(\sum_{i=1}^{n} a_i X_i \ge x\sqrt{n}(\sum_{i=1}^{n} a_i^2 X_i^2)^{1/2}\right)$$
$$= \lim_{n\to\infty} n^{-1} \sup_{b\ge 0} \inf_{t\ge 0} \sum_{i=1}^{n} \log E e^{t(ba_i X - x(a_i^2 X^2 + b^2)/2)}$$

hold for $0 < x < 1$?

6. Let $\{a_i\}$ and $\{b_i\}$ be two sequences of real numbers with $b_i > 0$. Under what conditions does a large deviation result hold for $\sum_{i=1}^{n} a_i X_i / (\sum_{i=1}^{n} b_i^2 X_i^2)^{1/2}$?

Chapter 4
Weak Convergence of Self-Normalized Sums

In this chapter we prove a self-normalized central limit theorem for i.i.d. random variables belonging to the domain of a normal distribution. We also prove a related functional central limit theorem for self-normalized sums and describe analogous results that the i.i.d. random variables belonging to the domain of attraction of a stable distribution with index $0 < \alpha < 2$.

4.1 Self-Normalized Central Limit Theorem

Let $X, X_1, X_2, \ldots$ be i.i.d. random variables. Set $S_n = \sum_{i=1}^n X_i$ and $V_n^2 = \sum_{i=1}^n X_i^2$. As noted in Sect. 2.3.1, X belongs to the domain of attraction of a normal distribution if there exist $a_n > 0$ and b_n such that

$$(S_n - b_n)/a_n \xrightarrow{D} N(0,1). \tag{4.1}$$

An equivalent condition that involves the tail of $|X|$ is

$$l(x) := EX^2 I(|X| \le x) \text{ is a slowly varying function;} \tag{4.2}$$

see Ibragimove and Linnik (1971). Recall the definition (2.36) of "slowly varying" (at ∞) and note that the normal distribution is stable with index $\alpha = 2$. When X is in the domain of attraction of a normal distribution, it is known that b_n in (4.1) can be taken as $nE(X)$ and a_n is so chosen that $a_n^{-2} EX^2 I(|X| \le a_n) \sim n^{-1}$.

The next theorem is due to Giné et al. (1997). It was conjectured by Logan et al. (1973) that if $EX = 0$, then "S_n/V_n is asymptotically normal if (and perhaps only if) X is in the domain of attraction of the normal law." The "if" part has by now been known for a long time (see, e.g., Maller, 1981), but the "only if" part has remained open until Giné et al. (1997).

V.H. de la Peña et al., *Self-Normalized Processes: Limit Theory and Statistical Applications*, Probability and its Applications,

© Springer-Verlag Berlin Heidelberg 2009

Theorem 4.1. $S_n/V_n \xrightarrow{D} N(0,1)$ *if and only if* $E(X)=0$ *and* X *is in the domain of attraction of a normal distribution.*

We start with some basic properties of a positive slowly varying function $l(x)$; see Bingham et al. (1987).

(P1) $l(x)$ is representable in the form $l(x)=c(x)\exp\left(\int_1^x \frac{a(y)}{y}dy\right)$, where $c(x)\to c>0$, for some c, and $a(x)\to 0$ as $x\to\infty$.

(P2) For $0<c<C<\infty$, $\lim_{x\to\infty}\frac{l(tx)}{l(x)}=1$ uniformly in $c\le t\le C$.

(P3) $\forall\,\varepsilon>0$, $\lim_{x\to\infty}x^{-\varepsilon}l(x)=0$ and $\lim_{x\to\infty}x^{\varepsilon}l(x)=\infty$.

(P4) For any $\varepsilon>0$, there exists x_0 such that for all x, $xt\ge x_0$

$$(1-\varepsilon)\left(t\vee\frac{1}{t}\right)^{-\varepsilon}\le\frac{l(tx)}{l(x)}\le(1+\varepsilon)\left(t\vee\frac{1}{t}\right)^{\varepsilon},$$

$$\left|\frac{l(tx)}{l(x)}-1\right|\le 2\left(\left(t\vee\frac{1}{t}\right)^{\varepsilon}-1\right).$$

(P5) For any $\theta>-1$, $\int_a^x y^\theta l(y)\,dy\sim\frac{x^{\theta+1}l(x)}{\theta+1}$ as $x\to\infty$.

(P6) For any $\theta<-1$, $\int_x^\infty y^\theta l(y)\,dy\sim\frac{x^{\theta+1}l(x)}{-\theta-1}$ as $x\to\infty$.

Lemma 4.2. *Assume that* $l(x):=EX^2I(|X|\le x)$ *is slowly varying. Then as* $x\to\infty$,

$$P(|X|\ge x)=o\left(l(x)/x^2\right), \tag{4.3}$$

$$E|X|I(|X|\ge x)=o(l(x)/x), \tag{4.4}$$

$$E|X|^pI(|X|\le x)=o\left(x^{p-2}l(x)\right)\qquad \textit{for } p>2. \tag{4.5}$$

Proof. Note that

$$\begin{aligned}
P(|X|\ge x)&=\sum_{k=0}^{\infty}P\left(2^kx\le|X|<2^{k+1}x\right)\\
&\le\sum_{k=0}^{\infty}(x2^k)^{-2}EX^2I\left(2^kx\le|X|\le 2^{k+1}x\right)\\
&=x^{-2}l(x)\sum_{k=0}^{\infty}\left(\frac{l(2^{k+1}x)}{l(2^kx)}-1\right)\frac{l(2^kx)}{2^{2k}l(x)}\\
&=x^{-2}l(x)\sum_{k=0}^{\infty}o(1)2^{-k}\qquad\text{by (P2) and (P4)}\\
&=o\left(l(x)/x^2\right).
\end{aligned}$$

This proves (4.3). To prove (4.4), application of (4.3) and (P6) yields

$$\begin{aligned} E|X|I(|X| \geq x) &= xP(|X| \geq x) + \int_x^\infty P(|X| \geq t)\,dt \\ &= o(l(x)/x) + o(1)\int_x^\infty l(t)/t^2 dt = o(l(x)/x) \end{aligned}$$

Similarly, for any $0 < \varepsilon < 1$,

$$\begin{aligned} E|X|^p I(|X| \leq x) &= E|X|^p I(|X| \leq \varepsilon x) + E|X|^p I(\varepsilon x < |X| \leq x) \\ &\leq (\varepsilon x)^{p-2} l(x) + x^{p-2}(l(x) - l(\varepsilon x)) \\ &= (\varepsilon x)^{p-2} l(x) + x^{p-2} o(l(x)) \end{aligned}$$

by (P2). This proves (4.5) since ε can be arbitrarily small. □

Proof (of the "if" part in Theorem 4.1). We follow the proof by Maller (1981) who uses the fact that if $E(X) = 0$, then $S_n/a_n \xrightarrow{D} N(0,1)$ for a sequence of positive constants a_n if and only if $V_n/a_n \xrightarrow{P} 1$. Let

$$\begin{aligned} l(x) &= EX^2 I(|X| \leq x), \qquad b = \inf\{x \geq 1 : l(x) > 0\}, \\ z_n &= \inf\left\{s : s \geq b+1, \tfrac{l(s)}{s^2} \leq \tfrac{1}{n}\right\}. \end{aligned} \tag{4.6}$$

Since $l(s)/s^2 \to 0$ as $s \to \infty$ and the function $l(x)$ is right continuous, we have $\lim_{n\to\infty} z_n = \infty$ and

$$n\,l(z_n) = z_n^2. \tag{4.7}$$

It follows from (4.3), (4.4) and (4.7) that

$$P\left(\max_{1\leq i\leq n} |X_i| > z_n\right) \leq nP(|X| \geq z_n) = o\left(nl(z_n)/z_n^2\right) = o(1), \tag{4.8}$$

$$\sum_{i=1}^n E|X_i|I(|X_i| > z_n) = o(nl(z_n)/z_n) = o\left(\sqrt{nl(z_n)}\right). \tag{4.9}$$

It suffices to show that

$$\frac{\sum_{i=1}^n X_i^2 I(|X_i| \leq z_n)}{\sqrt{nl(z_n)}} \xrightarrow{P} 1. \tag{4.10}$$

In fact, (4.8)–(4.10) imply that $S_n/V_n \xrightarrow{D} N(0,1)$ is equivalent to

$$\frac{\sum_{i=1}^n \{X_i I(|X_i| \leq z_n) - EX_i I(|X_i| \leq z_n)\}}{\sqrt{nl(z_n)}} \xrightarrow{D} N(0,1). \tag{4.11}$$

Observe that

$$\frac{1}{nl(z_n)} \mathrm{Var}\left(\sum_{i=1}^n \{X_i I(|X_i| \leq z_n) - EX_i I(|X_i| \leq z_n)\}\right) \to 1,$$

$nl(z_n) = z_n^2$, and for $0 < \varepsilon < 1$,

$$\frac{1}{nl(z_n)} \sum_{i=1}^{n} EX_i^2 I(\varepsilon z_n < |X_i| \le z_n) \to 0.$$

Hence the Lindeberg condition is satisfied and therefore (4.11) holds. Now (4.10) follows from

$$\mathrm{Var}\left(\sum_{i=1}^{n} X_i^2 I(|X_i| \le z_n)\right) \le nEX^4 I(|X| \le z_n) = o\left(nz_n^2 l(z_n)\right) = o\left((nl(z_n))^2\right),$$

by making use of (4.5). □

Proof (of the "only if" part in Theorem 4.1). The proof is considerably more complicated, and we refer the details to Giné et al. (1997). The main ingredient of the proof is

$$S_n/V_n = O_p(1) \iff \sup_{n \ge 1} E \exp\left(\gamma (S_n/V_n)^2\right) < \infty \qquad \text{for some } \gamma > 0.$$

An alternative proof is given by Mason (2005). □

As a generalization of the central limit theorem, the classical weak invariance principle (or functional central limit theorem) states that on some probability space,

$$\sup_{0 \le t \le 1} \left| \frac{1}{\sqrt{n}\sigma} S_{[nt]} - \frac{1}{\sqrt{n}} W(nt) \right| = o(1) \text{ in probability}$$

if and only if $EX = 0$ and $\mathrm{Var}(X) = \sigma^2$, where $\{W(t), t \ge 0\}$ is a standard Brownian motion; see Csörgő and Révész (1981). In view of the self-normalized central limit theorem, Csörgő et al. (2003a) proved a self-normalized version of the weak invariance principle. Let $D[0,1]$ denote the space of functions on $[0,1]$ that are right continuous and have left-hand limits.

Theorem 4.3. *As $n \to \infty$, the following statements are equivalent:*

(a) $EX = 0$ and X is in the domain of attraction of a normal distribution.

(b) $S_{[nt]}/V_n$ converges weakly to Brownian motion in $(D[0,1], \rho)$, where ρ is the sup-norm metric, $\rho(f,g) = \sup_{0 \le t \le 1} |f(t) - g(t)|$, for functions in $D[0,1]$.

(c) By redefining the random variables on some probability space, a standard Brownian motion $\{W(t), t \ge 0\}$ can be constructed such that

$$\sup_{0 \le t \le 1} |S_{[nt]}/V_n - W(nt)/\sqrt{n}| \xrightarrow{P} 0.$$

Proof. Clearly (c) implies (b). By Theorem 4.1, it suffices to show that (a) implies (c), which consists of two steps. Let z_j be defined as in (4.6) and put

$$X_i^* = X_i I(|X_i| \le z_i) - EX_i I(|X_i| \le z_i), \qquad S_k^* = \sum_{i=1}^{k} X_i^*.$$

The first step is to show

$$\max_{0 \le t \le 1} \left| \frac{S_{[nt]}}{V_n} - \frac{S^*_{[nt]}}{\sqrt{nl(z_n)}} \right| \xrightarrow{P} 0,$$

and the second step is to use strong approximation to show

$$\max_{0 \le t \le 1} \left| S^*_{[nt]} - B(nt) \right| = o\left(\sqrt{nl(n)}\right) \text{ in probability.}$$

The details can be found in Csörgő et al. (2003a). □

In view of (c), as $n \to \infty$, $h\{S_{[n\cdot]}/V_n\}$ converges in distribution to $h\{W(\cdot)\}$ for every measurable $h : D \to \mathbb{R}$ that is ρ-continuous except at points in a set of Winer measure zero on $(D, \mathscr{D})$, where $\mathscr{D}$ denotes the σ-field generated by the finite-dimensional subsets of D. In particular, if $EX = 0$ and X is in the domain of attraction of a normal distribution, then

$$P\left(\max_{1 \le i \le n} S_i/V_n \le x\right) \to P\left(\sup_{0 \le t \le 1} W(t) \le x\right),$$

$$P\left(\max_{1 \le i \le n} |S_i|/V_n \le x\right) \to P\left(\sup_{0 \le t \le 1} |W(t)| \le x\right),$$

$$P\left(n^{-1} \sum_{i=1}^n S_i^2/V_n^2 \le x\right) \to P\left(\int_0^1 W^2(t)dt \le x\right),$$

$$P\left(n^{-1} \sum_{i=1}^n |S_i|/V_n \le x\right) \to P\left(\int_0^1 |W(t)|dt \le x\right).$$

For further results that are related to Theorems 4.1 and 4.3, we refer to Csörgő et al. (2004, 2008).

4.2 Non-Normal Limiting Distributions for Self-Normalized Sums

Theorem 2.24 says that a necessary and sufficient condition for the existence of normalized constants $a_n > 0$ and b_n such that the normalized partial sums $(S_n - b_n)/a_n$ converge in distribution to a stable law with index α, $0 < \alpha < 2$, is

$$P(X \ge x) = \frac{(c_1 + o(1))\,h(x)}{x^\alpha}, \qquad P(X \le -x) = \frac{(c_2 + o(1))\,h(x)}{x^\alpha} \tag{4.12}$$

as $x \to \infty$, where $c_1 \ge 0, c_2 \ge 0, c_1 + c_2 > 0$ and $h(x)$ is a slowly varying function at ∞. The normalizing constants are determined by $n(c_1 + c_2)h(a_n) \sim a_n^\alpha$ and $b_n = nEXI(|X| \le a_n)$. The following theorem gives the limiting distributions of the self-normalized sum S_n/V_n under (4.12).

Theorem 4.4. *In addition to assumption* (4.12), *assume that* X *is symmetric when* $\alpha = 1$ *and* $E(X) = 0$ *when* $1 < \alpha < 2$. *Let* $\delta_1, \delta_2, \dots$ *be i.i.d. with* $P(\delta_i = 1) = c_1/(c_1 + c_2)$ *and* $P(\delta_i = -1) = c_2/(c_1 + c_2)$, $\eta_1, \eta_2, \dots$ *i.i.d. exponential random variables with mean* 1 *independent of* $\{\delta_i\}$. *Put*

$$U_k = \left(\sum_{i=1}^{k} \eta_i \right)^{-1/\alpha}$$

and

$$d_k = \begin{cases} 0 & \text{if } 0 < \alpha \le 1, \\ \frac{(c_1 - c_2)}{c_1 + c_2} E U_k I(U_k < 1) & \text{if } 1 < \alpha < 2. \end{cases}$$

Then

$$\frac{S_n}{V_n} \xrightarrow{D} \frac{\sum_{k=1}^{\infty} (\delta_k U_k - d_k)}{(\sum_{k=1}^{\infty} U_k^2)^{1/2}} \tag{4.13}$$

and the limiting density function $p(x)$ *of* S_n/V_n *exists and satisfies*

$$p(x) \sim \frac{1}{\alpha} \left(\frac{2}{\pi} \right)^{1/2} \sqrt{2\beta(\alpha, c_1, c_2)} e^{-x^2 \beta(\alpha, c_1, c_2)} \tag{4.14}$$

as $x \to \infty$, *where the constant* $\beta(\alpha, c_1, c_2)$ *is the solution to* (6.29) *in Chap. 6 on self-normalized moderate deviations.*

The characteristic function of the limiting distribution on the right-hand side of (4.13) was proved by Logan et al. (1973), whose method will be described in Sect. 15.1.1. The representation of the limiting distribution in (4.13) is due to LePage et al. (1981). The following theorem, due to Chistyakov and Götze (2004a), shows that it is necessary for X to be in the domain of attraction of a stable distribution in order that S_n/V_n has a non-degenerate limiting distribution. A random variable X is said to satisfy *Feller's condition* if $\lim_{n \to \infty} nE \sin(X/a_n)$ exists and is finite, where $a_n = \inf\{x > 0 : nx^{-2}[1 + EX^2 I(|X| \le x)] \le 1\}$.

Theorem 4.5. S_n/V_n *converges in distribution to a random variable* Z *such that* $P(|Z| = 1) < 1$ *if and only if:*

(a) X *is in the domain of attraction of a stable distribution with index* $\alpha \in (0, 2]$*;*
(b) $EX = 0$ *if* $1 < \alpha \le 2$*;*
(c) X *is in the domain of attraction of the Cauchy distribution and Feller's condition holds if* $\alpha = 1$.

4.3 Supplementary Results and Problems

1. Let X be a random variable and define $l(x) = EX^2 I(|X| \le x)$. Prove that if $P(|X| \ge x) = o(l(x)/x^2)$ as $x \to \infty$, then $l(x)$ is a slowly varying function at ∞.

2. Let $l(x)$ be defined as in Problem 1. If $l(x)$ is slowly varying at ∞, then $E|X|^p < \infty$ for all $0 < p < 2$.
3. Let X satisfy (4.12). Prove that $E|X|^p < \infty$ for $0 < p < \alpha$ and $E|X|^p = \infty$ for $p > \alpha$. What can be said about $E|X|^p$ when $p = \alpha$?
4. Let $X, X_1, X_2, \ldots$ be i.i.d. random variables with $E(X) = 0$. The following theorem on the limiting distribution of the maximum of the standardized sums is due to Darling and Erdős (1956): *If $E|X|^3 < \infty$, then for every $t \in \mathbb{R}$,*

$$\lim_{n\to\infty} P\left(a(n) \max_{1\le k\le n} S_k/(\sigma\sqrt{k}) \le t + b(n)\right) = \exp(-e^{-t}),$$

where $\sigma^2 = EX^2$, $a(n) = (2\log\log n)^{1/2}$ and

$$b(n) = 2\log\log n + \frac{1}{2}\log\log\log n - \frac{1}{2}\log(4\pi).$$

Einmahl (1989) proved that the Darling–Erdős theorem holds whenever

$$EX^2 I(|X| \ge x) = o\left((\log\log x)^{-1}\right) \qquad \text{as } x \to \infty,$$

which is also necessary. Csörgő et al. (2003b) proved the following self-normalized Darling–Erdős-type theorem: *Suppose that $EX = 0$ and that $l(x) := EX^2 I(|X| \le x)$ is a slowly varying function at ∞, satisfying $l(x^2) \le Cl(x)$ for some $C > 0$. Then, for every $t \in \mathbb{R}$*

$$\lim_{n\to\infty} P\left(a(n) \max_{1\le k\le n} S_k/V_k \le t + b(n)\right) = \exp(-e^{-t}).$$

Develop a self-normalized version of the Darling–Erdős-type theorem when X is in the domain of attraction of a stable distribution with exponent $0 < \alpha < 2$. Moreover, while Bertoin (1998) has extended the Darling–Erdős theorem to the case where X belongs to the domain of attraction of a stable distribution, it would be of interest to develop a self-normalized version.

Chapter 5
Stein's Method and Self-Normalized Berry–Esseen Inequality

The standard method to prove central limit theorems and Berry–Esseen inequalities is based on characteristic functions, as shown in Sect. 2.3. A different method to derive normal approximations was introduced by Stein (1972). Stein's method works well not only for independent random variables but also for dependent ones. It can also be applied to many other probability approximations, notably to Poisson, Poisson process, compound Poisson and binomial approximations. In this chapter we give an overview of the use of Stein's method for normal approximations. We start with basic results on the Stein equations and their solutions and then prove several classical limit theorems and the Berry–Esseen inequality for self-normalized sums.

5.1 Stein's Method

5.1.1 The Stein Equation

Let Z be a standard normally distributed random variable and let $\mathscr{C}$ be the set of continuous and piecewise continuously differentiable functions $f : \mathbb{R} \to \mathbb{R}$ with $E|f'(Z)| < \infty$. Stein's method rests on the following observation.

Lemma 5.1. *Let W be a real-valued random variable. If W has a standard normal distribution, then*

$$Ef'(W) = EWf(W) \tag{5.1}$$

for any absolutely continuous function f with $E|f'(W)| < \infty$. If (5.1) holds for any continuous and piecewise continuously differentiable functions $f : \mathbb{R} \to \mathbb{R}$ with $E|f'(Z)| < \infty$, then W has a standard normal distribution.

V.H. de la Peña et al., *Self-Normalized Processes: Limit Theory and Statistical Applications*, Probability and its Applications,

© Springer-Verlag Berlin Heidelberg 2009

Proof. Necessity: If W has a standard normal distribution, then

$$\begin{aligned}
Ef'(W) &= \frac{1}{\sqrt{2\pi}}\int_{-\infty}^{\infty} f'(w)e^{-w^2/2}dw \\
&= \frac{1}{\sqrt{2\pi}}\int_{-\infty}^{0} f'(w)\left(\int_{-\infty}^{w}(-x)e^{-x^2/2}dx\right)dw \\
&\quad +\frac{1}{\sqrt{2\pi}}\int_{0}^{\infty} f'(w)\left(\int_{w}^{\infty} xe^{-x^2/2}dx\right)dw \\
&= \frac{1}{\sqrt{2\pi}}\int_{-\infty}^{0}\left(\int_{x}^{0} f'(w)dw\right)(-x)e^{-x^2/2}dx \\
&\quad +\frac{1}{\sqrt{2\pi}}\int_{0}^{\infty}\left(\int_{0}^{x} f'(w)dw\right)xe^{-x^2/2}dx \\
&= \frac{1}{\sqrt{2\pi}}\int_{-\infty}^{\infty}[f(x)-f(0)]\,xe^{-x^2/2}dx \\
&= EWf(W).
\end{aligned}$$

Sufficiency: For fixed $z \in \mathbb{R}$, let $f(w) := f_z(w)$ be the solution of the following equation

$$f'(w) - wf(w) = I(w \le z) - \Phi(z). \tag{5.2}$$

Multiplying by $e^{-w^2/2}$ on both sides of (5.2) yields

$$\left(e^{-w^2/2}f(w)\right)' = e^{-w^2/2}\left[I(w \le z) - \Phi(z)\right].$$

Thus,

$$\begin{aligned}
f_z(w) &= e^{w^2/2}\int_{-\infty}^{w}[I(x\le z)-\Phi(z)]\,e^{-x^2/2}dx \\
&= -e^{w^2/2}\int_{w}^{\infty}[I(x\le z)-\Phi(z)]\,e^{-x^2/2}dx \\
&= \begin{cases} \sqrt{2\pi}e^{w^2/2}\Phi(w)\,[1-\Phi(z)] & \text{if } w\le z \\ \sqrt{2\pi}e^{w^2/2}\Phi(z)\,[1-\Phi(w)] & \text{if } w\ge z. \end{cases}
\end{aligned} \tag{5.3}$$

The solution f_z above is a bounded continuous and piecewise continuously differentiable function; see Lemma 5.2 below. Suppose that (5.1) holds for all $f \in \mathscr{C}$. Then it holds for f_z. By (5.2),

$$0 = E\left[f_z'(W) - Wf_z(W)\right] = E\left[I(W\le z)-\Phi(z)\right] = P(W \le z) - \Phi(z).$$

Thus, W has a standard normal distribution. □

When f is bounded and absolutely continuous, one can prove (5.1) by using integration by parts, noting that for $W \sim N(0,1)$,

$$\begin{aligned} E\{Wf(W)\} &= \frac{1}{\sqrt{2\pi}}\int_{-\infty}^{\infty} wf(w)e^{-w^2/2}dw \\ &= -\frac{1}{\sqrt{2\pi}}\int_{-\infty}^{\infty} f(w)d\left(e^{-w^2/2}\right) \\ &= \frac{1}{\sqrt{2\pi}}\int_{-\infty}^{\infty} f'(w)e^{-w^2/2}dw = Ef'(W). \end{aligned}$$

More generally, for a given real-valued measurable function h with $E|h(Z)| < \infty$, Stein's equation refers to

$$f'(w) - wf(w) = h(w) - Eh(Z). \tag{5.4}$$

Equation (5.2) is a special case of (5.4) with $h(w) = I(w \le z)$. Similarly to (5.3), the solution $f = f_h$ is given by

$$\begin{aligned} f_h(w) &= e^{w^2/2}\int_{-\infty}^{w} [h(x) - Eh(Z)]\, e^{-x^2/2}\, dx \\ &= -e^{w^2/2}\int_{w}^{\infty} [h(x) - Eh(Z)]\, e^{-x^2/2}\, dx. \end{aligned} \tag{5.5}$$

Below are some basic properties of solutions to the Stein equations, which can be used to derive error bounds for various approximations.

Lemma 5.2.

(a) For the function f_z defined by (5.3) *and all real w, u and v,*

$$0 < f_z(w) \le 2, \tag{5.6}$$

$$|wf_z(w)| \le 1, \tag{5.7}$$

$$|f_z'(w)| \le 2, \tag{5.8}$$

$$|(w+u)f_z(w+u) - (w+v)f_z(w+v)| \le 2\,(|w|+1)\,(|u|+|v|). \tag{5.9}$$

(b) For the function f_h defined in (5.5),

$$\sup_w |f_h(w)| \le 3\sup_w |h(w)|, \qquad \sup_w |wf_h(w)| \le 2\sup_w |h(w)|, \tag{5.10}$$

$$\sup_w |f_h'(w)| \le 4\sup_w |h(w)|. \tag{5.11}$$

If h is absolutely continuous, then

$$\sup_w |f_h(w)| \le 2\sup_w |h'(w)|, \tag{5.12}$$

$$\sup_w |f_h'(w)| \le 4\sup_w |h'(w)|, \tag{5.13}$$

$$\sup_w |f_h''(w)| \le 2\sup_w |h'(w)|. \tag{5.14}$$

Proof.

(a) The inequality

$$1-\Phi(w) \le \min\left(\frac{1}{2}, \frac{1}{w\sqrt{2\pi}}\right) e^{-w^2/2}, \qquad w>0, \tag{5.15}$$

can be used to derive (5.6) and (5.7). From (5.2) and (5.7), (5.8) follows. As to (5.9), we obtain by (5.8) and (5.6) that

$$\begin{aligned}
&|(w+u)f_z(w+u)-(w+v)f_z(w+v)| \\
&\quad \le |w||f_z(w+u)-f_z(w+v)| + |u|f_z(w+u) + |v|f_z(w+v) \\
&\quad \le 2|w|(|u|+|v|) + 2|u| + 2|v|.
\end{aligned}$$

(b) Let $c_0 = \sup_w |h(w)|$. Noting that $|h(x)-Eh(Z)| \le 2c_0$, (5.10) follows from (5.5) and (5.15), while (5.11) is a consequence of (5.10) and (5.4). The proofs of (5.12), (5.13) and (5.14) require much more lengthy arguments and are omitted here but can be found at the Web site for the book given in the Preface, where the properties (5.6)–(5.9) are further refined; see below. □

The properties (5.6)–(5.9) for f_z can be refined as follows: For all real w, u and v,

$$wf_z(w) \text{ is an increasing function of } w, \tag{5.16}$$

$$|wf_z(w)| \le 1, \qquad |wf_z(w)-uf_z(u)| \le 1, \tag{5.17}$$

$$|f_z'(w)| \le 1, \qquad |f_z'(w)-f_z'(v)| \le 1, \tag{5.18}$$

$$0 < f_z(w) \le \min(\sqrt{2\pi}/4, 1/|z|), \tag{5.19}$$

$$|(w+u)f_z(w+u)-(w+v)f_z(w+v)| \le (|w|+\sqrt{2\pi}/4)(|u|+|v|). \tag{5.20}$$

5.1.2 Stein's Method: Illustration of Main Ideas

Stein's equation (5.4) is the starting point for normal approximations. To illustrate the main ideas, let $\xi_1, \xi_2, \ldots, \xi_n$ be independent random variables such that $E\xi_i = 0$ for $1 \le i \le n$ and $\sum_{i=1}^n E\xi_i^2 = 1$. Put

$$W = \sum_{i=1}^n \xi_i, \qquad W^{(i)} = W - \xi_i \tag{5.21}$$

and

$$K_i(t) = E\xi_i\left[I(0 \le t \le \xi_i) - I(\xi_i \le t < 0)\right]. \tag{5.22}$$

It is easy to see that $K_i(t) \geq 0$ for all real t and that

$$\int_{-\infty}^{\infty} K_i(t)dt = E\xi_i^2, \qquad \int_{-\infty}^{\infty} |t|K_i(t)dt = E|\xi_i|^3/2. \tag{5.23}$$

Let h be a measurable function with $E|h(Z)| < \infty$, and $f = f_h$ be the solution of the Stein equation (5.4). Our goal is to estimate

$$Eh(W) - Eh(Z) = Ef'(W) - EWf(W). \tag{5.24}$$

The main idea of Stein's method is to rewrite $EWf(W)$ in terms of a functional of f'. Since ξ_i and $W^{(i)}$ are independent by (5.21) and $E\xi_i = 0$ for $1 \leq i \leq n$,

$$\begin{aligned}
EWf(W) &= \sum_{i=1}^{n} E\xi_i f(W) \\
&= \sum_{i=1}^{n} E\xi_i \left[f(W) - f\left(W^{(i)}\right)\right] \\
&= \sum_{i=1}^{n} E\xi_i \int_0^{\xi_i} f'\left(W^{(i)} + t\right) dt \\
&= \sum_{i=1}^{n} E \int_{-\infty}^{\infty} f'\left(W^{(i)} + t\right) \xi_i [I(0 \leq t \leq \xi_i) - I(\xi_i \leq t < 0)]dt \\
&= \sum_{i=1}^{n} E \int_{-\infty}^{\infty} f'\left(W^{(i)} + t\right) K_i(t)dt.
\end{aligned} \tag{5.25}$$

From $\sum_{i=1}^{n} \int_{-\infty}^{\infty} K_i(t)dt = \sum_{i=1}^{n} E\xi_i^2 = 1$, it follows that

$$Ef'(W) = \sum_{i=1}^{n} E \int_{-\infty}^{\infty} f'(W)K_i(t)dt. \tag{5.26}$$

Thus, by (5.25) and (5.26),

$$Ef'(W) - EWf(W) = \sum_{i=1}^{n} E \int_{-\infty}^{\infty} \left[f'(W) - f'\left(W^{(i)} + t\right)\right] K_i(t)dt. \tag{5.27}$$

Equations (5.25) and (5.27) play a key role in proving a Berry–Esseen type inequality. Since W and $W^{(i)}$ are close, one expects that $f'(W) - f'(W^{(i)} + t)$ is also small. This becomes clear when f has a bounded second derivative, which is the case when h has a bounded derivative.

5.1.3 Normal Approximation for Smooth Functions

Let $\xi_1, \xi_2, \ldots, \xi_n$ be independent random variables satisfying $E\xi_i = 0$ and $E|\xi_i|^3 < \infty$ for $1 \le i \le n$, and such that $\sum_{i=1}^n E\xi_i^2 = 1$. Let

$$\beta = \sum_{i=1}^n E|\xi_i|^3. \tag{5.28}$$

Theorem 5.3. *Assume that h is a smooth function satisfying*

$$||h'|| := \sup_w |h'(w)| < \infty. \tag{5.29}$$

Then

$$|Eh(W) - Eh(Z)| \le 3\beta ||h'||. \tag{5.30}$$

In particular, we have

$$\left| E|W| - \sqrt{\frac{2}{\pi}} \right| \le 3\beta.$$

Proof. It follows from (5.14) that $||f_h''|| \le 2||h'||$. Therefore, by (5.27) and the mean value theorem,

$$\begin{aligned} |E\{f_h'(W) - Wf_h(W)\}| &\le \sum_{i=1}^n \int_{-\infty}^{\infty} E\left|f_h'(W) - f_h'\left(W^{(i)} + t\right)\right| K_i(t)\,dt \\ &\le 2||h'|| \sum_{i=1}^n \int_{-\infty}^{\infty} E\left(|t| + |\xi_i|\right) K_i(t)\,dt. \end{aligned}$$

Using (5.23), it then follows that

$$\begin{aligned} |E\{f_h'(W) - Wf_h(W)\}| &\le 2||h'|| \sum_{i=1}^n \left(E|\xi_i|^3/2 + E|\xi_i| E\xi_i^2\right) \\ &\le 3||h'|| \sum_{i=1}^n E|\xi_i|^3. \end{aligned} \tag{5.31}$$

□

The following theorem removes the assumption $E|\xi_i|^3 < \infty$ in Theorem 5.3. Theorem 5.5 then shows how results of the type (5.30) can be used to bound $\sup_z |P(W \le z) - \Phi(Z)|$.

Theorem 5.4. *Let $\xi_1, \xi_2, \ldots, \xi_n$ be independent random variables satisfying $E\xi_i = 0$ for $1 \le i \le n$ and such that $\sum_{i=1}^n E\xi_i^2 = 1$. Then for h satisfying* (5.29),

$$|Eh(W) - Eh(Z)| \le 16(\beta_2 + \beta_3)||h'||, \tag{5.32}$$

where

$$\beta_2 = \sum_{i=1}^{n} E\xi_i^2 I(|\xi_i| > 1) \quad and \quad \beta_3 = \sum_{i=1}^{n} E|\xi_i|^3 I(|\xi_i| \le 1). \tag{5.33}$$

Proof. Defining $W^{(i)}$ by (5.21), we use (5.13) and (5.14) to show

$$\left|f_h'(W) - f_h'\left(W^{(i)} + t\right)\right| \le \|h'\| \min(8, 2(|t| + |\xi_i|)) \le 8\|h'\| (|t| \wedge 1 + |\xi_i| \wedge 1),$$

where $a \wedge b$ denotes $\min(a,b)$. Substituting this bound into (5.27), we obtain

$$|Eh(W) - Eh(Z)| \le 8\|h'\| \sum_{i=1}^{n} \int_{-\infty}^{\infty} E(|t| \wedge 1 + |\xi_i| \wedge 1) K_i(t)\, dt. \tag{5.34}$$

Making use of

$$x \int_{-\infty}^{\infty} (|t| \wedge 1) [I(0 \le t \le x) - I(x \le t < 0)]\, dt = \begin{cases} x^2 - |x|/2 & \text{if } |x| > 1, \\ \frac{1}{2}|x|^3 & \text{if } |x| \le 1, \end{cases}$$

we obtain

$$\int_{-\infty}^{\infty} E(|t| \wedge 1 + |\xi_i| \wedge 1) K_i(t)\, dt = E\left\{\xi_i^2 I(|\xi_i| > 1)\right\} - \frac{1}{2} E\left\{|\xi_i| I(|\xi_i| > 1)\right\}$$
$$+ \frac{1}{2} E\left\{|\xi_i|^3 I(|\xi_i| \le 1)\right\} + E\left\{\xi_i^2 E(|\xi_i| \wedge 1)\right\}.$$

It then follows from (5.34) that

$$|Eh(W) - Eh(Z)| \le 8\|h'\| \left(\beta_2 + \beta_3 + \sum_{i=1}^{n} E\xi_i^2 E(|\xi_i| \wedge 1)\right). \tag{5.35}$$

Since both x^2 and $(x \wedge 1)$ are increasing functions of $x \ge 0$, it follows that for any random variable ξ,

$$E\xi^2 E(|\xi| \wedge 1) \le E\left\{\xi^2 (|\xi| \wedge 1)\right\} = E|\xi|^3 I(|\xi| \le 1) + E\xi^2 I(|\xi| > 1), \tag{5.36}$$

and therefore the sum in (5.35) is no greater than $\beta_3 + \beta_2$, proving (5.32). □

Although we cannot derive a sharp Berry–Esseen bound from Theorem 5.3 or 5.4, the following result still provides a partial rate of convergence.

Theorem 5.5. *Assume that there exists* δ *such that for any* h *satisfying* (5.29),

$$|Eh(W) - Eh(Z)| \le \delta \|h'\|. \tag{5.37}$$

Then

$$\sup_z |P(W \le z) - \Phi(z)| \le 2\delta^{1/2}. \tag{5.38}$$

Proof. We can assume that $\delta \le 1/4$, since otherwise (5.38) is trivial. Let $\alpha = \delta^{1/2}(2\pi)^{1/4}$, and define for fixed z

$$h_\alpha(w) = \begin{cases} 1 & \text{if } w \le z, \\ 0 & \text{if } w \ge z+\alpha, \\ \text{linear} & \text{if } z \le w \le z+\alpha. \end{cases}$$

Then $\|h'\| = 1/\alpha$ and hence by (5.37),

$$\begin{aligned} P(W \le z) - \Phi(z) &\le Eh_\alpha(W) - Eh_\alpha(Z) + Eh_\alpha(Z) - \Phi(z) \\ &\le \frac{\delta}{\alpha} + P\{z \le Z \le z+\alpha\} \\ &\le \frac{\delta}{\alpha} + \frac{\alpha}{\sqrt{2\pi}}. \end{aligned}$$

Therefore

$$P(W \le z) - \Phi(z) \le 2(2\pi)^{-1/4}\delta^{1/2} \le 2\delta^{1/2}. \tag{5.39}$$

Similarly, we have

$$P(W \le z) - \Phi(z) \ge -2\delta^{1/2}, \tag{5.40}$$

proving (5.38). □

Theorems 5.4 and 5.5 together yield the Lindeberg central limit theorem.

Corollary 5.6. *Let $X_1, \ldots, X_n$ be independent random variables with $EX_i = 0$ and $EX_i^2 < \infty$ for $1 \le i \le n$. Put $S_n = \sum_{i=1}^n X_i$ and $B_n^2 = \sum_{i=1}^n EX_i^2$. If*

$$B_n^{-2} \sum_{i=1}^n EX_i^2 I(|X_i| > \varepsilon B_n) \to 0 \qquad \textit{for all } \varepsilon > 0, \tag{5.41}$$

then

$$S_n/B_n \xrightarrow{D} N(0,1). \tag{5.42}$$

Proof. To apply Theorems 5.4 and 5.5, let

$$\xi_i = X_i/B_n \quad \text{and} \quad W = S_n/B_n. \tag{5.43}$$

Clearly, the Lindeberg condition (5.41) is equivalent to

$$\lim_{n\to\infty} \sum_{i=1}^n E\xi_i^2 I(|\xi_i| > \varepsilon) = 0 \qquad \text{for all } \varepsilon > 0. \tag{5.44}$$

Define β_2 and β_3 as in (5.33), and observe that for any $0 < \varepsilon < 1$,

$$\begin{aligned}
\beta_2+\beta_3 &= \sum_{i=1}^n E\xi_i^2 I(|\xi_i|>1)+\sum_{i=1}^n E|\xi_i|^3 I(|\xi_i|\le 1) \\
&= \sum_{i=1}^n E\xi_i^2 I(|\xi_i|>1)+\sum_{i=1}^n E|\xi_i|^3 I(|\xi_i|\le \varepsilon)+\sum_{i=1}^n E|\xi_i|^3 I(\varepsilon<|\xi_i|\le 1) \\
&\le \sum_{i=1}^n E\xi_i^2 I(|\xi_i|>1)+\varepsilon\sum_{i=1}^n E|\xi_i|^2+\sum_{i=1}^n E|\xi_i|^2 I(\varepsilon<|\xi_i|\le 1) \\
&\le 2\sum_{i=1}^n E\xi_i^2 I(|\xi_i|>\varepsilon)+\varepsilon. \qquad (5.45)
\end{aligned}$$

Then (5.44) and (5.45) imply $\beta_2+\beta_3 \to 0$ as $n\to\infty$, since ε is arbitrary. Hence, by Theorems 5.4 and 5.5,

$$\sup_z |P(S_n/B_n \le z)-\Phi(z)| \le 8(\beta_2+\beta_3)^{1/2} \to 0 \quad \text{as } n\to\infty.$$

□

5.2 Concentration Inequality and Classical Berry–Esseen Bound

We use (5.27) and a concentration inequality in Lemma 5.8 below, which provides a key tool for overcoming the non-smoothness of the indicator function $I(W \le z)$, to derive the classical Berry–Esseen bound.

Theorem 5.7. *Let $\xi_1, \xi_2, \ldots, \xi_n$ be independent random variables satisfying $E\xi_i = 0$, $E|\xi_i|^3 < \infty$ for $1 \le i \le n$ and such that $\sum_{i=1}^n E\xi_i^2 = 1$. Then*

$$|P(W \le z) - \Phi(z)| \le 12\beta, \qquad (5.46)$$

where $W = \sum_{i=1}^n \xi_i$ and $\beta = \sum_{i=1}^n E|\xi_i|^3$.

Proof. Let f_z be the solution of the Stein equation (5.2) and define $W^{(i)}$ by (5.21). To apply (5.27), rewrite

$$\begin{aligned}
f_z'(W)-f_z'\left(W^{(i)}+t\right) &= \left(W^{(i)}+\xi_i\right)f_z\left(W^{(i)}+\xi_i\right)-\left(W^{(i)}+t\right)f_z\left(W^{(i)}+t\right) \\
&\quad +I\left(W^{(i)}\le z-\xi_i\right)-I\left(W^{(i)}\le z-t\right).
\end{aligned}$$

Thus, by (5.9),

$$\begin{aligned}
&\left|E\left\{f_z'(W)-f_z'(W^{(i)}+t)\right\}\right| \\
&\quad \le E\left|\left(W^{(i)}+\xi_i\right)f_z\left(W^{(i)}+\xi_i\right)-\left(W^{(i)}+t\right)f_z\left(W^{(i)}+t\right)\right| \\
&\qquad +\left|P\left(W^{(i)}\le z-\xi_i\right)-P\left(W^{(i)}\le z-t\right)\right|
\end{aligned}$$

$$\leq 2E\left\{\left(\left|W^{(i)}\right|+1\right)(|t|+|\xi_i|)\right\}+\left|P\left(W^{(i)}\leq z-\xi_i\right)-P\left(W^{(i)}\leq z-t\right)\right|$$
$$\leq 4(|t|+E|\xi_i|)+\left|P\left(W^{(i)}\leq z-\xi_i\right)-P\left(W^{(i)}\leq z-t\right)\right|.$$

From (5.27), it follows that

$$\begin{aligned}|P(W\leq z)-\Phi(z)| &= |Ef_z'(W)-EWf_z(W)|\\ &\leq \sum_{i=1}^n 4\int_{-\infty}^{\infty}(|t|+E|\xi_i|)K_i(t)dt\\ &\quad+\sum_{i=1}^n\int_{-\infty}^{\infty}\left|P\left(W^{(i)}\leq z-\xi_i\right)-P\left(W^{(i)}\leq z-t\right)\right|K_i(t)dt\\ &\leq 4\sum_{i=1}^n(E|\xi_i|^3/2+E|\xi_i|E\xi_i^2)\\ &\quad+\sum_{i=1}^n\int_{-\infty}^{\infty}(2|t|+2E|\xi_i|+3\beta)K_i(t)dt\leq 12\beta,\end{aligned}$$

where in the second inequality, we have used (5.23) to obtain the first sum, and Lemma 5.8 below to obtain the second sum, noting that

$$\begin{aligned}\left|P\left(W^{(i)}\leq z-\xi_i\right)-P\left(W^{(i)}\leq z-t\right)\right| &\leq P\left(z-\max(t,\xi_i)\leq W^{(i)}\leq z-\min(t,\xi_i)\right)\\ &\leq 2(|t|+E|\xi_i|)+3\beta.\quad\square\end{aligned}$$

Lemma 5.8. *With the same notation and assumptions as in Theorem 5.7,*

$$P\left(a\leq W^{(i)}\leq b\right)\leq 2(b-a)+3\beta \tag{5.47}$$

for all $a<b$ and $1\leq i\leq n$, where $W^{(i)}=W-\xi_i$.

Proof. Define $\delta=\beta/2$ and let

$$f(w)=\begin{cases}-\frac{1}{2}(b-a)-\delta & \text{if } w<a-\delta,\\ w-\frac{1}{2}(b+a) & \text{if } a-\delta\leq w\leq b+\delta,\\ \frac{1}{2}(b-a)+\delta & \text{if } w>b+\delta,\end{cases} \tag{5.48}$$

so that $f'(x)=I(a-\delta<x<b+\delta)$ and $\|f\|=(b-a)/2+\delta$. Since ξ_j and $W^{(i)}-\xi_j$ are independent for $j\neq i$, and since ξ_i is independent of $W^{(i)}$ and $E\xi_j=0$ for all j,

$$\begin{aligned}&E\left\{W^{(i)}f\left(W^{(i)}\right)\right\}-E\left\{\xi_i f\left(W^{(i)}-\xi_i\right)\right\}\\ &\qquad=\sum_{j=1}^n E\left\{\xi_j\left[f(W^{(i)})-f(W^{(i)}-\xi_j)\right]\right\}\end{aligned}$$

$$= \sum_{j=1}^{n} E\xi_j \int_{-\xi_j}^{0} f'\left(W^{(i)}+t\right) dt$$

$$= \sum_{j=1}^{n} E \int_{-\infty}^{\infty} f'\left(W^{(i)}+t\right) \hat{M}_j(t) dt, \tag{5.49}$$

where $\hat{M}_j(t) = \xi_j\{I(-\xi_j \le t \le 0) - I(0 < t \le -\xi_j)\}$. Noting that $\hat{M}_j \ge 0$ and $f' \ge 0$,

$$\begin{aligned}
\sum_{j=1}^{n} E \int_{-\infty}^{\infty} f'\left(W^{(i)}+t\right) \hat{M}_j(t) dt &\ge \sum_{j=1}^{n} E \int_{|t|\le\delta} f'\left(W^{(i)}+t\right) \hat{M}_j(t) dt \\
&\ge \sum_{j=1}^{n} EI\left(a \le W^{(i)} \le b\right) \int_{|t|\le\delta} \hat{M}_j(t) dt \\
&= E\left\{I\left(a \le W^{(i)} \le b\right) \sum_{j=1}^{n} |\xi_j| \min(\delta, |\xi_j|)\right\} \\
&\ge H_{1,1} - H_{1,2},
\end{aligned} \tag{5.50}$$

where

$$H_{1,1} = P\left(a \le W^{(i)} \le b\right) \sum_{j=1}^{n} E|\xi_j| \min(\delta, |\xi_j|),$$

$$H_{1,2} = E\left|\sum_{j=1}^{n} \{|\xi_j| \min(\delta, |\xi_j|) - E|\xi_j| \min(\delta, |\xi_j|)\}\right|.$$

Simple algebra yields

$$\min(x,y) \ge x - x^2/(4y) \qquad x > 0,\ y > 0,$$

implying that

$$\sum_{j=1}^{n} E|\xi_j| \min(\delta, |\xi_j|) \ge \sum_{j=1}^{n} \left\{E\xi_j^2 - \frac{E|\xi_j|^3}{4\delta}\right\} = \frac{1}{2} \tag{5.51}$$

since $\delta = \beta/2$, and therefore

$$H_{1,1} \ge \frac{1}{2} P\left(a \le W^{(i)} \le b\right). \tag{5.52}$$

By the Hölder inequality,

$$\begin{aligned}
H_{1,2} &\le \left(\operatorname{Var}\left\{\sum_{j=1}^{n} |\xi_j| \min(\delta, |\xi_j|)\right\}\right)^{1/2} \\
&\le \left(\sum_{j=1}^{n} E\xi_j^2 \min(\delta, |\xi_j|)^2\right)^{1/2} \\
&\le \delta \left(\sum_{j=1}^{n} E\xi_j^2\right)^{1/2} = \delta.
\end{aligned} \tag{5.53}$$

On the other hand, recalling that $\|f\| \le \frac{1}{2}(b-a)+\delta$, we have

$$\begin{aligned} E\left\{W^{(i)} f\left(W^{(i)}\right)\right\} - E\left\{\xi_i f\left(W^{(i)}-\xi_i\right)\right\} & \\ &\le \{(b-a)/2+\delta\}\left(E\left|W^{(i)}\right| + E|\xi_i|\right) \\ &\le (b-a)+2\delta = (b-a)+\beta. \end{aligned} \tag{5.54}$$

Combining (5.49), (5.50) and (5.52)–(5.54) yields

$$P\left(a \le W^{(i)} \le b\right) \le 2\{(b-a)+\beta+\delta\} = 2(b-a)+3\beta.$$

□

Following the lines of the previous proof, one can prove that

$$|P(W \le z) - \Phi(z)| \le 20(\beta_2+\beta_3), \tag{5.55}$$

dispensing with the third moment assumption, where β_2 and β_3 are as in (5.33). With a more refined concentration inequality, the constant 20 can be reduced to 4.1; see Chen and Shao (2001).

5.3 A Self-Normalized Berry–Esseen Inequality

Let $X_1,\ldots,X_n$ be independent random variables with $EX_i = 0$ and $EX_i^2 < \infty$. Put

$$S_n = \sum_{i=1}^n X_i, \qquad V_n^2 = \sum_{i=1}^n X_i^2, \qquad B_n^2 = \sum_{i=1}^n EX_i^2. \tag{5.56}$$

The study of the Berry–Esseen bound for the self-normalized sum S_n/V_n has a long history. The first general result is due to Bentkus and Götze (1996) for the i.i.d. case, which is extended to the non-i.i.d. case by Bentkus et al. (1996). In particular, they have shown that

$$\sup_z |P(S_n/V_n \le z) - \Phi(z)| \le C(\beta_2+\beta_3), \tag{5.57}$$

where C is an absolute constant and

$$\beta_2 = B_n^{-2}\sum_{i=1}^n EX_i^2 I(|X_i| > B_n), \qquad \beta_3 = B_n^{-3}\sum_{i=1}^n E|X_i|^3 I(|X_i| \le B_n). \tag{5.58}$$

The bound, therefore, coincides with the classical Berry–Esseen bound for the standardized mean S_n/B_n up to an absolute constant. Their proof is based on the traditional characteristic function approach. In this section, we give a direct proof of (5.57) by using Stein's method, which has been used by Shao (2005) to obtain a more explicit bound.

Theorem 5.9. *Let $X_1,\ldots,X_n$ be independent random variables with $EX_i = 0$ and $EX_i^2 < \infty$ for $1 \le i \le n$. Define S_n, V_n and B_n by (5.56). Then*

$$\sup_z |P(S_n/V_n \le z) - \Phi(z)| \le 11B_n^{-2} \sum_{i=1}^n EX_i^2 I(|X_i| > B_n/2) + B_n^{-3} \sum_{i=1}^n E|X_i|^3 I(|X_i| \le B_n/2).$$

In particular, for $2 < p \le 3$,

$$\sup_z |P(S_n/V_n \le z) - \Phi(z)| \le 25B_n^{-p} \sum_{i=1}^n E|X_i|^p.$$

5.3.1 Proof: Outline of Main Ideas

Without loss of generality, assume that $B_n = 1$. Let

$$\xi_i = X_i/V_n \quad \text{and} \quad W = \sum_{i=1}^n \xi_i. \tag{5.59}$$

A key observation is that for any absolutely continuous function f,

$$\begin{aligned} Wf(W) - \sum_{i=1}^n \xi_i f(W-\xi_i) &= \sum_{i=1}^n \xi_i (f(W) - f(W-\xi_i)) = \sum_{i=1}^n \xi_i \int_{-\xi_i}^0 f'(W+t)dt \\ &= \sum_{i=1}^n \xi_i \int_{-1}^1 f'(W+t)\,[I(-\xi_i \le t \le 0) - I(0 < t \le -\xi_i)]\,dt \\ &= \sum_{i=1}^n \int_{-1}^1 f'(W+t)\hat{m}_i(t)dt \\ &= \int_{-1}^1 f'(W+t)\hat{m}(t)dt, \end{aligned} \tag{5.60}$$

where

$$\hat{m}_i(t) = \xi_i\,[I(-\xi_i \le t \le 0) - I(0 < t \le -\xi_i)], \qquad \hat{m}(t) = \sum_{i=1}^n \hat{m}_i(t).$$

Noting that

$$\int_{-1}^1 \hat{m}(t)dt = \sum_{i=1}^n \xi_i^2 = 1,$$

we have

$$f'(W) - Wf(W) = -\sum_{i=1}^{n} \xi_i f(W - \xi_i) + \int_{-1}^{1} \left[f'(W) - f'(W+t) \right] \hat{m}(t) dt. \tag{5.61}$$

Let $f = f_z$ be the solution (5.3) to the Stein equation (5.2). Then

$$\begin{aligned} I(W \le z) - \Phi(z) &= \int_{-1}^{1} (W f_z(W) - (W+t) f_z(W+t)) \hat{m}(t) dt \\ &\quad + \int_{-1}^{1} [I(W \le z) - I(W+t \le z)] \hat{m}(t) dt - \sum_{i=1}^{n} \xi_i f_z(W - \xi_i) \\ &= R_1 + R_2 - R_3, \end{aligned} \tag{5.62}$$

where

$$R_1 = \int_{-1}^{1} (W f_z(W) - (W+t) f_z(W+t)) \hat{m}(t) dt, \tag{5.63}$$

$$R_2 = I(W \le z) - \int_{-1}^{1} I(W+t \le z) \hat{m}(t) dt, \tag{5.64}$$

$$R_3 = \sum_{i=1}^{n} \xi_i f_z(W - \xi_i). \tag{5.65}$$

It is easy to see that

$$|R_1| \le \int_{-1}^{1} (|W|+1) |t| \hat{m}(t) dt \le (1/2) \sum_{i=1}^{n} (|W|+1) |\xi_i|^3. \tag{5.66}$$

Since $B_n = 1$, we expect that V_n is close to 1 with high probability, so ξ_i is close X_i, which can be used to show that

$$E|R_1| I(V_n \ge 1/2) = O(1)(\beta_2 + \beta_3). \tag{5.67}$$

As to R_2, since $\hat{m}(t) \ge 0$ and $\int_{-1}^{1} \hat{m}(t) dt = 1$, we can view $\hat{m}(t)$ as a conditional density function given $\mathscr{X}_n := (X_i, 1 \le i \le n)$. Let T be a random variable such that the conditional density function of T given $\mathscr{X}_n$ is $\hat{m}(t)$. Then we can rewrite

$$\int_{-1}^{1} I(W+t \le z) \hat{m}(t) dt = E\left[I(W+T \le z) | \mathscr{X}_n \right]$$

and

$$R_2 = I(W \le z) - E\left[I(W+T \le z) | \mathscr{X}_n \right]. \tag{5.68}$$

Now $E(R_2) = P(W \le z) - P(W + T \le z)$. Similarly to Lemma 5.8, we expect the following randomized concentration inequality to hold:

$$|P(W \le z) - P(W + T \le z)| = O(1)(\beta_2 + \beta_3). \tag{5.69}$$

Noting that ξ_i and $W - \xi_i$ are almost independent, we also expect that

$$|ER_3 I(V_n \ge 1/2)| = O(1)(\beta_2 + \beta_3). \tag{5.70}$$

5.3.2 Proof: Details

We now give a detailed proof of (5.57). Without loss of generality, assume $B_n = 1$, $z \ge 0$, and use the notation in Sect. 5.3.1. It follows from (5.62) that

$$[I(W \le z) - \Phi(z)]I(V_n \ge 1/2) = (R_1 + R_2 - R_3)I(V_n \ge 1/2)$$

and therefore

$$\begin{aligned} &|P(W \le z) - \Phi(z)| &(5.71)\\ &\quad = |E[I(W \le z) - \Phi(z)]I(V_n < 1/2)| + |E[I(W \le z) - \Phi(z)]I(V_n \ge 1/2)| \\ &\quad \le P(V_n < 1/2) + |ER_1 I(V_n \ge 1/2)| \\ &\qquad + |ER_2 I(V_n \ge 1/2)| + |ER_3 I(V_n \ge 1/2)|. &(5.72)\end{aligned}$$

Since (5.57) is trivial when $\beta_2 \ge 0.1$ or $\beta_3 \ge 0.1$, we assume

$$\beta_2 < 0.1 \quad \text{and} \quad \beta_3 < 0.1, \tag{5.73}$$

and divide the proof of (5.57) into four steps.

Step 1. Show that

$$P(V_n \le 1/2) \le 0.4\beta_3. \tag{5.74}$$

Application of Theorem 2.19 yields

$$\begin{aligned} P(V_n \le 1/2) &\le P\{\textstyle\sum_{i=1}^n X_i^2 I(|X_i| \le 1) \le 1/4\} \\ &\le \exp\left(-\frac{\left[\sum_{i=1}^n EX_i^2 I(|X_i| \le 1) - 1/4\right]^2}{2\sum_{i=1}^n EX_i^4 I(|X_i| \le 1)}\right) \\ &\le \exp\left[-(1 - \beta_2 - 0.25)^2/(2\beta_3)\right] \\ &\le \exp(-0.65^2/2\beta_3) \le 0.4\beta_3 \qquad \text{by (5.73).} \end{aligned}$$

Step 2. Bound $ER_1 I(V_n \ge 1/2)$ by

$$|ER_1 I(V_n \ge 1/2)| \le 2\beta_2 + 16\beta_3. \tag{5.75}$$

Noting that $|\xi_i| \le V_n^{-1}|X_i|I(|X_i| \le 1) + I(|X_i| > 1)$, we have by (5.66),

$$\begin{aligned}
&|ER_1 I(V_n \ge 1/2)| \\
&\le \frac{1}{2}\sum_{i=1}^n E|\xi_i|^3 I(V_n \ge 1/2) + \frac{1}{2}\sum_{i=1}^n E|\xi_i|^3|W|I(V_n \ge 1/2) \\
&\le \frac{1}{2}\sum_{i=1}^n E\left(|\xi_i|^3 + |\xi_i|^4\right) I(V_n \ge 1/2) + \frac{1}{2}\sum_{i=1}^n E|\xi_i|^3|W - \xi_i| I(V_n \ge 1/2) \\
&\le \sum_{i=1}^n E|\xi_i|^3 I(V_n \ge 1/2) + \frac{1}{2}\sum_{i=1}^n E|\xi_i|^3|W - \xi_i| I(V_n \ge 1/2) \\
&\le \sum_{i=1}^n E\left[V_n^{-3}|X_i|^3 I(|X_i| \le 1) + I(|X_i| > 1)\right] I(V_n \ge 1/2) \\
&\quad + \frac{1}{2}\sum_{i=1}^n E\left[V_n^{-3}|X_i|^3 I(|X_i| \le 1) + I(|X_i| > 1)\right] |S_n - X_i| V_n^{-1} I(V_n \ge 1/2) \\
&\le \sum_{i=1}^n \left\{8E|X_i|^3 I(|X_i| \le 1) + P(|X_i| > 1)\right\} \\
&\quad + \sum_{i=1}^n \left\{E\left[8|X_i|^3 I(|X_i| \le 1) + I(|X_i| > 1)\right] |S_n - X_i|\right\} \\
&\le 8\beta_3 + \beta_2 + 8\beta_3 + \beta_2 = 2\beta_2 + 16\beta_3.
\end{aligned} \tag{5.76}$$

Step 3. Bound $|ER_2 I(V_n \ge 1/2)|$ by

$$|ER_2 I(V_n \ge 1/2)| \le A\beta_2 + A\beta_3. \tag{5.77}$$

By (5.68), $ER_2 I(V_n \ge 1/2) = E[I(W \le z) - I(W + T \le z)]I(V_n \ge 1/2)$. Therefore

$$ER_2 I(V_n \ge 1/2) \begin{cases} \le EI(z - |T| \le W \le z) I(V_n \ge 1/2), \\ \ge -EI(z \le W \le z + |T|) I(V_n \ge 1/2). \end{cases}$$

We now develop a randomized concentration inequality for $P(z - |T| \le W \le z, V_n \ge 1/2)$. Following the proof of Lemma 5.8, let $\delta = (1/2)\sum_{i=1}^n |\xi_i|^3$,

$$f_{T,\delta}(w) = \begin{cases} -|T|/2 - \delta & \text{if } w \le z - |T| - \delta, \\ (w - z - |T|/2) & \text{if } z - |T| - \delta \le w \le z + \delta, \\ |T|/2 + \delta & \text{if } w \ge z + \delta; \end{cases}$$

$$h(w) = \begin{cases} 0 & \text{if } w \le 1/4, \\ \text{linear} & \text{if } 1/4 < w < 1/2, \\ 1 & \text{if } w \ge 1/2. \end{cases}$$

By (5.60) and the fact that $f'_{T,\delta} \geq 0$,

$$\begin{aligned}
W f_{T,\delta}(W) &- \sum_{i=1}^{n} \xi_i f_{T,\delta}(W - \xi_i) \\
&= \int_{-1}^{1} f'_{T,\delta}(W+t)\hat{m}(t)dt \\
&\geq \int_{|t|\leq\delta} f'_{T,\delta}(W+t)I(z-|T| \leq W \leq z)\hat{m}(t)dt \\
&= I(z-|T| \leq W \leq z)\sum_{i=1}^{n} |\xi_i| \min(|\xi_i|, \delta) \\
&\geq I(z-|T| \leq W \leq z)\sum_{i=1}^{n} \{\xi_i^2 - |\xi_i|^3/(4\delta)\} \\
&= (1/2)I(z-|T| \leq W \leq z),
\end{aligned}$$

in which the last inequality follow from $\min(x,y) \geq x - x^2/(4y)$ for $x \geq 0$ and $y > 0$. Therefore

$$\begin{aligned}
EI&(z-|T| \leq W \leq z)I(V_n > 1/2) \\
&\leq EI(z-|T| \leq W \leq z)h(V_n) \\
&\leq 2EWf_{T,\delta}(W)h(V_n) - 2\sum_{i=1}^{n} E\xi_i f_{T,\delta}(W-\xi_i)h(V_n) \\
&:= \Delta_1 + \Delta_2.
\end{aligned} \tag{5.78}$$

Recalling that the conditional density function of T given $\mathscr{X}_n$ is $\hat{m}(t)$, we obtain

$$\begin{aligned}
|\Delta_1| &\leq 2E|W|\left(|T|/2+\delta\right)h(V_n) \\
&\leq 2E|W|\delta h(V_n) + E\left(|W|h(V_n)E\left(|T|\,|\mathscr{X}_n\right)\right) \\
&= 2E|W|\delta h(V_n) + E|W|h(V_n)\int_{-1}^{1} |t|\hat{m}(t)dt \\
&= 2E|W|\delta h(V_n) + \frac{1}{2}\sum_{i=1}^{n} E|W|h(V_n)|\xi_i|^3 \\
&\leq \frac{3}{2}\sum_{i=1}^{n} E|W||\xi_i|^3 I(V_n > 1/4) \\
&\leq A(\beta_2+\beta_3),
\end{aligned} \tag{5.79}$$

following the proof of (5.76). To estimate Δ_2, let

$$V_n^* = \max(V_n, 1/4), \qquad V_{(i)}^* = \max\left(\left(\Sigma_{j\neq i} X_j^2\right)^{1/2}, 1/4\right).$$

The main idea is to replace V_n in Δ_2 by $V^*_{(i)}$ and then use the independence of X_i and $\{X_j, j \neq i\}$. First note that

$$\left|\frac{d}{dx}\left(xf_{T,\delta}(ax)\right)\right| \leq |f_{T,\delta}(xa)| + |axf'_{T,\delta}(ax)| \leq A\left(|a|+1\right)$$

for $0 \leq x \leq 4$, and that

$$\begin{aligned}|h(x)-h(y)| &\leq 2\min\left(1,|x-y|\right)\\ &= 2\min\left(1,|x^2-y^2|/(|x|+|y|)\right) \leq A\min\left(1,|x^2-y^2|\right)\end{aligned}$$

for $x \geq 1/4, y \geq 1/4$. Moreover,

$$\begin{aligned}0 \leq \frac{1}{V^*_{(i)}} - \frac{1}{V^*_n} &= \frac{V_n^{*2}-V_{(i)}^{*2}}{V^*_n V^*_{(i)}(V^*_n+V^*_{(i)})}\\ &\leq \frac{|X_i|^2}{V^*_n V^*_{(i)}(V^*_n+V^*_{(i)})} \leq A\min(X_i^2,1).\end{aligned} \tag{5.80}$$

Also note that $g(V_n)h(V_n) = g(V^*_n)h(V^*_n)$ for any measurable function g because $h(V_n) = h(V^*_n) = 0$ when $V_n < 1/4$. Hence, with $\delta^* = \sum_{i=1}^n |X_i|^3/V_n^{*3}$, we have

$$\begin{aligned}&E\xi_i f_{T,\delta}(W-\xi_i)h(V_n)\\ &\quad= E\frac{X_i}{V^*_n} f_{T,\delta^*}\left(\frac{S_n-X_i}{V^*_n}\right)h\left(V^*_n\right)\\ &\quad\leq E\frac{X_i}{V^*_{(i)}} f_{T,\delta^*}\left(\frac{S_n-X_i}{V^*_{(i)}}\right)h\left(V^*_{(i)}\right) + AE|X_i|\left(|S_n-X_i|+1\right)\min(1,X_i^2)\\ &\quad= E\frac{X_i}{V^*_{(i)}} f_{T,\delta^*}\left(\frac{S_n-X_i}{V^*_{(i)}}\right)h\left(V^*_{(i)}\right) + AE|X_i|\min(1,X_i^2)E\left(|S_n-X_i|+1\right)\\ &\quad\leq E\frac{X_i}{V^*_{(i)}} f_{T,\delta^*}\left(\frac{S_n-X_i}{V^*_{(i)}}\right)h\left(V^*_{(i)}\right) + AEX_i^2 I\left(|X_i| \geq 1\right) + AE|X_i|^3 I(|X_i| \leq 1).\end{aligned}$$

We can replace δ^* by $\delta^*_{(i)}$, where $\delta^*_{(i)} = (1/2)\sum_{j\neq i}|X_j|^3/V_{(i)}^{*3}$, because

$$\begin{aligned}\left|\delta^*-\delta^*_{(i)}\right| &\leq \frac{|X_i|^3}{V_n^{*3}} + \sum_{j\neq i}|X_j|^3\left(\frac{1}{V_{(i)}^{*3}}-\frac{1}{V_n^{*3}}\right)\\ &\leq A\min\left(|X_i|^3,1\right) + A\sum_{j\neq i}\frac{|X_j|^3}{V_n^{*3}}\min(1,X_i^2) \leq A\min(1,X_i^2),\end{aligned}$$

which implies that

$$\begin{aligned}
&E\frac{X_i}{V_{(i)}^*} f_{T,\delta^*}\left(\frac{S_n - X_i}{V_{(i)}^*}\right) h\left(V_{(i)}^*\right) \\
&\quad \le E\frac{X_i}{V_{(i)}^*} f_{T,\delta_{(i)}^*}\left(\frac{S_n - X_i}{V_{(i)}^*}\right) h\left(V_{(i)}^*\right) + AE|X_i|\min(1, X_i^2) \\
&\quad \le E\frac{X_i}{V_{(i)}^*} f_{T,\delta_{(i)}^*}\left(\frac{S_n - X_i}{V_{(i)}^*}\right) h\left(V_{(i)}^*\right) + AEX_i^2 I(|X_i| \ge 1) + AE|X_i|^3 I(|X_i| < 1) \\
&\quad = E\left\{\frac{X_i}{V_{(i)}^*} h\left(V_{(i)}^*\right) E\left[f_{T,\delta_{(i)}^*}\left(\frac{S_n - X_i}{V_{(i)}^*}\right)\Big|\mathscr{X}_n\right]\right\} \\
&\qquad + AEX_i^2 I(|X_i| \ge 1) + AE|X_i|^3 I(|X_i| < 1).
\end{aligned}$$

We next compute the conditional expected value of $f_{T,\delta_{(i)}}$ given $\mathscr{X}_n$:

$$\begin{aligned}
E\left(f_{T,\delta_{(i)}^*}(S_n - X_i)\Big|\mathscr{X}_n\right) &= \int_{-1}^{1} f_{t,\delta_{(i)}^*}(S_n - X_i)\hat{m}(t)dt \\
&= \sum_{j=1}^{n}\int_{-1}^{1} f_{t,\delta_{(i)}^*}(S_n - X_i)\hat{m}_j(t)dt \\
&= \sum_{j=1}^{n}\xi_j\int_{-\xi_j}^{0} f_{t,\delta_{(i)}^*}(S_n - X_i)dt \\
&= \xi_i\int_{-\xi_i}^{0} f_{t,\delta_{(i)}^*}(S_n - X_i)dt + \sum_{j\ne i}\xi_j\int_{-\xi_j}^{0} f_{t,\delta_{(i)}^*}(S_n - X_i)dt.
\end{aligned}$$

As before, we show that ξ_j above can be replaced by $X_j/V_{(i)}^*$. Since $|f_{t,\delta_{(i)}^*}| \le 2$ for $-1 \le t \le 1$ and

$$\left|\frac{d}{dx}\left(x\int_{-x}^{0} f_{t,\delta_{(i)}^*}(S_n - X_i)dt\right)\right| \le 2|x|,$$

we have

$$\begin{aligned}
&\sum_{j\ne i}\xi_j\int_{-\xi_j}^{0} f_{t,\delta_{(i)}^*}(S_n - X_i)dt \\
&\qquad \le \sum_{j\ne i}\frac{X_j}{V_{(i)}^*}\int_{-X_j/V_{(i)}^*}^{0} f_{t,\delta_{(i)}^*}(S_n - X_i)dt + 2\sum_{j\ne i}|X_j|\left(\frac{1}{V_{(i)}^*} - \frac{1}{V_n^*}\right)\frac{|X_j|}{V_{(i)}^*} \\
&\qquad \le \sum_{j\ne i}\frac{X_j}{V_{(i)}^*}\int_{-X_j/V_{(i)}^*}^{0} f_{t,\delta_{(i)}^*}(S_n - X_i)dt + A\sum_{j\ne i}\frac{|X_j|^2}{V_{(i)}^{*2}}\frac{X_i^2}{V_n^{2*}} \\
&\qquad \le \sum_{j\ne i}\frac{X_j}{V_{(i)}^*}\int_{-X_j/V_{(i)}^*}^{0} f_{t,\delta_{(i)}^*}(S_n - X_i)dt + A\min(1, X_i^2).
\end{aligned}$$

Because $E(X_i)=0$, X_i and $\{X_j, j\neq i\}$ are independent, we have

$$\begin{aligned}E\frac{X_i}{V^*_{(i)}}f_{T,\delta^*_{(i)}}\left(\frac{S_n-X_i}{V^*_{(i)}}\right)h\left(V^*_{(i)}\right) &\leq E\frac{|X_i|^3}{V_n^{*2}}+AE|X_i|\min(1,X_i^2)\\ &\quad+\sum_{j\neq i}E\frac{X_i}{V^*_{(i)}}\frac{X_j}{V^*_{(i)}}h\left(V^*_{(i)}\right)\int_{-X_j/V^*_{(i)}}^{0}f_{t,\delta^*_{(i)}}(S_n-X_i)dt\\ &\leq AE|X_i|\min(1,X_i^2).\end{aligned}$$

Putting the above inequalities together gives

$$|\Delta_2|\leq A(\beta_2+\beta_3). \tag{5.81}$$

Therefore

$$EI(z-|T|\leq W\leq z)I(V_n\geq 1/2)\leq A(\beta_2+\beta_3).$$

Similarly, $EI(z\leq W\leq z+|T|)I(V_n\geq 1/2)\leq A(\beta_2+\beta_3)$. This proves (5.77).

Step 4. Following the proof of (5.81), it is readily seen that

$$|ER_3I(V_n\geq 1/2)|\leq A(\beta_2+\beta_3).$$

Completing the proof of (5.57). □

5.4 Supplementary Results and Problems

1. Prove (5.15) and show how it can be used to derive (5.6) and (5.7).
2. Let Y be a random variable with density function p with respect to Lebesgue measure. Assume that $p(y)>0$ for all $y\in\mathbb{R}$ and $p(-\infty)=p(\infty)=0$. Let $f:\mathbb{R}\longrightarrow\mathbb{R}$ be bounded and absolutely continuous:

   (a) Prove that for $f\in\mathscr{C}$,

   $$E\left((p(Y)f(Y))'/p(Y)\right)=0. \tag{5.82}$$

   (b) Let h be a measurable function such that $E|h(Y)|<\infty$. Solve $f=f_h$ for the Stein equation

   $$f'(y)+f(y)p'(y)/p(y)=h(y)-Eh(Y).$$

3. Modify the proof of Lemma 5.8 to derive (5.55).
4. Stein's method can also be applied to prove the following non-uniform Berry–Esseen bound (Chen and Shao, 2001): *Let $\xi_1,\xi_2,\ldots,\xi_n$ be independent random variables satisfying $E\xi_i=0$ and $E|\xi_i|^3<\infty$ for each $1\leq i\leq n$ and such that $\sum_{i=1}^n E\xi_i^2=1$. Put $W=\sum_{i=1}^n\xi_i$. Then*

$$|P(W \le z) - \Phi(z)| \le A(1+|z|)^{-3}\gamma, \tag{5.83}$$

where A is an absolute constant and $\gamma = \sum_{i=1}^n E|\xi_i|^3$.

A key step in the proof of (5.83) is the following non-uniform concentration inequality

$$P(a \le W \le b) \le 10e^{-a/2}(b-a+\gamma) \tag{5.84}$$

for all real $b > a$. Prove (5.84).

5. Chen and Shao (2007) have proved the following "randomized concentration inequality": *Let* $\xi_1, \xi_2, \dots, \xi_n$ *be independent random variables satisfying* $E\xi_i = 0$ *and* $E|\xi_i|^3 < \infty$ *for each* $1 \le i \le n$ *and such that* $\sum_{i=1}^n E\xi_i^2 = 1$. *Put* $W = \sum_{i=1}^n \xi_i$. *Let* Δ_1 *and* Δ_2 *be measurable functions of* $\{\xi_i, 1 \le i \le n\}$. *Then*

$$\begin{aligned} P(\Delta_1 \le W \le \Delta_2) \le\ & E|W(\Delta_2 - \Delta_1)| + 2\gamma \\ & + \sum_{i=1}^n \{E|\xi_i(\Delta_1 - \Delta_{1,i})| + E|\xi_i(\Delta_2 - \Delta_{2,i})|\}, \end{aligned} \tag{5.85}$$

where $\Delta_{1,i}$ *and* $\Delta_{2,i}$ *are Borel measurable functions of* $(\xi_j, 1 \le j \le n,\ j \ne i)$.

Making use of (5.85), one can obtain Berry–Esseen bounds for many non-linear statistics. In particular, carry this out for U-statistics; see (8.5) and Sect. 8.2.1 for an introduction to U-statistics.

6. It would be of interest to investigate if Stein's method can be used to prove the following results that have been proved by other methods and to derive more precise bounds.

   (a) Hall and Wang (2004): Let $X, X_1, X_2, \dots$ be i.i.d. random variables in the domain of attraction of the normal law with $E(X) = 0$. Then

$$\sup_x |P(S_n/V_n \le x) - \Phi(x) - L_n(x)| = o(\delta_n) + O(n^{-1/2}), \tag{5.86}$$

   where

$$\begin{aligned} L_n(x) &= nE\left(\Phi\left[x(1+X^2/b_n^2)^{1/2} - (X/b_n)\right] - \Phi(x)\right), \\ \delta_n &= nP(|X| > b_n) + nb_n^{-1}|EXI(|X| \le b_n)| \\ &\quad + nb_n^{-3}|EX^3I(|X| \le b_n)| + nb_n^{-4}EX^4I(|X| \le b_n), \\ b_n &= \sup\{x:\ nx^{-2}EX^2I(|X| \le x) \ge 1\}. \end{aligned}$$

   If, in addition, Cramér's condition (2.31) is satisfied, then $O(n^{-1/2})$ on the right-hand side of (5.86) can be replaced by $O(n^{-1})$.

   (b) Hall and Wang (2004): Let $X, X_1, X_2, \dots$ be i.i.d. random variables with $E(X) = 0$, $\sigma^2 = E(X^2)$ and $E|X|^3 < \infty$. Assume that the distribution of X is nonlattice, then

$$\sup_x |P(S_n/V_n \le x) - \Phi(x) - F_n(x)| = o(n^{-1/2}), \tag{5.87}$$

   where $F_n(x) = \frac{EX^3}{6\sqrt{n}\sigma^3}(2x^2+1)\phi(x)$.

Chapter 6
Self-Normalized Moderate Deviations and Laws of the Iterated Logarithm

Let $X, X_1, \ldots, X_n$ be i.i.d. random variables. Shao (1997) has developed a theory of moderate deviations for the self-normalized sum of the X_i when X belongs to the domain of attraction of a stable distribution with index α $(0 < \alpha \leq 2)$. In this chapter, Sect. 6.1 describes this theory when X is attracted to a normal distribution $(\alpha = 2)$, and Sect. 6.2 describes the theory for the case $0 < \alpha < 2$. Section 6.3 applies the theory to self-normalized laws of the iterated logarithm.

6.1 Self-Normalized Moderate Deviations: Normal Case

Throughout this chapter we let $X, X_1, X_2, \ldots$ be i.i.d. random variables and set

$$S_n = \sum_{i=1}^{n} X_i \quad \text{and} \quad V_n^2 = \sum_{i=1}^{n} X_i^2.$$

Let $\{x_n, n \geq 1\}$ be a sequence of positive numbers with $x_n \to \infty$ as $n \to \infty$. It is known that

$$\lim_{n\to\infty} x_n^{-2} \log P\left(\frac{|S_n|}{\sqrt{n}} \geq x_n\right) = -\frac{1}{2}$$

holds for any sequence $\{x_n\}$ with $x_n \to \infty$ and $x_n = o(\sqrt{n})$ if and only if $EX = 0$, $EX^2 = 1$ and $Ee^{t_0|X|} < \infty$ for some $t_0 > 0$. The "if" part follows from the theory of large deviations in Sect. 3.1. For the "only if" part, see Problem 6.1. While we have given a treatment of the self-normalized large deviation probability $P(S_n \geq x_n V_n)$ with $x_n \asymp \sqrt{n}$ in Sect. 3.2, we now consider the case $x_n = o(\sqrt{n})$ and show that $\log P(S_n \geq x_n V_n)$ is asymptotically distribution-free if X belongs to the domain of attraction of a normal law.

Theorem 6.1. *Let $\{x_n, n \geq 1\}$ be a sequence of positive numbers with $x_n \to \infty$ and $x_n = o(\sqrt{n})$ as $n \to \infty$. If $EX = 0$ and $EX^2 I(|X| \leq x)$ is slowly varying as $x \to \infty$, then*

V.H. de la Peña et al., *Self-Normalized Processes: Limit Theory and Statistical Applications*, Probability and its Applications,

© Springer-Verlag Berlin Heidelberg 2009

$$\lim_{n\to\infty} x_n^{-2} \log P\left(\frac{S_n}{V_n} \ge x_n\right) = -\frac{1}{2}. \tag{6.1}$$

The proof is divided into two parts. The first part proves the upper bound

$$\limsup_{n\to\infty} x_n^{-2} \log P\left(\frac{S_n}{V_n} \ge x_n\right) \le -\frac{1}{2}, \tag{6.2}$$

and the second part proves the lower bound

$$\liminf_{n\to\infty} x_n^{-2} \log P\left(\frac{S_n}{V_n} \ge x_n\right) \ge -\frac{1}{2}. \tag{6.3}$$

6.1.1 Proof of the Upper Bound

Let

$$\begin{aligned} l(x) &= EX^2 I(|X| \le x), \qquad b = \inf\{x \ge 1 : l(x) > 0\}, \\ z_n &= \inf\left\{s : s \ge b+1, \tfrac{l(s)}{s^2} \le \tfrac{x_n^2}{n}\right\}. \end{aligned} \tag{6.4}$$

By Lemma 4.2, we have that similar to (4.7),

$$z_n \to \infty \quad \text{and} \quad n\,l(z_n) = x_n^2 z_n^2 \qquad \text{for every } n \text{ sufficiently large,} \tag{6.5}$$

$$P(|X| \ge x) = o\left(l(x)/x^2\right), \qquad E|X|I(|X| \ge x) = o\left(l(x)/x\right), \tag{6.6}$$

and

$$E|X|^k I(|X| \le x) = o\left(x^{k-2} l(x)\right) \qquad \text{for each } k > 2 \tag{6.7}$$

as $x \to \infty$. Let $\bar{X}_i = X_i I(|X_i| \le z_n)$. For any $0 < \varepsilon < 1/4$,

$$\begin{aligned} P(S_n \ge x_n V_n) &\le P\left(\sum_{i=1}^n \bar{X}_i \ge (1-\varepsilon) x_n V_n\right) \\ &\quad + P\left(\sum_{i=1}^n X_i I(|X_i| > z_n) \ge \varepsilon x_n V_n\right) \\ &\le P\left(\sum_{i=1}^n \bar{X}_i \ge (1-\varepsilon)^2 x_n \sqrt{n l(z_n)}\right) \\ &\quad + P\left(V_n \le (1-\varepsilon)\sqrt{n l(z_n)}\right) \\ &\quad + P\left(\sum_{i=1}^n I(|X_i| > z_n) \ge \varepsilon^2 x_n^2\right) \\ &:= J_1 + J_2 + J_3. \end{aligned} \tag{6.8}$$

We next apply the exponential inequality (2.22) to the truncated variables $\bar{X}_i$, noting that

$$\sum_{i=1}^{n} E\bar{X}_i = o\left(nl(z_n)/z_n\right) = o\left(x_n\sqrt{nl(z_n)}\right),$$

$$\sum_{i=1}^{n} \mathrm{Var}(\bar{X}_i) \le \sum_{i=1}^{n} E\bar{X}_i^2 = nl(z_n) := B_n^2,$$

$$\beta_n := \sum_{i=1}^{n} E|\bar{X}_i - E\bar{X}_i|^3 \le 8nE|\bar{X}_i|^3 = o\left(nz_n l(z_n)\right) \qquad \text{by (6.7)},$$

$$\frac{\left(x_n\sqrt{nl(z_n)}\right)^3}{B_n^6}\beta_n e^{x_n\sqrt{nl(z_n)}z_n/B_n^2} = o(x_n^2) \qquad \text{by (6.5)}.$$

From (2.22), it then follows that for sufficiently large n,

$$\begin{aligned} J_1 &\le P\left(\sum_{i=1}^{n}(\bar{X}_i - E\bar{X}_i) \ge (1-\varepsilon)^3 x_n\sqrt{nl(z_n)}\right) \\ &\le \exp\left(-\frac{(1-\varepsilon)^6 x_n^2}{2}\right). \end{aligned} \tag{6.9}$$

To bound J_2, application of (2.26) yields

$$\begin{aligned} J_2 &\le P\left(\sum_{i=1}^{n}\bar{X}_i^2 \le (1-\varepsilon)^2 nl(z_n)\right) \\ &\le \exp\left(-\frac{(1-(1-\varepsilon)^2)^2(nl(z_n))^2}{2nEX^4 I(|X|\le z_n)}\right) \\ &\le \exp\left(-\frac{\varepsilon^2\left(nl(z_n)\right)^2}{o\left(nz_n^2 l(z_n)\right)}\right) \\ &\le \exp\left(-x_n^2\varepsilon^2/o(1)\right) \le \exp(-2x_n^2) \end{aligned} \tag{6.10}$$

by (6.5) and (6.7). We next consider J_3. Recalling that $\sum_{i=1}^{n} I(|X_i| > z_n)$ has a binomial distribution and applying (2.27), we obtain from (6.6) and (6.5) that

$$\begin{aligned} J_3 &\le \left(\frac{3nP(|X|>z_n)}{\varepsilon^2 x_n^2}\right)^{\varepsilon^2 x_n^2} \\ &= \left(o\left(\frac{l(z_n)}{z_n^2}\right)\cdot\frac{n}{\varepsilon^2 x_n^2}\right)^{\varepsilon^2 x_n^2} = \left(\frac{o(1)}{\varepsilon^2}\right)^{\varepsilon^2 x_n^2} \le \exp(-2x_n^2). \end{aligned} \tag{6.11}$$

Since ε is arbitrary, (6.2) follows from (6.8)–(6.11). □

6.1.2 Proof of the Lower Bound

Define z_n as in (6.4). The proof is based on the following observation:

$$x_n V_n \leq \frac{1}{2b}\left(b^2 V_n^2 + x_n^2\right) \tag{6.12}$$

for any $b > 0$ and equality holds when $b = x_n/V_n$. From the proof of (6.2), we can see that V_n is close to $(nl(z_n))^{1/2}$. Thus we can choose $b = 1/z_n$ in (6.12). To carry out this idea, we need the following two lemmas.

Lemma 6.2. *Let $\{\xi, \xi_n, n \geq 1\}$ be a sequence of independent random variables, having the same non-degenerate distribution function $F(x)$. Assume that*

$$H := \sup\left\{h : Ee^{h\xi} < \infty\right\} > 0.$$

For $0 < h < H$, put

$$m(h) = E\xi e^{h\xi}/Ee^{h\xi}, \qquad \sigma^2(h) = E\xi^2 e^{h\xi}/Ee^{h\xi} - m^2(h).$$

Then

$$P\left(\sum_{i=1}^n \xi_i \geq nx\right) \geq \frac{3}{4}\left(Ee^{h\xi}\right)^n e^{-nhm(h) - 2h\sigma(h)\sqrt{n}} \tag{6.13}$$

provided that

$$m(h) \geq x + 2\sigma(h)/\sqrt{n}. \tag{6.14}$$

Proof. Let

$$V(x) = \frac{1}{Ee^{h\xi}}\int_{-\infty}^{x} e^{hy}\, dF(y).$$

Consider the sequence of independent random variables $\{\eta, \eta_n, n \geq 1\}$, having the same distribution function $V(x)$. Denote by $F_n(x)$ the distribution function of the random variable $(\sum_{i=1}^n(\eta_i - E\eta_i))/\sqrt{n\mathrm{Var}\,\eta}$. By the conjugate method in Petrov (1965), which we have explained with another notation in Sect. 3.1,

$$P\left(\sum_{i=1}^n \xi_i \geq nx\right) = \left(Ee^{h\xi}\right)^n e^{-nhm(h)} \int_{-(m(h)-x)\sqrt{n}/\sigma(h)}^{\infty} e^{-h\sigma(h)t\sqrt{n}}\, dF_n(t).$$

Since $m(h) \geq x + 2\sigma(h)/\sqrt{n}$,

$$\begin{aligned}
\int_{-(m(h)-x)\sqrt{n}/\sigma(h)}^{\infty} e^{-h\sigma(h)t\sqrt{n}}\, dF_n(t) &\geq \int_{-2}^{2} e^{-h\sigma(h)t\sqrt{n}}\, dF_n(t) \\
&\geq e^{-2h\sigma(h)\sqrt{n}} P\left(\left|\sum_{i=1}^n (\eta_i - E\eta_i)\right| \leq 2\sqrt{n\mathrm{Var}\,\eta}\right) \\
&\geq \frac{3}{4} e^{-2h\sigma(h)\sqrt{n}},
\end{aligned}$$

proving (6.13). □

Lemma 6.3. *Let* $0 < \varepsilon < 1/4$,

$$b_n = 1/z_n, \quad \xi := \xi_n = 2b_nX - b_n^2X^2, \quad \text{and} \quad h := h_\varepsilon = (1+\varepsilon)/2.$$

Then, under the condition of Theorem 6.1, as $n \to \infty$,

$$Ee^{h\xi} = 1 + \varepsilon(1+\varepsilon)x_n^2/(2n) + o\left(x_n^2/n\right), \tag{6.15}$$

$$E\xi e^{h\xi} = (1+2\varepsilon)x_n^2/n + o\left(x_n^2/n\right), \tag{6.16}$$

$$E\xi^2 e^{h\xi} = 4x_n^2/n + o\left(x_n^2/n\right). \tag{6.17}$$

Proof. Note that

$$h\xi = h\left(1 - (b_nX - 1)^2\right) \le h \le 1. \tag{6.18}$$

In view of (6.6), we have

$$\begin{aligned} Ee^{h\xi} &= Ee^{h\xi}I(|X| > z_n) + Ee^{h\xi}I(|X| \le z_n) \\ &= o\left(l(z_n)/z_n^2\right) + E\left(1 + h\xi + \frac{(h\xi)^2}{2}\right)I(|X| \le z_n) \\ &\quad + E\left(e^{h\xi} - 1 - h\xi - \frac{(h\xi)^2}{2}\right)I(|X| \le z_n). \end{aligned} \tag{6.19}$$

From (6.5)–(6.7), it follows that

$$\begin{aligned} &E\left(1 + h\xi + \frac{(h\xi)^2}{2}\right)I(|X| \le z_n) \\ &= 1 - P(|X| > z_n) - 2hb_nEXI(|X| > z_n) - hb_n^2 l(z_n) \\ &\quad + 2h^2b_n^2 l(z_n) - 2h^2b_n^3EX^3I(|X| \le z_n) + h^2b_n^4EX^4I(|X| \le z_n)/2 \\ &= 1 - hb_n^2l(z_n) + 2h^2b_n^2l(z_n) \\ &\quad + o\left(l(z_n)/z_n^2\right) + hb_n o\left(l(z_n)/z_n\right) + h^2b_n^3 o\left(z_nl(z_n)\right) + h^2b_n^4 o\left(z_n^2l(z_n)\right) \\ &= 1 + \varepsilon(1+\varepsilon)b_n^2l(z_n)/2 + o\left(b_n^2l(z_n)\right) \\ &= 1 + \varepsilon(1+\varepsilon)x_n^2/(2n) + o(x_n^2/n). \end{aligned} \tag{6.20}$$

Similarly, by using the inequality $|e^x - 1 - x - x^2/2| \le |x|^3e^{|x|}$,

$$\begin{aligned} &\left|E\left(e^{h\xi} - 1 - h\xi - \frac{(h\xi)^2}{2}\right)I(|X| \le z_n)\right| \\ &\qquad \le E|h\xi|^3e^{h|\xi|}I(|X| \le z_n) \\ &\qquad \le 4h^3Ee^{h(1+(b_nX-1)^2)}\left(8b_n^3|X|^3 + b_n^6X^6\right)I(|X| \le z_n) \\ &\qquad \le 4h^3e^3E\left(8b_n^3|X|^3 + b_n^6X^6\right)I(|X| \le z_n) \\ &\qquad \le 4h^3e^3\left(b_n^3 o\left(z_nl(z_n)\right) + b_n^6 o\left(z_n^4l(z_n)\right)\right) \\ &\qquad = o\left(b_n^2 l(z_n)\right) = o(x_n^2/n). \end{aligned} \tag{6.21}$$

From (6.19)–(6.21), (6.15) follows.

To estimate $E\xi e^{h\xi}$, write

$$E\xi e^{h\xi} = E\xi e^{h\xi} I(|X| > z_n) + E\xi(1+h\xi)I(|X| \le z_n) + E\xi(e^{h\xi} - 1 - h\xi)I(|X| \le z_n).$$

Noting that $\sup_{-\infty < x \le 1} |x|e^x = e$, we have

$$\begin{aligned}\left|E\xi e^{h\xi} I(|X| > z_n)\right| &\le h^{-1} Eh|\xi| e^{h\xi} I(|X| > z_n) \\ &\le h^{-1} e P(|X| > z_n) \\ &= h^{-1} o\left(l(z_n)/z_n^2\right) \\ &= o\left(x_n^2/n\right)\end{aligned}$$

by (6.18) and (6.6). Similar to (6.20),

$$E\xi(1+h\xi)I(|X| \le z_n) = (1+2\varepsilon)x_n^2/n + o\left(x_n^2/n\right).$$

Using the inequality $|e^x - 1 - x| \le x^2 e^{|x|}$, we can proceed along the lines of the proof of (6.21) to show

$$E\xi(e^{h\xi} - 1 - h\xi)I(|X| \le z_n) = o\left(x_n^2/n\right).$$

Combining these bounds yields (6.16). The proof of (6.17) is similar. □

Proof (of the lower bound (6.3)*).* Let b_n, h and ξ be the same as in Lemma 6.3. Put

$$\xi_i = 2b_n X_i - b_n^2 X_i^2, \qquad i = 1, 2, \ldots.$$

By (6.12),

$$\begin{aligned}P(S_n \ge x_n V_n) &\ge P\left(S_n \ge \frac{1}{2b_n}\left(b_n^2 V_n^2 + x_n^2\right)\right) \\ &= P\left(\sum_{i=1}^n \xi_i \ge x_n^2\right).\end{aligned} \tag{6.22}$$

As in Lemma 6.2, let

$$m(h) = E\xi e^{h\xi}/Ee^{h\xi}, \qquad \sigma^2(h) = E\xi^2 e^{h\xi}/Ee^{h\xi} - m^2(h), \qquad x = x_n^2/n.$$

From Lemma 6.3, it follows that

$$m(h) = (1+2\varepsilon)x_n^2/n + o\left(x_n^2/n\right),$$

$$E\xi e^{h\xi} - (x_n^2/n)Ee^{h\xi} = 2\varepsilon x_n^2/n + o\left(x_n^2/n\right),$$

$$\frac{\sigma(h)\left(Ee^{h\xi}\right)^{1/2}}{\sqrt{n}} = \frac{2(1+o(1))x_n/\sqrt{n}}{\sqrt{n}} = o\left(x_n^2/n\right).$$

Therefore, (6.14) is satisfied for every sufficiently large n. By Lemma 6.2 and (6.15),

$$\begin{aligned} P\left(\sum_{i=1}^{n} \xi_i \geq x_n^2\right) &\geq \frac{3}{4}\left(Ee^{h\xi}\right)^n e^{-nhm(h)-2h\sigma(h)\sqrt{n}} \\ &\geq \frac{3}{4}e^{\varepsilon(1+\varepsilon)x_n^2/2-h(1+2\varepsilon)x_n^2+o(x_n^2)} \\ &= \frac{3}{4}e^{-(1+\varepsilon)^2x_n^2/2+o(x_n^2)}. \end{aligned} \tag{6.23}$$

Since ε is arbitrary, (6.3) follows from (6.22) and (6.23). □

Remark 6.4. The above proof of Theorem 6.1 has shown actually that the convergence in (6.1) is uniform: For arbitrary $0 < \varepsilon < 1/4$, there exist $0 < \delta < 1$, $x_0 > 1$ and n_0 such that for any $n \geq n_0$ and $x_0 < x < \delta\sqrt{n}$,

$$e^{-(1+\varepsilon)x^2/2} \leq P\left(\frac{S_n}{V_n} \geq x\right) \leq e^{-(1-\varepsilon)x^2/2}. \tag{6.24}$$

Remark 6.5. Following the proof of (6.2) and using the Ottaviani maximal inequality (2.28) to bound J_1, one can obtain the following result under the conditions of Theorem 6.1: For any $0 < \varepsilon < 1/2$, there exist $\theta > 1$, $0 < \delta < 1$, $x_0 > 1$ and n_0 such that for any $n \geq n_0$ and $x_0 < x < \delta\sqrt{n}$,

$$P\left(\max_{n\leq k\leq \theta n} \frac{S_k}{V_k} \geq x\right) \leq e^{-(1-\varepsilon)x^2/2}. \tag{6.25}$$

6.2 Self-Normalized Moderate Deviations: Stable Case

Let X be in the domain of attraction of a stable distribution with exponent α $(0 < \alpha < 2)$ and x_n be a sequence of constants satisfying

$$x_n \to \infty \quad \text{and} \quad x_n = o(\sqrt{n})$$

as $n \to \infty$. In this section we prove that the tail probability of the self-normalized sum $P(S_n \geq x_n V_n)$ is also Gaussian-like. Specifically, we have

Theorem 6.6. *Assume that there exist* $0 < \alpha < 2$, $c_1 \geq 0$, $c_2 \geq 0$, $c_1 + c_2 > 0$ *and a slowly varying function* $h(x)$ *such that*

$$P(X \geq x) = \frac{c_1 + o(1)}{x^\alpha}h(x) \quad \text{and} \quad P(X \leq -x) = \frac{c_2 + o(1)}{x^\alpha}h(x) \tag{6.26}$$

as $x \to \infty$. *Moreover, assume that* $EX = 0$ *if* $1 < \alpha < 2$, X *is symmetric if* $\alpha = 1$ *and that* $c_1 > 0$ *if* $0 < \alpha < 1$. *Then*

$$\lim_{n\to\infty} x_n^{-2} \log P(S_n \geq x_n V_n) = -\beta(\alpha, c_1, c_2), \tag{6.27}$$

where $\beta(\alpha, c_1, c_2)$ is the solution of

$$\Gamma(\beta, \alpha, c_1, c_2) = 0, \tag{6.28}$$

in which $\Gamma(\beta, \alpha, c_1, c_2)$ is given by

$$\begin{cases} c_1 \int_0^\infty \frac{1+2x-e^{2x-x^2/\beta}}{x^{\alpha+1}} dx + c_2 \int_0^\infty \frac{1-2x-e^{-2x-x^2/\beta}}{x^{\alpha+1}} dx & \text{if } 1 < \alpha < 2, \\ c_1 \int_0^\infty \frac{2-e^{2x-x^2/\beta}-e^{-2x-x^2/\beta}}{x^2} dx & \text{if } \alpha = 1, \\ c_1 \int_0^\infty \frac{1-e^{2x-x^2/\beta}}{x^{\alpha+1}} dx + c_2 \int_0^\infty \frac{1-e^{-2x-x^2/\beta}}{x^{\alpha+1}} dx & \text{if } 0 < \alpha < 1. \end{cases} \tag{6.29}$$

In particular, if X is symmetric, then

$$\lim_{n\to\infty} x_n^{-2} \log P(S_n \geq x_n V_n) = -\beta(\alpha), \tag{6.30}$$

where $\beta(\alpha)$ is the solution of

$$\int_0^\infty \frac{2 - e^{2x-x^2/\beta} - e^{-2x-x^2/\beta}}{x^{\alpha+1}} dx = 0.$$

Remark 6.7. It is easy to see that $\Gamma(\beta, \alpha, c_1, c_2)$ is strictly decreasing and continuous on $(0, \infty)$ and, by the L'Hôpital rule, that

$$\lim_{\beta \downarrow 0} \Gamma(\beta, \alpha, c_1, c_2) = \infty \quad \text{and} \quad \lim_{\beta \uparrow \infty} \Gamma(\beta, \alpha, c_1, c_2) = -\infty.$$

Therefore the solution of $\Gamma(\beta, \alpha, c_1, c_2) = 0$ indeed exists and is unique.

6.2.1 Preliminary Lemmas

To prove Theorem 6.6, we start with some preliminary lemmas. Statements below are understood to hold for every sufficiently large n. Let

$$y_n = x_n^2/n \tag{6.31}$$

and let z_n be a sequence of positive numbers such that

$$h(z_n) z_n^{-\alpha} \sim y_n \qquad \text{as } n \to \infty. \tag{6.32}$$

Lemma 6.8. *Under the conditions of Theorem 6.6, we have as $x \to \infty$,*

$$P(|X| \geq x) \sim \frac{c_1 + c_2}{x^\alpha} h(x), \tag{6.33}$$

$$E|X|^2 I(|X| \le x) \sim \frac{\alpha(c_1+c_2)}{2-\alpha} x^{2-\alpha} h(x), \tag{6.34}$$

$$E|X| I(|X| \ge x) \sim \frac{\alpha(c_1+c_2)}{\alpha-1} x^{1-\alpha} h(x) \qquad \text{if } 1 < \alpha < 2, \tag{6.35}$$

$$E|X| I(|X| \le x) \sim \frac{\alpha(c_1+c_2)}{1-\alpha} x^{1-\alpha} h(x) \qquad \text{if } 0 < \alpha < 1. \tag{6.36}$$

Proof. Equation (6.33) follows from the assumption (6.26). For (6.34), write

$$x^2 = \int_0^\infty 2t I(t \le |x|) dt$$

and

$$\begin{aligned} E|X|^2 I(|X| \le x) &= E \int_0^\infty 2t I(t \le |X|) I(|X| \le x) dt \\ &= \int_0^x 2t \{P(|X| \ge t) - P(|X| > x)\} dt \\ &\sim \frac{2(c_1+c_2)x^{2-\alpha}h(x)}{2-\alpha} - (c_1+c_2)x^{2-\alpha}h(x) \\ &= \frac{\alpha(c_1+c_2)}{2-\alpha} x^{2-\alpha} h(x) \end{aligned}$$

by (6.33) and (P5) in Sect. 4.1. This proves (6.34), and the proofs of (6.35) and (6.36) are similar. □

Lemma 6.9. *Under the conditions of Theorem 6.6,*

$$Ee^{-b^2X^2} = 1 - 2(c_1+c_2)b^\alpha h(1/b) \int_0^\infty x^{1-\alpha} e^{-x^2} dx + o(b^\alpha h(1/b)) \tag{6.37}$$

as $b \downarrow 0$.

Proof. Observe that

$$1 - e^{-y^2} = \int_0^{|y|} 2x e^{-x^2} dx = \int_0^\infty 2x e^{-x^2} I(|y| \ge x) dx.$$

We have

$$\begin{aligned} 1 - Ee^{-b^2X^2} &= 2 \int_0^\infty x e^{-x^2} P(|X| \ge x/b) dx \\ &= 2 \int_0^\infty x e^{-x^2} \frac{(c_1+c_2+o(1))h(x/b)}{(x/b)^\alpha} dx \qquad \text{by (6.33)} \\ &= 2(c_1+c_2)b^\alpha h(1/b) \int_0^\infty x^{1-\alpha} e^{-x^2} dx + o(b^\alpha h(1/b)) \end{aligned}$$

by (P2) and (P4) in Sect. 4.1. □

For $t > 0$, put

$$\gamma(t) = \begin{cases} c_1\alpha\int_0^\infty \frac{1+2tx-e^{t(2x-x^2)}}{x^{\alpha+1}}dx + c_2\alpha\int_0^\infty \frac{1-2tx-e^{t(-2x-x^2)}}{x^{\alpha+1}}dx & \text{if } 1<\alpha<2, \\ c_1\int_0^\infty \frac{2-e^{t(2x-x^2)}-e^{t(-2x-x^2)}}{x^2}dx & \text{if } \alpha=1, \\ c_1\alpha\int_0^\infty \frac{1-e^{t(2x-x^2)}}{x^{\alpha+1}}dx + c_2\alpha\int_0^\infty \frac{1-e^{t(-2x-x^2)}}{x^{\alpha+1}}dx & \text{if } 0<\alpha<1. \end{cases} \tag{6.38}$$

Note that

$$\gamma'(t) = \begin{cases} c_1\alpha\int_0^\infty \frac{2-(2-x)e^{t(2x-x^2)}}{x^{\alpha}}dx + c_2\alpha\int_0^\infty \frac{(2+x)e^{t(-2x-x^2)}-2}{x^{\alpha}}dx & \text{if } 1<\alpha<2, \\ c_1\int_0^\infty \frac{(x-2)e^{t(2x-x^2)}+(2+x)e^{t(-2x-x^2)}}{x}dx & \text{if } \alpha=1, \\ c_1\alpha\int_0^\infty \frac{(x-2)e^{t(2x-x^2)}}{x^{\alpha}}dx + c_2\alpha\int_0^\infty \frac{(2+x)e^{t(-2x-x^2)}}{x^{\alpha}}dx & \text{if } 0<\alpha<1; \end{cases} \tag{6.39}$$

and

$$\gamma''(t) = \begin{cases} -c_1\alpha\int_0^\infty \frac{(2-x)^2e^{t(2x-x^2)}}{x^{\alpha-1}}dx - c_2\alpha\int_0^\infty \frac{(2+x)^2e^{t(-2x-x^2)}}{x^{\alpha-1}}dx & \text{if } 1<\alpha<2, \\ -c_1\int_0^\infty (x-2)^2e^{t(2x-x^2)} + (2+x)^2e^{t(-2x-x^2)}dx & \text{if } \alpha=1, \\ -c_1\alpha\int_0^\infty \frac{(x-2)^2e^{t(2x-x^2)}}{x^{\alpha-1}}dx - c_2\alpha\int_0^\infty \frac{(2+x)^2e^{t(-2x-x^2)}}{x^{\alpha}}dx & \text{if } 0<\alpha<1. \end{cases} \tag{6.40}$$

The next two lemmas play a key role in the proof of Theorem 6.6.

Lemma 6.10. *Let*

$$\xi := \xi_b = 2bX - (bX)^2, \qquad b>0$$

and let $0<d<D<\infty$. *Under the conditions of Theorem 6.6, as* $b\downarrow 0$,

$$1 - Ee^{t\xi} = \gamma(t)b^\alpha h(1/b) + o(b^\alpha h(1/b)), \tag{6.41}$$

$$E\xi e^{t\xi} = -b^\alpha h(1/b)\gamma'(t) + o(b^\alpha h(1/b)), \tag{6.42}$$

and

$$E\xi^2 e^{t\xi} = -b^\alpha h(1/b)\gamma''(t) + o(b^\alpha h(1/b)) \tag{6.43}$$

for any $d\le t\le D$, *where* $\gamma(t)$ *is defined as in* (6.38) *and the constants implied in* $o(\cdot)$ *do not depend on* t.

Proof. Let $'$ denote derivative with respect to x. In the case $1<\alpha<2$, we use integration by parts and $EX=0$ to obtain

$$
\begin{aligned}
1-Ee^{t\xi} &= 2t\int_0^\infty P(X\ge x/b)\left(1-(1-x)e^{t(2x-x^2)}\right)dx\\
&\quad +2t\int_0^\infty P(X\le -x/b)\left(-1+(1+x)e^{t(-2x-x^2)}\right)dx\\
&= 2t\int_0^\infty \frac{(c_1+o(1))h(x/b)}{(x/b)^\alpha}\left(1-(1-x)e^{t(2x-x^2)}\right)dx\\
&\quad +2t\int_0^\infty \frac{(c_2+o(1))\,h(x/b)}{(x/b)^\alpha}\left(-1+(1+x)e^{t(-2x-x^2)}\right)dx\\
&= (c_1+o(1))\,b^\alpha h(1/b)\,(1+o(1))\,2t\int_0^\infty \frac{1}{x^\alpha}\left(1-(1-x)e^{t(2x-x^2)}\right)dx\\
&\quad +(c_2+o(1))\,b^\alpha h(1/b)\,(1+o(1))\,2t\int_0^\infty \frac{1}{x^\alpha}\left(-1+(1+x)e^{t(-2x-x^2)}\right)dx\\
&= \gamma(t)b^\alpha h(1/b)+o\left(b^\alpha h(1/b)\right),
\end{aligned} \tag{6.44}
$$

where (P2) is used for the third equality and integration by parts is used for the last equality. Moreover,

$$
\begin{aligned}
E\xi e^{t\xi} &= 2\int_0^\infty P(X\ge x/b)\left((1-x)\left(t(2x-x^2)+1\right)e^{t(2x-x^2)}-1\right)dx\\
&\quad +2\int_0^\infty P(X\le -x/b)\left((1+x^2)\left(t(2x+x^2)-1\right)e^{t(-2x-x^2)}+1\right)dx\\
&= 2\int_0^\infty \frac{(c_1+o(1))\,h(x/b)}{(x/b)^\alpha}\left((1-x)\left(t(2x-x^2)+1\right)e^{t(2x-x^2)}-1\right)dx\\
&\quad +2\int_0^\infty \frac{(c_2+o(1))\,h(x/b)}{(x/b)^\alpha}\left((1+x^2)\left(t(2x+x^2)-1\right)e^{t(-2x-x^2)}+1\right)dx\\
&= 2c_1b^\alpha h(1/b)\int_0^\infty \frac{1}{x^\alpha}\left((1-x)\left(t(2x-x^2)+1\right)e^{t(2x-x^2)}-1\right)dx\\
&\quad +2c_2b^\alpha h(1/b)\int_0^\infty \frac{1}{x^\alpha}\left((1+x^2)\left(t(2x+x^2)-1\right)e^{t(-2x-x^2)}+1\right)dx\\
&\quad +o\left(b^\alpha h(1/b)\right)\\
&= -\gamma'(t)b^\alpha h(1/b)+o\left(b^\alpha h(1/b)\right),
\end{aligned} \tag{6.45}
$$

proving (6.41) and (6.42). To prove (6.43), we proceed similarly to obtain

$$
\begin{aligned}
E\xi^2e^{t\xi} &= \int_0^\infty P(X\ge x/b)\left((2x-x^2)^2e^{t(2x-x^2)}\right)'dx\\
&\quad +\int_0^\infty P(X\le -x/b)\left((2x+x^2)^2e^{t(-2x-x^2)}\right)'dx\\
&= c_1b^\alpha h(1/b)\int_0^\infty x^{-\alpha}\left((2x-x^2)^2e^{t(2x-x^2)}\right)'dx\\
&\quad +c_2b^\alpha h(1/b)\int_0^\infty x^{-\alpha}\left((2x+x^2)^2e^{t(-2x-x^2)}\right)'dx+o\left(b^\alpha h(1/b)\right)
\end{aligned}
$$

$$\begin{aligned}
&= \alpha c_1 b^\alpha h(1/b) \int_0^\infty \frac{(2-x)^2 e^{t(2x-x^2)}}{x^{\alpha-1}} dx \\
&\quad + \alpha c_2 b^\alpha h(1/b) \int_0^\infty \frac{(2+x)^2 e^{t(2x-x^2)}}{x^{\alpha-1}} dx + o(b^\alpha h(1/b)) \\
&= -b^\alpha h(1/b)\gamma''(t) + o(b^\alpha h(1/b)). \qquad (6.46)
\end{aligned}$$

For the case $\alpha = 1$, since X is symmetric,

$$\begin{aligned}
1 - Ee^{t\xi} &= -\int_0^\infty \left(1 - e^{t(2x-x^2)}\right) dP(X \geq x/b) \\
&\quad - \int_0^\infty \left(1 - e^{t(-2x-x^2)}\right) dP(X \leq -x/b) \\
&= -\int_0^\infty \left(2 - e^{t(2x-x^2)} - e^{t(-2x-x^2)}\right) dP(X \geq x/b) \\
&= \int_0^\infty P(X \geq x/b) \left(2 - e^{t(2x-x^2)} - e^{t(-2x-x^2)}\right)' dx.
\end{aligned}$$

In the case $0 < \alpha < 1$, we do not have this simplification and work directly with

$$\begin{aligned}
1 - Ee^{t\xi} &= \int_0^\infty P(X \geq x/b) \left(1 - e^{t(2x-x^2)}\right)' dx \\
&\quad + \int_0^\infty P(X \leq -x/b) \left(1 - e^{t(-2x-x^2)}\right)' dx.
\end{aligned}$$

We can then proceed as in (6.44)–(6.46) to complete the proof. □

Lemma 6.11. *Let $0 < d \leq D < \infty$. Then, under the conditions of Theorem 6.6,*

$$\sup_{0<b\leq D/z_n} \inf_{t>0} e^{-tcy_n} Ee^{t(2bX-|bX|^2)} \leq e^{-\beta c y_n + o(y_n)}$$

for every $d \leq c \leq D$, where $\beta := \beta_p(\alpha, c_1, c_2)$ is defined as in Theorem 6.6, z_n and y_n are as in (6.31) and (6.32), and the constant implied by $o(y_n)$ is uniform in $c \in [d, D]$.

Proof. Let $0 < \delta < d$ and divide $0 < b < D/z_n$ into two parts: $0 < b < \delta/z_n$ and $\delta/z_n \leq b \leq D/z_n$. From (6.41) it follows that for $0 < b < \delta/z_n$,

$$\begin{aligned}
Ee^{3\beta(2bX-|bX|^2)} &\leq 1 - \gamma(3\beta) b^\alpha h(1/b) + o(b^\alpha h(1/b)) \\
&\leq \exp\left((|\gamma(3\beta)| + 1) b^\alpha h(1/b)\right) \\
&\leq \exp\left(K(\delta/z_n)^\alpha h(z_n/\delta)\right) \\
&\leq \exp\left(K_1 \delta^{\alpha/2} z_n^{-\alpha} h(z_n)\right) \\
&\leq \exp\left(K_2 \delta^{\alpha/2} y_n\right) \leq \exp(c\beta y_n),
\end{aligned}$$

provided that δ is chosen to be sufficiently small, and that n is large enough; here and in the sequel, K and $K_1, K_2, \ldots$ denote positive constants which depend only on α and other given constants, but may be different from line to line. Hence there exists $\delta > 0$ such that

$$\begin{aligned}
&\sup_{0<b\le\delta/z_n} \inf_{t>0} e^{-tcy_n} E e^{t(2bX-|bX|^2)} \\
&\qquad \le \sup_{0<b\le\delta/z_n} e^{-3\beta c y_n} E e^{3\beta(2bX-|bX|^2)} \le e^{-2\beta c y_n}. \qquad (6.47)
\end{aligned}$$

Next estimate $\sup_{\delta/z_n\le b\le D/z_n} \inf_{t>0} e^{-tcy_n} E e^{t(pbX-|bX|^p)}$. Let $\gamma(t)$, $\gamma'(t)$ and $\gamma''(t)$ be defined as in (6.38), (6.39) and (6.40) respectively. In view of (6.40) and the fact that

$$\gamma''(t) < 0 \text{ for } t > 0, \quad \lim_{t\downarrow 0} \gamma'(t) = \infty \quad \text{and} \quad \lim_{t\uparrow\infty} \gamma'(t) = -\infty,$$

there exists a unique t_b such that

$$\gamma'(t_b) = -\frac{y_n c}{b^\alpha h(z_n)}. \qquad (6.48)$$

Since

$$0 < K_1 \le \frac{d\, y_n z_n^\alpha}{D^\alpha h(z_n)} \le \frac{y_n c}{b^\alpha h(z_n)} \le \frac{D y_n z_n^\alpha}{\delta^\alpha h(z_n)} \le K_2 < \infty$$

for $\delta/z_n \le b \le D/z_n$, we have

$$K_3 \le t_b \le K_4.$$

Applying (P2) in Sect. 4.1 and (6.41) again, we obtain

$$\begin{aligned}
&\sup_{\delta/z_n\le b\le D/z_n} \inf_{t>0} e^{-tcy_n} E e^{t(2bX-|bX|^2)} \\
&\quad \le \sup_{\delta/z_n\le b\le D/z_n} e^{-t_b c y_n} E e^{t_b(2bX-|bX|^2)} \\
&\quad \le \sup_{\delta/z_n\le b\le D/z_n} \exp\left(-t_b c y_n - \gamma(t_b) b^\alpha h(1/b) + o\left(b^\alpha h(1/b)\right)\right) \\
&\quad \le \sup_{\delta/z_n\le b\le D/z_n} \exp\left(-t_b c y_n - \gamma(t_b) b^\alpha h(z_n) + \gamma(t_b) b^\alpha h(z_n) o(1) + o(y_n)\right) \\
&\quad \le \sup_{\delta/z_n\le b\le D/z_n} \exp\left(-t_b c y_n - \gamma(t_b) b^\alpha h(z_n) + o(y_n)\right).
\end{aligned}$$

Let

$$g(b) = -t_b c y_n - \gamma(t_b) b^\alpha h(z_n)$$

and b_0 be such that $t_{b_0} = \beta$. Noting that $\gamma(t) = \alpha t^\alpha \Gamma_p(t, \alpha, c_1, c_2)$, we have

$$\gamma(t_{b_0}) = 0.$$

By (6.48),

$$g'(b) = -\gamma(t_b)\alpha b^{\alpha-1}h(z_n)\begin{cases} > 0 & \text{if } b < b_0, \\ = 0 & \text{if } b = b_0, \\ < 0 & \text{if } b > b_0 \end{cases}$$

for t_b a decreasing function of b, and $\gamma(t)/t^\alpha$ is a decreasing function of t. Thus, $g(b)$ achieves the maximum at $b = b_0$ and $g(b_0) = -\beta c y_n$. Consequently,

$$\sup_{\delta/z_n \le b \le D/z_n} \inf_{t>0} e^{-tcy_n} E e^{t(2bX-|bX|^2)} \le \exp\left(-\beta c y_n + o(y_n)\right). \tag{6.49}$$

From (6.47) and (6.49), the desired conclusion follows. □

6.2.2 *Proof of Theorem 6.6*

Let $\beta = \beta(\alpha, c_1, c_2)$. We first show that for any $0 < \varepsilon < 1/2$,

$$P\left(S_n \ge x_n V_n\right) \le \exp\left(-(1-\varepsilon)\beta x_n^2\right) \tag{6.50}$$

provided that n is sufficiently large. Define y_n and z_n as in (6.31) and (6.32) and let $0 < \delta < A < \infty$. The values of δ and A will be specified later, with δ sufficiently small and A sufficiently large. Similar to (6.8),

$$\begin{aligned} P(S_n \ge x_n V_n) \le\ & P\left(S_n \ge x_n V_n, \delta x_n z_n < V_n < A x_n z_n\right) \\ & + P\left(S_n \ge x_n V_n, V_n \ge A x_n z_n\right) + P(V_n \le \delta x_n z_n) \\ \le\ & P\left(S_n \ge \inf_{b = x_n/V_n} \left((bV_n)^2 + x_n^2\right)/(2b), \delta x_n z_n \le V_n \le A x_n z_n\right) \\ & + P\left(\sum_{i=1}^n X_i I\left(|X_i| \le \sqrt{A} z_n\right) \ge A x_n^2 z_n/2\right) \\ & + P\left(\sum_{i=1}^n X_i I\left(|X_i| > \sqrt{A} z_n\right) \ge x_n V_n/2\right) + P(V_n \le \delta x_n z_n) \\ \le\ & P\left(S_n \ge \inf_{1/(Az_n) \le b \le 1/(\delta z_n)} \left((bV_n)^2 + x_n^2\right)/(2b)\right) \\ & + P\left(\sum_{i=1}^n X_i I(|X_i| \le A z_n) \ge A x_n^2 z_n/2\right) \\ & + P\left(\sum_{i=1}^n I\left(|X_i| > \sqrt{A} z_n\right) \ge (x_n/2)^2\right) + P(V_n^2 \le \delta^2 x_n^2 z_n^2) \\ :=\ & T_1 + T_2 + T_3 + T_4. \end{aligned} \tag{6.51}$$

From (2.27), (P4), (6.33), and (6.32), it follows that

$$\begin{aligned}
T_3 &\le \left(\frac{4enP(|X|>\sqrt{A}z_n)}{x_n 2}\right)^{x_n^2/4} \le \left(\frac{16(c_1+c_2)h(Az_n)}{A^{\alpha/2}z_n^\alpha y_n}\right)^{x_n^2/4} \\
&\le \left(\frac{20(c_1+c_2)h(z_n)}{A^{\alpha/4}z_n^\alpha y_n}\right)^{x_n^2/4} \le \left(25(c_1+c_2)/A^{\alpha/4}\right)^{x_n^2/4} \le e^{-2\beta x_n^2}, \quad (6.52)
\end{aligned}$$

provided that A is large enough. Let $t = 1/(\delta z_n)$ and $c_0 = \int_0^\infty x^{1-\alpha}e^{-x^2}dx$. It follows from (6.37) that for δ sufficiently small,

$$\begin{aligned}
T_4 &\le e^{t\delta^2 x_n^2 z_n^2} E e^{-tV_n^2} = e^{\delta x_n^2}\left(Ee^{-tX^2}\right)^n \\
&\le \exp\left(\delta x_n^2 - (c_1+c_2)c_0 n t^\alpha h(t)\right) \\
&\le \exp\left(\delta x_n^2 - 0.6(c_1+c_2)c_0 n\,(\delta z_n)^{-\alpha}h(z_n)\right) \\
&\le \exp\left(\delta x_n^2 - 0.5(c_1+c_2)c_0\delta^{-\delta}x_n^2\right) \le \exp(-2\beta x_n^2). \quad (6.53)
\end{aligned}$$

To bound T_2, we apply Lemma 6.8 to obtain

$$\begin{aligned}
&\sum_{i=1}^n \left|EX_i I\left(|X_i| \le \sqrt{A}z_n\right)\right| = n\left|EXI\left(|X| \le \sqrt{A}z_n\right)\right| \\
&\le \begin{cases} nE|X|I\left(|X| > \sqrt{A}z_n\right) & \text{if } 1<\alpha<2 \\ 0 & \text{if } \alpha=1 \\ nE|X|I\left(|X| \le \sqrt{A}z_n\right) & \text{if } 0<\alpha<1 \end{cases} \\
&\le \begin{cases} 2n\alpha(c_1+c_2)(\sqrt{A}z_n)^{1-\alpha}h(\sqrt{A}z_n)/(\alpha-1) & \text{if } 1<\alpha<2 \\ 0 & \text{if } \alpha=1 \\ 2n\alpha(c_1+c_2)(\sqrt{A}z_n)^{1-\alpha}h(\sqrt{A}z_n)/(1-\alpha) & \text{if } 0<\alpha<1 \end{cases} \\
&\le \begin{cases} 2n\alpha(c_1+c_2)A^{1-\alpha/2}z_n^{1-\alpha}h(z_n)/(\alpha-1) & \text{if } 1<\alpha<2 \\ 0 & \text{if } \alpha=1 \\ 2n\alpha(c_1+c_2)A^{1-\alpha/2}z_n^{1-\alpha}h(z_n)/(1-\alpha) & \text{if } 0<\alpha<1 \end{cases} \\
&\le Ax_n^2 z_n/4, \qquad (6.54)
\end{aligned}$$

$$nEX^2 I\left(|X| \le \sqrt{A}z_n\right) \le \frac{4(c_1+c_2)}{2-\alpha} n(\sqrt{A}z_n)^{2-\alpha}h(\sqrt{A}z_n) \le Az_n^2 x_n^2. \qquad (6.55)$$

Therefore, by the Bernstein inequality (2.17), we have for all sufficiently large A,

$$\begin{aligned}
T_2 &\le P\left(\sum_{i=1}^{n}\{X_i I\left(|X_i|\le\sqrt{A}z_n\right)-EX_i I\left(|X_i|\le\sqrt{A}z_n\right)\ge Ax_n^2 z_n/4\right)\\
&\le \exp\left(-\frac{A^2x_n^4z_n^2}{32(Az_n^2x_n^2+A^{3/2}z_n^2x_n^2)}\right)\\
&\le \exp\left(-\frac{A^{1/2}x_n^2}{64}\right)\le\exp(-2\beta x_n^2). \qquad (6.56)
\end{aligned}$$

To bound T_1, let $\theta=(1-\varepsilon/2)^{-1/2}$ and $b_j=\theta^j/(Az_n)$, $j=0,1,2,\ldots$. It follows from Lemma 6.11 that

$$\begin{aligned}
T_1 &= P\left(\sup_{1/(Az_n)\le b\le 1/(\delta z_n)}(2bS_n-b^2V_n^2)\ge x_n^2\right)\\
&\le P\left(\max_{0\le j\le\log_\theta(A/\delta)}\sup_{b_j\le b\le b_{j+1}}(2bS_n-b^2V_n^2)\ge x_n^2\right)\\
&\le P\left(\max_{0\le j\le\log_\theta(A/\delta)}(2b_{j+1}S_n-b_j^2V_2^2)\ge x_n^2\right)\\
&\le \sum_{0\le j\le\log_\theta(A/\delta)} P\left(2\theta b_jS_n-b_j^2V_n^2\ge x_n^2\right)\\
&= \sum_{0\le j\le\log_\theta(A/\delta)} P\left(2(b_j/\theta S_n-(b_j/\theta)^p2V_n^2\ge (x_n/\theta)^2\right)\\
&\le (1+\log_\theta(A/\delta))\sup_{0<b\le 1/(\delta z_n)}P\left(2bS_n-b^2V_n^2\ge(x_n/\theta)^2\right)\\
&\le (1+\log_\theta(A/\delta))\sup_{0<b\le 1/(\delta z_n)}\inf_{t>0}e^{-t(x_n/\theta)^2}Ee^{t\left(2bS_n-b^2V_n^2\right)}\\
&\le (1+\log_\theta(A/\delta))\left(\sup_{0<b\le 1/(\delta z_n)}\inf_{t>0}e^{-ty_n/\theta^2}Ee^{t\left(2bX-|bX|^2\right)}\right)^n\\
&\le (1+\log_\theta(A/\delta))\exp(-\beta ny_n/\theta^2+o(y_n)n)\\
&= (1+\log_\theta(A/\delta))\exp\left(-(p-1)\beta x_n^2/\theta^2+o(x_n^2)\right)\\
&= (1+\log_\theta(A/\delta))\exp\left(-\beta(1-\varepsilon/2)x_n^2+o(x_n^2)\right). \qquad (6.57)
\end{aligned}$$

From (6.52), (6.53), (6.56) and (6.57), (6.50) follows.

We next use the same idea as that in the proof of (6.3) to show

$$P(S_n\ge x_nV_n)\ge\exp\left(-(1+\varepsilon)\beta x_n^2\right). \qquad (6.58)$$

Recalling that $\gamma(t)=\alpha t^\alpha\Gamma_p(t,\alpha,c_1,c_2)$, we have $\gamma(\beta)=0$. Since $\gamma(t)$ is concave on $(0,\infty)$ and $\lim_{t\downarrow 0}\gamma(t)=0$, it follows from $\gamma(\beta)=0$ that $\gamma'(\beta)<0$. Let $\delta=\varepsilon/3$ and $\gamma'(t)$ be as in (6.39). Put

$$b := b_{n,\delta} = \left(-\frac{(1+\delta)y_n}{\gamma'(\beta)h(z_n)} \right)^{1/\alpha},$$

$$\xi = 2bX - |bX|^2 \quad \text{and} \quad \xi_i = 2bX_i - |bX_i|^2, \qquad i = 1, 2, \ldots.$$

Application of (6.12) yields

$$P(S_n \geq x_n V_n) \geq P\left(S_n \geq \left(b^2 V_n^2 + x_n^2\right)/(2b)\right) = P\left(\sum_{i=1}^{n} \xi_i \geq n(p-1)y_n \right).$$

To verify condition (6.14), let $m(\cdot)$ and $\sigma(\cdot)$ be the same as in Lemma 6.2. From (6.32), it follows that

$$b \sim \frac{1}{z_n} \left(-\frac{(1+\delta)}{\gamma'(\beta)} \right)^{1/\alpha}.$$

By Lemma 6.10, (P2) and (6.32), we obtain

$$Ee^{\beta\xi} = 1 + o(y_n),$$

$$E\xi\, e^{\beta\xi} = (1+\delta)y_n + o(y_n),$$

$$\left(E\xi^2 e^{\beta\xi}\right)^{1/2} \Big/ \sqrt{n} = O\left(\sqrt{y_n}/\sqrt{n}\right) = o(y_n),$$

and hence

$$m(\beta) = (1+\delta)y_n + o(y_n),$$

$$\sigma(\beta)/\sqrt{n} = o(y_n).$$

Thus, the condition (6.14) is satisfied with $h = \beta$. Therefore, by Lemma 6.2,

$$\begin{aligned} P\left(\sum_{i=1}^{n} \xi_i \geq n y_n \right) &\geq \frac{3}{4} \left(Ee^{\beta\xi}\right)^n \exp\left(-n\beta m(\beta) - 2\beta\, \sigma(\beta)\sqrt{n}\right) \\ &\geq \frac{3}{4} \exp\left(o(y_n)n - n(1+\delta)\beta y_n\right) \\ &\geq \exp\left(-(1+\varepsilon)\beta x_n^q\right), \end{aligned}$$

as desired, proving Theorem 6.6. □

Remark 6.12. Analogous to Remark 6.4, the convergence in (6.27) is uniform: For arbitrary $0 < \varepsilon < 1/2$, there exist $0 < \delta < 1, x_0 > 1$ and n_0 such that for any $n \geq n_0$ and $x_0 < x < \delta\sqrt{n}$,

$$e^{-(1+\varepsilon)\beta(\alpha,c_1,c_2)x^2} \leq P(S_n \geq xV_n) \leq e^{-(1-\varepsilon)\beta(\alpha,c_1,c_2)x^2}.$$

Moreover, analogous to Remark 6.5, we have a strong version of (6.50).

Theorem 6.13. *Under the conditions of Theorem 6.6, for any* $0 < \varepsilon < 1/2$ *there exists* $\theta > 1$ *such that*

$$P\left(\max_{n\le k\le \theta n}\frac{S_k}{V_k}\ge x_n\right)\le \exp\left(-(1-\varepsilon)\beta(\alpha,c_1,c_2)x_n^2\right) \tag{6.59}$$

for every n sufficiently large.

Proof. Let $\eta=(1-(1-\varepsilon/2)^{1/4})/3$. Clearly,

$$P\left(\max_{n\le k\le \theta n}\frac{S_k}{V_k}\ge x_n\right)\le P\left(\frac{S_n}{V_n}\ge (1-3\eta)x_n\right) \\ +P\left(\max_{n<k\le \theta n}\frac{S_k-S_n}{V_k}\ge 3\eta x_n\right). \tag{6.60}$$

By Theorem 6.6, if n is sufficiently large,

$$P\left(\frac{S_n}{V_n}\ge (1-3\eta)x_n\right)\le \exp\left(-(1-\varepsilon/2)\beta(\alpha,c_1,c_2)x_n^2\right). \tag{6.61}$$

Next we show that the second term on the right hand side of (6.60) is bounded by $\exp(-2\beta(\alpha,c_1,c_2)x_n^2)$. Let z_n be as in (6.32) and let $0<\delta<1/4$. Write

$$\begin{aligned}P\left(\max_{n<k\le \theta n}\frac{S_k-S_n}{V_k}\ge 3\eta x_n\right) &\le P\left(\max_{n<k\le \theta n}\frac{\sum_{i=n+1}^k X_i I\left(|X_i|\le (\eta\delta)^2 z_n\right)}{V_k}\ge 2\eta x_n\right)\\ &\quad +P\left(\max_{n<k\le \theta n}\frac{\sum_{i=n+1}^k |X_i| I\left(|X_i|\ge (\eta\delta)^2 z_n\right)}{V_k}\ge \eta x_n\right)\\ &\le P\left(\max_{n<k\le \theta n}\sum_{i=n+1}^k X_i I\left(|X_i|\le (\eta\delta)^2 z_n\right)\ge 2\eta\delta x_n^2 z_n\right)\\ &\quad +P(V_n\le \delta x_n z_n)+P\left(\sum_{i=n+1}^{[\theta n]} I\left(|X_i|\ge (\eta\delta)^2 z_n\right)\ge (\eta x_n)^2\right).\end{aligned}$$

By (6.53), there exists $\delta>0$ such that

$$P(V_n\le \delta x_n z_n)\le \exp(-2\beta(\alpha,c_1,c_2)x_n^2).$$

Similar to (6.52), we have

$$\begin{aligned}P\left(\sum_{i=n+1}^{[\theta n]} I\left(|X_i|\ge (\eta\delta)^2 z_n\right)\ge (\eta x_n)^2\right) &\le \left(\frac{3(\theta-1)nP\left(|X|\ge (\eta\delta)^2 z_n\right)}{(\eta x_n)^2}\right)^{(\eta x_n)^2}\\ &\le \left(\frac{6(c_1+c_2)(\theta-1)nh(z_n)}{(\eta x_n)^2(\eta\delta)^{2\alpha}z_n^\alpha}\right)^{(\eta x_n)^2}\\ &\le \left(\frac{8(\theta-1)(c_1+c_2)}{\eta^6\delta^4}\right)^{\eta^2 x_n^2}\\ &\le \exp\left(-2\beta(\alpha,c_1,c_2)x_n^2\right),\end{aligned}$$

provided that θ is sufficiently near 1. In view of the proof of (6.54), we can choose $\theta - 1 > 0$ sufficiently small so that

$$\sum_{i=n+1}^{[\theta n]} \left| EX_i I\left(|X_i| \leq (\eta\delta)^2 z_n\right) \right| \leq K(\theta - 1)x_n^2 z_n \leq \frac{1}{2}\eta\delta x_n^2 z_n,$$

$$\begin{aligned} \sum_{i=n+1}^{[\theta n]} \operatorname{Var}\left(X_i I\left(|X_i| \leq (\eta\delta)^2 z_n\right)\right) &\leq (\theta - 1)nEX^2 I\left(|X| \leq (\eta\delta)^2 z_n\right) \\ &\leq \frac{2(\theta-1)n\alpha(c_1+c_2)}{2-\alpha}(\eta\delta)^{4-2\alpha} z_n^{2-\alpha} h(z_n) \\ &\leq K(\theta-1)x_n^2 z_n^2 \leq \eta^4\delta^4 x_n^2 z_n^2, \end{aligned}$$

where K is a constant depending only on $\alpha, c_1, c_2, \eta, \delta$. Therefore, by the Ottaviani inequality (2.28) and the Bernstein inequality (2.17),

$$\begin{aligned} &P\left(\max_{n<k\leq\theta n} \sum_{i=n+1}^{k} X_i I\left(|X_i| \leq (\eta\delta)^2 z_n\right) \geq 2\eta\delta x_n^2 z_n\right) \\ &\leq 2P\left(\sum_{i=n+1}^{[\theta n]} \left\{X_i I\left(|X_i| \leq (\eta\delta)^2 z_n\right) - EX_i I\left(|X_i| \leq (\eta\delta)^2 z_n\right)\right\} \geq \eta\delta x_n^2 z_n\right) \\ &\leq \exp\left(-\frac{(\eta\delta x_n^2 z_n)^2}{2\eta^4\delta^4 x_n^2 z_n^2 + 4(\eta\delta)^2 z_n(\eta\delta x_n^2 z_n)}\right) \\ &= \exp\left(-\frac{x_n^2}{6\eta^2\delta^2}\right) \leq \exp\left(-2\beta(\alpha, c_1, c_2)x_n^2\right), \end{aligned}$$

provided that δ is small. Putting together the above inequalities yields

$$P\left(\max_{n<k\leq\theta n} \frac{S_k - S_n}{V_k} \geq 3\eta x_n\right) \leq 4\exp\left(-2\beta(\alpha, c_1, c_2)x_n^2\right). \tag{6.62}$$

From (6.61), (6.60) and (6.62), (6.59) follows. □

6.3 Self-Normalized Laws of the Iterated Logarithm

Let $X, X_1, X_2, \ldots$ be i.i.d. random variables. Finiteness of the second moment is necessary for the classical law of the iterated logarithm to hold; see Theorem 2.4. Moreover, if X is symmetric and in the domain of attraction of a stable law with index α $(0 < \alpha < 2)$, then

$$\limsup_{n\to\infty} \frac{S_n}{a_n} = 0 \quad \text{or} \quad \infty \ a.s. \tag{6.63}$$

for any sequence $\{a_n, n \geq 1\}$ of positive numbers with $a_n \to \infty$; see Feller (1946). In contrast, Griffin and Kuelbs (1989) have proved that a self-normalized law of the iterated logarithm holds for all distributions in the domain of attraction of a normal or stable law. This is the content of the following theorem, in which the constant specified in (6.65) is due to Shao (1997).

Theorem 6.14.

(a) If $EX = 0$ *and* $EX^2 I(|X| \leq x)$ *is slowly varying as* $x \to \infty$*, then*

$$\limsup_{n\to\infty} \frac{S_n}{V_n(2\log\log n)^{1/2}} = 1 \ \ a.s. \tag{6.64}$$

(b) Under the conditions of Theorem 6.6, we have

$$\limsup_{n\to\infty} \frac{S_n}{V_n(\log\log n)^{1/2}} = (\beta(\alpha, c_1, c_2))^{-1/2} \ \ a.s. \tag{6.65}$$

In particular, if X *is symmetric, then*

$$\limsup_{n\to\infty} \frac{S_n}{V_n(\log\log n)^{1/2}} = (\beta(\alpha))^{-1/2} \ \ a.s., \tag{6.66}$$

where $\beta(\alpha, c_1, c_2)$ *and* $\beta(\alpha)$ *are defined as in Theorem 6.6.*

Proof. We only prove part (b) since part (a) is similar. We first show that

$$\limsup_{n\to\infty} \frac{S_n}{V_n(\log\log n)^{1/2}} \leq (\beta(\alpha, c_1, c_2))^{-1/2} \ \ a.s. \tag{6.67}$$

For any $0 < \varepsilon < 1/4$, let $\theta > 1$ be given in (6.59). Then

$$\limsup_{n\to\infty} \frac{S_n}{V_n(\log\log n)^{1/2}} = \limsup_{k\to\infty} \max_{\theta^k \leq n \leq \theta^{k+1}} \frac{S_n}{V_n(\log k)^{1/2}}. \tag{6.68}$$

By (6.59),

$$\begin{aligned} &\sum_{k=1}^{\infty} P\left(\max_{\theta^k \leq n \leq \theta^{k+1}} \frac{S_n}{V_n(\log k)^{1/2}} \geq (1+\varepsilon)\beta(\alpha, c_1, c_2)^{-1/2} \right) \\ &\qquad \leq K \sum_{k=1}^{\infty} \exp\left(-(1-\varepsilon)(1+\varepsilon)^2 \log k\right) < \infty. \end{aligned}$$

In view of the Borel–Cantelli lemma and (6.68), (6.67) follows.

To prove the lower bound of the lim sup, let $\tau > 1$ and $n_k = [e^{k^\tau}]$, $k = 1, 2, \ldots$. Note that

$$\begin{aligned}
\limsup_{n\to\infty}\frac{S_n}{V_n(\log\log n)^{1/2}} &\geq \limsup_{k\to\infty}\frac{S_{n_k}}{V_{n_k}(\log\log n_k)^{1/2}} \\
&\geq \limsup_{k\to\infty}\frac{S_{n_k}-S_{n_{k-1}}}{V_{n_k}(\log\log n_k)^{1/2}}+\liminf_{k\to\infty}\frac{S_{n_{k-1}}}{V_{n_k}(\log\log n_k)^{1/2}} \\
&= \limsup_{k\to\infty}\frac{(V_{n_k}^2-V_{n_{k-1}}^2)^{1/2}}{V_{n_k}}\frac{S_{n_k}-S_{n_{k-1}}}{(V_{n_k}^2-V_{n_{k-1}}^2)^{1/2}(\log\log n_k)^{1/2}} \\
&\quad +\liminf_{k\to\infty}\frac{V_{n_{k-1}}}{V_{n_k}}\frac{S_{n_{k-1}}}{V_{n_{k-1}}(\log\log n_k)^{1/2}}.
\end{aligned} \tag{6.69}$$

Since $(S_{n_k}-S_{n_{k-1}})/(V_{n_k}^2-V_{n_{k-1}}^2)^{1/2})$, $k\geq 1$, are independent, it follows from Theorem 6.6 and the Borel–Cantelli lemma that

$$\limsup_{k\to\infty}\frac{S_{n_k}-S_{n_{k-1}}}{(V_{n_k}^2-V_{n_{k-1}}^2)^{1/2}(\log\log n_k)^{1/2}}\geq\frac{1}{\tau^2\beta(\alpha,c_1,c_2)^{1/2}}\quad a.s. \tag{6.70}$$

We shall show that

$$\lim_{k\to\infty}\frac{V_{n_k}}{V_{n_{k-1}}}=\infty\ a.s. \tag{6.71}$$

By (6.69)–(6.71) and (6.67),

$$\limsup_{n\to\infty}\frac{S_n}{V_n(\log\log n)^{1/2}}\geq\frac{1}{\tau^2\beta(\alpha,c_1,c_2)^{1/2}}\quad a.s. \tag{6.72}$$

Since $\tau>1$ is arbitrary, (6.65) follows from (6.67) and (6.72).

To prove (6.71), let $x_{n_k}=k$ and define z_{n_k} as in (6.32). Then, by (6.53) and the Borel–Cantelli lemma,

$$\liminf_{k\to\infty}\frac{V_{n_k}}{x_{n_k}z_{n_k}}\geq\delta>0\ a.s. \tag{6.73}$$

From Lemma 6.8, it follows that

$$\begin{aligned}
P(V_{n_{k-1}}>z_{n_k}) &\leq n_{k-1}P(|X|>z_{n_k})+P\left(\sum_{i=1}^{n_{k-1}}X_i^2I(|X_i|\leq z_n)>z_{n_k}^2\right) \\
&\leq\frac{O(1)n_{k-1}h(z_{n_k})}{z_{n_k}^{\alpha}}+\frac{n_{k-1}EX^2I(|X_i|\leq z_n)}{z_{n_k}^2} \\
&=\frac{O(1)n_{k-1}h(z_{n_k})}{z_{n_k}^{\alpha}}=\frac{O(1)n_{k-1}x_{n_k}^2}{n_k} \\
&=O(k^{-2})\qquad\text{by (6.32).}
\end{aligned}$$

Hence, by the Borel–Cantelli lemma again,

$$\limsup_{k\to\infty}\frac{V_{n_{k-1}}}{z_{n_k}}\leq 1\ a.s. \tag{6.74}$$

From (6.73) and (6.74), (6.71) follows. □

6.4 Supplementary Results and Problems

Let $X, X_1, X_2, \ldots$ be i.i.d. random variables. Put

$$S_n = \sum_{i=1}^n X_i, \quad V_n^2 = \sum_{i=1}^n X_i^2, \quad V_{n,p}^p = \sum_{i=1}^n |X_i|^p, \qquad p > 1.$$

1. Assume that $E(X) = 0$ and $E(X^2) = 1$.
   (a) If $Ee^{t_0|X|} < \infty$ for some $t_0 > 0$, show that
   $$\log P(|S_n| \geq x_n\sqrt{n}) \sim -x_n^2/2 \tag{6.75}$$
   for $x_n \to \infty$ and $x_n = o(n^{1/2})$.
   (b) If (6.75) holds for any $x_n \to \infty$ and $x_n = o(n^{1/2})$, show that there exists $t_0 > 0$ such that $Ee^{t_0|X|} < \infty$ (see Shao, 1989).
2. Assume that $E(X) = 0$ and $E(X^2) < \infty$. Let a_i, $1 \leq i \leq n$, be a sequence of real numbers. Under what condition does the result
   $$\log P\left(\sum_{i=1}^n a_i X_i \geq x_n \left(\sum_{i=1}^n a_i^2 X_i^2\right)^{1/2}\right) \sim -x_n^2/2$$
   hold for $x_n \to \infty$ and $x_n = o(n^{1/2})$?
3. Prove (6.25).
4. *Moderate deviation normalized by* $V_{n,p}$ (Shao, 1997): Assume that the conditions in Theorem 6.6 are satisfied. Let $p > \max(1, \alpha)$, and let $\{x_n, n \geq 1\}$ be a sequence of positive numbers with $x_n \to \infty$ and $x_n = o(n^{(p-1)/p})$ as $n \to \infty$. Then
   $$\lim_{n\to\infty} x_n^{-p/(p-1)} \log P\left(\frac{S_n}{V_{n,p}} \geq x_n\right) = -(p-1)\beta_p(\alpha, c_1, c_2), \tag{6.76}$$
   where $\beta_p(\alpha, c_1, c_2)$ is the solution of $\Gamma_p(\beta, \alpha, c_1, c_2) = 0$ and $\Gamma_p(\beta, \alpha, c_1, c_2) =$
   $$\begin{cases} c_1 \displaystyle\int_0^\infty \frac{1 + px - e^{px - x^p/\beta^{p-1}}}{x^{\alpha+1}}dx + c_2 \int_0^\infty \frac{1 - px - e^{-px - x^p/\beta^{p-1}}}{x^{\alpha+1}}dx & \text{if } 1 < \alpha < 2, \\ c_1 \displaystyle\int_0^\infty \frac{2 - e^{px - x^p/\beta^{p-1}} - e^{-px - x^p/\beta^{p-1}}}{x^2}dx & \text{if } \alpha = 1, \\ c_1 \displaystyle\int_0^\infty \frac{1 - e^{px - x^p/\beta^{p-1}}}{x^{\alpha+1}}dx + c_2 \int_0^\infty \frac{1 - e^{-px - x^p/\beta^{p-1}}}{x^{\alpha+1}}dx & \text{if } 0 < \alpha < 1. \end{cases}$$
5. *Moderate deviation normalized by* $\max_{1\leq k\leq n}|X_k|$ (Horváth and Shao, 1996): Assume that the conditions of Theorem 6.6 are satisfied. If $\{x_n, 1 \leq n < \infty\}$ is a sequence of positive numbers satisfying $x_n \to \infty$ and $x_n = o(n)$, then we have
   $$\lim_{n\to\infty} \frac{1}{x_n} \log P\left(S_n \geq x_n \max_{1\leq k\leq n}|X_k|\right) = -\tau(\alpha, c_1, c_2),$$

where $\tau = \tau(\alpha, c_1, c_2) > 0$ is the solution of $f(\tau) = c_1 + c_2$, and $f(t) =$

$$\begin{cases} \frac{t\alpha(c_2 - c_1)}{\alpha - 1} + c_1\alpha \int_0^1 \frac{e^{tx} - 1 - tx}{x^{\alpha+1}} dx + c_2\alpha \int_0^1 \frac{e^{-tx} - 1 + tx}{x^{\alpha+1}} dx & \text{if } 1 < \alpha < 2, \\ c_1 \int_0^1 \frac{e^{tx} + e^{-tx} - 2}{x^2} dx & \text{if } \alpha = 1, \\ c_1\alpha \int_0^1 \frac{e^{tx} - 1}{x^{\alpha+1}} dx + c_2\alpha \int_0^1 \frac{e^{-tx} - 1}{x^{\alpha+1}} dx & \text{if } 0 < \alpha < 1. \end{cases}$$

6. *Universal self-normalized moderate deviation for centered Feller class* (Jing et al., 2008): Let C_s denote the support of X, that is,

$$C_s = \{x : P(X \in (x - \varepsilon, x + \varepsilon)) > 0, \quad \text{for any } \varepsilon > 0\}.$$

Assume that

$$C_s \cap \mathbb{R}^+ \neq \emptyset \quad \text{and} \quad C_s \cap \mathbb{R}^- \neq \emptyset, \quad \text{where } \mathbb{R}^+ = \{x : x > 0\}, \ \mathbb{R}^- = \{x : x < 0\}$$

and that either $EX = 0$ or $EX^2 = \infty$. If X is in the centered Feller class, i.e.,

$$\limsup_{a \to \infty} \frac{a^2 \{P(|X| > a) + a^{-1}|EXI(|X| \le a)\}}{EX^2 I(|X| \le a)} < \infty,$$

then

$$\log P(S_n/V_n \ge x_n) \sim -n\lambda(x_n^2/n)$$

for any sequence $\{x_n, n \ge 1\}$ with $x_n \to \infty$ and $x_n = o(\sqrt{n})$, where

$$\lambda(x) = \inf_{b \ge 0} \sup_{t \ge 0} \left(tx - \log E \exp\{t(2bX - b^2X^2)\}\right).$$

If in addition, $\text{Card}(C_s) \ge 3$, then

$$\lim_{n \to \infty} x_n^{-2} \log P(S_n/V_n \ge x_n) = -t_0, \tag{6.77}$$

where $t_0 = \lim_{x \to 0^+} t_x$, and (t_x, b_x) is the solution (t, b) of the equations

$$Eb(2X - bX^2)\exp\{tb(2X - bX^2)\} = xE\exp\{tb(2X - bX^2)\}, \tag{6.78}$$
$$E(X - bX^2)\exp\{tb(2X - bX^2)\} = 0. \tag{6.79}$$

7. Let X satisfy the conditions in Theorem 6.6. Prove that X is in the centered Feller class. Also verify that $t_0 = \beta(\alpha, c_1, c_2)$, where t_0 is as in (6.77) and $\beta(\alpha, c_1, c_2)$ in (6.27).
8. Assume that X is symmetric. Prove that

$$\limsup_{n \to \infty} \frac{|S_n|}{V_n(2\log\log n)^{1/2}} \le 1 \ \ a.s. \tag{6.80}$$

Under what condition can the inequality ≤ 1 be changed to $= 1$?
Hint: Use (2.11).

Chapter 7
Cramér-Type Moderate Deviations for Self-Normalized Sums

Let $X_1, X_2, \ldots, X_n$ be a sequence of independent random variables with zero means and finite variances. In Sect. 2.1.3, we have described Cramér's moderate deviation results for $(\sum_{i=1}^n X_i)/(\sum_{i=1}^n EX_i^2)^{1/2}$. In this chapter we show that similar to self-normalized large and moderate deviation theorems in Chaps. 3 and 6, Cramér-type moderate deviation results again hold for self-normalized sums under minimal moment conditions.

7.1 Self-Normalized Cramér-Type Moderate Deviations

Let $X_1, X_2, \ldots, X_n$ be independent random variables with $EX_i = 0$ and $0 < EX_i^2 < \infty$. Let

$$S_n = \sum_{i=1}^n X_i, \qquad B_n^2 = \sum_{i=1}^n EX_i^2, \qquad V_n^2 = \sum_{i=1}^n X_i^2, \tag{7.1}$$

$$\begin{aligned} \Delta_{n,x} = {} & \frac{(1+x)^2}{B_n^2} \sum_{i=1}^n EX_i^2 I\left(|X_i| > B_n/(1+x)\right) \\ & + \frac{(1+x)^3}{B_n^3} \sum_{i=1}^n E|X_i|^3 I\left(|X_i| \le B_n/(1+x)\right) \end{aligned} \tag{7.2}$$

for $x \ge 0$. Jing et al. (2003) have proved the following theorem.

Theorem 7.1. *There exists an absolute constant A (> 1) such that*

$$\frac{P(S_n \ge xV_n)}{1 - \Phi(x)} = e^{O(1)\Delta_{n,x}} \quad \textit{and} \quad \frac{P(S_n \le -xV_n)}{\Phi(-x)} = e^{O(1)\Delta_{n,x}} \tag{7.3}$$

for all $x \ge 0$ satisfying

$$x^2 \max_{1 \le i \le n} EX_i^2 \le B_n^2 \tag{7.4}$$

V.H. de la Peña et al., *Self-Normalized Processes: Limit Theory and Statistical Applications*, Probability and its Applications,

© Springer-Verlag Berlin Heidelberg 2009

and

$$\Delta_{n,x} \le (1+x)^2/A, \qquad |O(1)| \le A\,. \tag{7.5}$$

Theorem 7.1 provides a very general framework and the following results are its direct consequences.

Theorem 7.2. *Let $\{a_n, n \ge 1\}$ be a sequence of positive numbers. Assume that*

$$a_n^2 \le B_n^2 / \max_{1\le i\le n} EX_i^2 \tag{7.6}$$

and

$$\forall\, \varepsilon > 0,\ B_n^{-2} \sum_{i=1}^n EX_i^2 I(|X_i| > \varepsilon B_n/(1+a_n)) \to 0 \quad \text{as } n\to\infty. \tag{7.7}$$

Then

$$\frac{\log P(S_n/V_n \ge x)}{\log(1-\Phi(x))} \to 1, \qquad \frac{\log P(S_n/V_n \le -x)}{\log \Phi(-x)} \to 1 \tag{7.8}$$

holds uniformly for $x \in (0, a_n)$.

The next corollary is a special case of Theorem 7.2 and may be of independent interest.

Corollary 7.3. *Suppose that $B_n \ge c\sqrt{n}$ for some $c > 0$ and that $\{X_i^2, i \ge 1\}$ is uniformly integrable. Then, for any sequence of real numbers x_n satisfying $x_n \to \infty$ and $x_n = o(\sqrt{n})$,*

$$\log P(S_n/V_n \ge x_n) \sim -x_n^2/2. \tag{7.9}$$

When the X_i's have finite $(2+\delta)$th absolute moments for $0 < \delta \le 1$, we have

Theorem 7.4. *Let $0 < \delta \le 1$ and set*

$$L_{n,\delta} = \sum_{i=1}^n E|X_i|^{2+\delta}, \qquad d_{n,\delta} = B_n / L_{n,\delta}^{1/(2+\delta)}.$$

Then for $0 \le x \le d_{n,\delta}$,

$$\frac{P(S_n/V_n \ge x)}{1-\Phi(x)} = 1 + O(1)\left(\frac{1+x}{d_{n,\delta}}\right)^{2+\delta}, \tag{7.10}$$

$$\frac{P(S_n/V_n \le -x)}{\Phi(-x)} = 1 + O(1)\left(\frac{1+x}{d_{n,\delta}}\right)^{2+\delta}, \tag{7.11}$$

where $O(1)$ is bounded by an absolute constant. In particular, if $d_{n,\delta} \to \infty$ as $n \to \infty$, we have

$$\frac{P(S_n \ge xV_n)}{1-\Phi(x)} \to 1, \qquad \frac{P(S_n \le -xV_n)}{\Phi(-x)} \to 1 \tag{7.12}$$

uniformly in $0 \le x \le o(d_{n,\delta})$.

Results (7.10) and (7.11) are useful because they provide not only the relative error but also a Berry–Esseen rate of convergence. By the fact that $1-\Phi(x) \le 2e^{-x^2/2}/(1+x)$ for $x \ge 0$, it follows from (7.10) that the following exponential non-uniform Berry–Esseen bound holds for $0 \le x \le d_{n,\delta}$:

$$|P(S_n/V_n \ge x) - (1-\Phi(x))| \le A(1+x)^{1+\delta} e^{-x^2/2}/d_{n,\delta}^{2+\delta}. \tag{7.13}$$

The next corollary specifies $d_{n,\delta}$ under certain circumstances and especially for i.i.d. cases.

Corollary 7.5. *Let* $0 < \delta \le 1$. *Assume that* $\{|X_i|^{2+\delta}, i \ge 1\}$ *is uniformly integrable and that* $B_n \ge c n^{1/2}$ *for some constant* $c > 0$. *Then* (7.12) *holds uniformly for* $x \in [0, o(n^{\delta/(4+2\delta)}))$.

For i.i.d. random variables, Theorem 7.1 reduces to

Corollary 7.6. *Let* $X, X_1, X_2, \dots$ *be i.i.d. random variables with* $EX = 0$ *and* $\sigma^2 = EX^2 < \infty$. *Then there exists an absolute constant* $A > 2$ *such that*

$$\frac{P(S_n \ge xV_n)}{1-\Phi(x)} = e^{O(1)\Delta_{n,x}} \quad \text{and} \quad \frac{P(S_n \le -xV_n)}{\Phi(-x)} = e^{O(1)\Delta_{n,x}}$$

for all $x \ge 0$ *satisfying* $\Delta_{n,x} \le (1+x)^2/A$, *where* $|O(1)| \le A$ *and*

$$\begin{aligned}\Delta_{n,x} = &(1+x)^2\sigma^{-2}EX_1^2 I\left(|X_1| > \sqrt{n}\sigma/(1+x)\right)\\ &+(1+x)^3\sigma^{-3}n^{-1/2}E|X_1|^3 I\left(|X_1| \le \sqrt{n}\sigma/(1+x)\right).\end{aligned}$$

Remark 7.7. If $X_1, X_2, \dots$ are i.i.d. random variables with $\sigma^2 = EX_1^2 < \infty$, then condition (7.4) reduces to $x \le \sqrt{n}$ while (7.5) reduces to

$$\frac{1}{\sigma^2}E\left\{X_1^2 I\left(|X_1| > \frac{\sqrt{n}\sigma}{1+x}\right)\right\} + \frac{1+x}{\sqrt{n}}\frac{E|X_1|^3}{\sigma^3} I\left(|X_1| \le \frac{\sqrt{n}\sigma}{1+x}\right) \le \frac{1}{A},$$

which in turn implies $(1+x) \le \sqrt{n}$. Hence, (7.5) implies (7.4) in the i.i.d. case. However, (7.5) does not imply (7.4) in general. On the other hand, it would be of interest to find out if condition (7.4) in Theorem 7.1 or condition (7.6) in Theorem 7.2 can be removed.

Remark 7.8. An example given in Shao (1999) shows that in the i.i.d. case, the condition $E|X_1|^{2+\delta} < \infty$ cannot be replaced by $E|X_1|^r < \infty$ for some $r < 2+\delta$ for (7.12) to hold.

Remark 7.9. When $X_1, X_2, \dots$ are i.i.d. random variables, $d_{n,\delta}$ is simply equal to $n^{\delta/(4+2\delta)}(EX_1^2)^{1/2}/(E|X_1|^{2+\delta})^{1/(2+\delta)}$.

7.2 Proof of Theorems

Throughout the remainder of this chapter, we use A to denote an absolute constant, which may assume different values at different places. We first prove Theorems 7.2 and 7.4 by making use of Theorem 7.1 and then give the proof of Theorem 7.1.

7.2.1 Proof of Theorems 7.2, 7.4 and Corollaries

Proof (of Theorem 7.2). Note that for $0 < \varepsilon \le 1$ and $0 \le x \le a_n$,

$$\begin{aligned}
&B_n^{-2}\sum_{i=1}^n EX_i^2 I(|X_i| > B_n/(1+x)) + (1+x)B_n^{-3}\sum_{i=1}^n E|X_i|^3 I(|X_i| \le B_n/(1+x)) \\
&= B_n^{-2}\sum_{i=1}^n EX_i^2 I(|X_i| > B_n/(1+x)) + (1+x)B_n^{-3}\sum_{i=1}^n E|X_i|^3 I(|X_i| \le \varepsilon B_n/(1+a_n)) \\
&\quad +(1+x)B_n^{-3}\sum_{i=1}^n E|X_i|^3 I(\varepsilon B_n/(1+a_n) < |X_i| \le B_n/(1+x)) \\
&\le B_n^{-2}\sum_{i=1}^n EX_i^2 I(|X_i| > B_n/(1+x)) + \varepsilon(1+x)B_n^{-2}/(1+a_n)\sum_{i=1}^n E|X_i|^2 \\
&\quad +B_n^{-2}\sum_{i=1}^n E|X_i|^2 I(\varepsilon B_n/(1+a_n) < |X_i| \le B_n/(1+x)) \\
&\le \varepsilon + B_n^{-2}\sum_{i=1}^n E|X_i|^2 I(|X_i| > \varepsilon B_n/(1+a_n)).
\end{aligned}$$

Therefore by (7.7),

$$\Delta_{n,x} = o\left((1+x)^2\right) \quad \text{as } n \to \infty$$

uniformly for $0 \le x \le a_n$. Hence Theorem 7.2 follows from Theorem 7.1. □

Proof (of Corollary 7.3). For any a_n satisfying $a_n \to \infty$ and $a_n = o(B_n)$, the uniform integrability assumption implies that (7.6) and (7.7) are satisfied and hence the corollary follows from Theorem 7.2. □

Proof (of Theorem 7.4). Equations (7.10) and (7.11) follow from Theorem 5.9 on the Berry–Esseen bound for $0 \le x \le A$. When $x > A$, it is easy to see that

$$\Delta_{n,x} \le (1+x)^{2+\delta} L_{n,\delta}/B_n^{2+\delta} = \left(\frac{1+x}{d_{n,\delta}}\right)^{2+\delta} \le (1+x)^2/A$$

and that $d_{n,\delta}^2 \max_{i\le n} EX_i^2 \le B_n^2$. Thus, conditions (7.4) and (7.5) are satisfied and the result follows from Theorem 7.1. □

Proof (of Corollary 7.5). Let $d > 0$ and $x_n = d\,n^{\delta/(4+2\delta)}$. It suffices to show that

$$\Delta_{n,x_n} = o(1) \quad \text{as } n \to \infty. \tag{7.14}$$

Similar to the proof of Theorem 7.2, we have, for any $0 < \varepsilon < 1$,

$$\begin{aligned}
\Delta_{n,x_n} &\le (1+x_n)^2 B_n^{-2} \sum_{i=1}^n EX_i^2 I(|X_i| > B_n/(1+x_n)) \\
&\quad + \varepsilon^{1-\delta}(1+x_n)^{2+\delta} B_n^{-(2+\delta)} \sum_{i=1}^n E|X_i|^{2+\delta} I(|X_i| \le \varepsilon B_n/(1+x_n)) \\
&\quad + (1+x_n)^{2+\delta} B_n^{-(2+\delta)} \sum_{i=1}^n E|X_i|^{2+\delta} I(\varepsilon B_n/(1+x_n) < |X_i| \le B_n/(1+x_n)) \\
&\le (1+x_n)^{2+\delta} B_n^{-(2+\delta)} \sum_{i=1}^n E|X_i|^{2+\delta} I(|X_i| > \varepsilon B_n/(1+x_n)) + O(1)\varepsilon^{1-\delta} \\
&= o(1) + O(1)\varepsilon^{1-\delta},
\end{aligned}$$

since $\{|X_i|^{2+\delta},\ i \ge 1\}$ is uniformly integrable. This proves (7.14) because ε can be arbitrarily small and hence the corollary. □

7.2.2 Proof of Theorem 7.1

We use the same notation as before and only prove the first part in (7.3) since the second part can be easily obtained by changing x to $-x$ in the first part. The main idea of the proof is to reduce the problem to a one-dimensional large deviation result. It suffices to show that

$$P(S_n \ge xV_n) \ge (1-\Phi(x))\, e^{-A\Delta_{n,x}} \tag{7.15}$$

and

$$P(S_n \ge xV_n) \le (1-\Phi(x))\, e^{A\Delta_{n,x}} \tag{7.16}$$

for all $x > 0$ satisfying (7.4) and (7.5). Let

$$b := b_x = x/B_n. \tag{7.17}$$

Since $xV_n \le (x^2 + b^2V_n^2)/(2b)$, it follows that

$$P(S_n \ge xV_n) \ge P\left(S_n \ge (x^2 + b^2V_n^2)/(2b)\right) = P(2bS_n - b^2V_n^2 \ge x^2).$$

Therefore, the lower bound (7.15) follows from the following proposition.

Proposition 7.10. *There exists an absolute constant $A > 1$ such that*

$$P(2bS_n - b^2V_n^2 \ge x^2) = (1-\Phi(x))\, e^{O(1)\Delta_{n,x}} \tag{7.18}$$

for all $x > 0$ satisfying (7.4) and (7.5), where $|O(1)| \le A$.

As for the upper bound (7.16), when $0 < x \le 2$, this bound is a direct consequence of the Berry–Esseen bound in Theorem 5.9. For $x > 2$, let

$$\theta := \theta_{n,x} = B_n/(1+x) \tag{7.19}$$

and define

$$\bar{X}_i = X_i I(|X_i| \le \theta), \qquad \bar{S}_n = \sum_{i=1}^n \bar{X}_i, \qquad \bar{V}_n^2 = \sum_{i=1}^n \bar{X}_i^2,$$
$$S_n^{(i)} = S_n - X_i, \qquad V_n^{(i)} = (V_n^2 - X_i^2)^{1/2}, \qquad \bar{B}_n^2 = \sum_{i=1}^n E\bar{X}_i^2.$$

Noting that for any $s,t \in R^1$, $c \ge 0$ and $x \ge 1$,

$$\begin{aligned} x\sqrt{c+t^2} &= \sqrt{(x^2-1)c+t^2+c+(x^2-1)t^2} \\ &\ge \sqrt{(x^2-1)c+t^2+2t\sqrt{(x^2-1)c}} \\ &= t+\sqrt{(x^2-1)c}, \end{aligned}$$

we have

$$\{s+t \ge x\sqrt{c+t^2}\} \subset \{s \ge (x^2-1)^{1/2}\sqrt{c}\}. \tag{7.20}$$

Hence,

$$\begin{aligned} P(S_n \ge xV_n) &\le P(\bar{S}_n \ge x\bar{V}_n) + P(S_n \ge xV_n, \max_{1\le i\le n} |X_i| > \theta) \qquad (7.21) \\ &\le P(\bar{S}_n \ge x\bar{V}_n) + \sum_{i=1}^n P(S_n \ge xV_n, |X_i| > \theta) \\ &\le P(\bar{S}_n \ge x\bar{V}_n) + \sum_{i=1}^n P\left(S_n^{(i)} \ge (x^2-1)^{1/2} V_n^{(i)}, |X_i| > \theta\right) \\ &\le P(\bar{S}_n \ge x\bar{V}_n) + \sum_{i=1}^n P\left(S_n^{(i)} \ge (x^2-1)^{1/2} V_n^{(i)}\right) P(|X_i| > \theta). \end{aligned}$$

Moreover, $P(\bar{S}_n \ge x\bar{V}_n)$ is equal to

$$\begin{aligned} &P\left\{\bar{S}_n \ge x\left[\bar{B}_n^2 + \textstyle\sum_{i=1}^n (\bar{X}_i^2 - E\bar{X}_i^2)\right]^{1/2}\right\} \qquad (7.22) \\ &\le P\left\{\bar{S}_n \ge x\bar{B}_n\left[1+\frac{1}{2\bar{B}_n^2}\textstyle\sum_{i=1}^n (\bar{X}_i^2 - E\bar{X}_i^2) - \frac{1}{\bar{B}_n^4}\left(\sum_{i=1}^n (\bar{X}_i^2 - E\bar{X}_i^2)\right)^2\right]\right\} := K_n, \end{aligned}$$

where the inequality follows from $(1+y)^{1/2} \ge 1+y/2-y^2$ for any $y \ge -1$. Hence the upper bound (7.16) follows from the next three propositions.

Proposition 7.11. *There exists an absolute constant A such that*

$$P(S_n^{(i)} \geq x V_n^{(i)}) \leq (1+x^{-1}) \frac{1}{\sqrt{2\pi}x} \exp(-x^2/2 + A\Delta_{n,x}) \tag{7.23}$$

for any $x > 2$ satisfying (7.4) *and* (7.5).

Proposition 7.12. *There exists an absolute constant A such that*

$$K_n \leq (1 - \Phi(x)) e^{A\Delta_{n,x}} + Ae^{-3x^2} \tag{7.24}$$

for all $x > 2$ satisfying (7.4) *and* (7.5).

Proposition 7.13. *There exists an absolute constant A such that*

$$K_n \leq (1 - \Phi(x)) e^{A\Delta_{n,x}} + A\left(\Delta_{n,x}/(1+x)^2\right)^{4/3} \tag{7.25}$$

for $x > 2$ with $\Delta_{n,x}/(1+x)^2 \leq 1/128$.

To complete the proof of (7.16), we first use Propositions 7.12 and 7.13 to show that

$$K_n \leq (1 - \Phi(x)) e^{A\Delta_{n,x}} \tag{7.26}$$

for all $x > 2$ satisfying conditions (7.4) and (7.5). We consider two cases. If $\Delta_{n,x}/(1+x)^2 \leq (1-\Phi(x))^3/128$, then by (7.25),

$$\begin{aligned}
K_n &\leq (1 - \Phi(x)) e^{A\Delta_{n,x}} \left(1 + A(1+x)^{-2}\Delta_{n,x}\left(\Delta_{n,x}/(1+x)^2\right)^{1/3}/(1-\Phi(x))\right) \\
&\leq (1 - \Phi(x)) e^{A\Delta_{n,x}} \left(1 + A\Delta_{n,x}/(1+x)^2\right) \\
&\leq (1 - \Phi(x)) e^{2A\Delta_{n,x}}.
\end{aligned}$$

When $\Delta_{n,x}/(1+x)^2 > (1-\Phi(x))^3/128$, by (7.24),

$$\begin{aligned}
K_n &\leq (1 - \Phi(x)) e^{A\Delta_{n,x}} \left(1 + Ae^{-3x^2}/(1-\Phi(x))\right) \\
&\leq (1 - \Phi(x)) e^{A\Delta_{n,x}} \left(1 + A(1-\Phi(x))^3\right) \\
&\leq (1 - \Phi(x)) e^{A\Delta_{n,x}} \left(1 + 128A\Delta_{n,x}/(1+x)^2\right) \\
&\leq (1 - \Phi(x)) e^{129A\Delta_{n,x}}.
\end{aligned}$$

For $x > 2$, we next use Proposition 7.11 and the fact that $(2\pi)^{-1/2}(x^{-1} - x^{-3})e^{-x^2/2} \leq 1 - \Phi(x)$ for $x > 0$ to obtain

$$\begin{aligned}
& P\left(S_n^{(i)} \geq (x^2-1)^{1/2} V_n^{(i)}\right) \\
& \leq \left(1+(x^2-1)^{-1/2}\right)(2\pi)^{-1/2}(x^2-1)^{-1/2}\exp(-x^2/2+A\Delta_{n,x}) \\
& \leq (2\pi)^{-1/2}\frac{A}{x}\exp(-x^2/2+A\Delta_{n,x}) \\
& \leq (2\pi)^{-1/2}A(x^{-1}-x^{-3})\exp(-x^2/2+A\Delta_{n,x}) \\
& \leq A\,(1-\Phi(x))\exp(A\Delta_{n,x}). \qquad (7.27)
\end{aligned}$$

It follows from (7.21), (7.22), (7.26) and (7.27) that

$$\begin{aligned}
P(S_n \geq xV_n) &\leq P(\bar{S}_n \geq x\bar{V}_n) + \sum_{i=1}^n P\left(S_n^{(i)} \geq (x^2-1)^{1/2}V_n^{(i)}\right)P(|X_i|>\theta) \\
&\leq (1-\Phi(x))\,e^{A\Delta_{n,x}} + \sum_{i=1}^n A\,(1-\Phi(x))\exp(A\Delta_{n,x})P(|X_i|>\theta) \\
&\leq (1-\Phi(x))\,e^{A\Delta_{n,x}}\left(1+A\sum_{i=1}^n P(|X_i|>\theta)\right) \\
&\leq (1-\Phi(x))\,e^{A\Delta_{n,x}}\left(1+A\sum_{i=1}^n \theta^{-2}EX_i^2 I(|X_i|>\theta)\right) \\
&\leq (1-\Phi(x))\,e^{A\Delta_{n,x}}(1+A\Delta_{n,x}) \\
&\leq (1-\Phi(x))\,e^{2A\Delta_{n,x}}.
\end{aligned}$$

This completes the proof of (7.16) and therefore also that of Theorem 7.1.

7.2.3 Proof of Propositions

A key ingredient in the proofs of the propositions an appropriately chosen expansion for $E\{(\lambda bX-\theta(bX)^2)^k e^{\lambda bX-\theta(bX)^2}\}$ as $b \downarrow 0$, $k=0,1,2,3$. This is provided by the following lemmas whose proofs can be found in Jing et al. (2003).

Lemma 7.14. *Let X be a random variable with $EX=0$ and $EX^2<\infty$. Then, for any $0<b<\infty$, $\lambda>0$ and $\theta>0$,*

$$Ee^{\lambda bX-\theta(bX)^2} = 1+(\lambda^2/2-\theta)b^2EX^2+O_{\lambda,\theta}\,\delta_b, \qquad (7.28)$$

where $\delta_b = b^2EX^2I(|bX|>1)+b^3E|X|^3I(|bX|\leq 1)$ and $O_{\lambda,\theta}$ denotes a quantity that is bounded by a finite constant depending only on λ and θ. In (7.28), $|O_{\lambda,\theta}| \leq \max(\lambda+|\lambda^2/2-\theta|+e^{\lambda^2/(4\theta)},\ \lambda\theta+\theta^2/2+(\lambda+\theta)^3e^\lambda/6)$.

Lemma 7.15. *Let X be a random variable with $EX=0$ and $EX^2<\infty$. For $0<b<\infty$, let $\xi := \xi_b = 2bX-(bX)^2$. Then, for $\lambda>0$,*

$$
\begin{aligned}
Ee^{\lambda\xi} &= 1+(2\lambda^2-\lambda)b^2EX^2+O_{\lambda,0}\,\delta_b, && (7.29)\\
E\xi e^{\lambda\xi} &= (4\lambda-1)b^2EX^2+O_{\lambda,1}\,\delta_b, && (7.30)\\
E\xi^2e^{\lambda\xi} &= 4b^2EX^2+O_{\lambda,2}\,\delta_b, && (7.31)\\
E|\xi|^3e^{\lambda\xi} &= O_{\lambda,3}\,\delta_b, && (7.32)\\
(E\xi e^{\lambda\xi})^2 &= O_{\lambda,4}\delta_b, && (7.33)
\end{aligned}
$$

where δ_b is defined as in Lemma 7.14 and

$$
\begin{aligned}
|O_{\lambda,0}| &\le \max(2\lambda+|2\lambda^2-\lambda|+e^{\lambda},2.5\lambda^2+4\lambda^3e^{\lambda}/3),\\
|O_{\lambda,1}| &\le \max\left(2+|4\lambda-1|+\max(e^{\lambda},e/\lambda),5\lambda+13.5\lambda^2e^{\lambda}\right),\\
|O_{\lambda,2}| &\le \max\left(4+\max\left(e^{\lambda},(e/(2\lambda))^2\right),5+27\lambda e^{\lambda}\right),\\
|O_{\lambda,3}| &\le 27e^{\lambda},\\
|O_{\lambda,4}| &\le 2\max\left(\left(\max(e^{\lambda},e/\lambda)+2\right)^2,(1+9\lambda e^{\lambda})^2\right).
\end{aligned}
$$

In particular, when $\lambda=1/2$, $|O_{\lambda,0}|\le 2.65$, $|O_{\lambda,1}|\le 8.1$, $|O_{\lambda,2}|\le 28$, $|O_{\lambda,3}|\le 45$, $|O_{\lambda,4}|\le 150$, and

$$Ee^{\xi/2}=e^{O_5\delta_b},\qquad \text{where } |O_5|\le 5.5. \tag{7.34}$$

Lemma 7.16. *Let $\{\xi_i,\ 1\le i\le n\}$ be a sequence of independent random variables with $Ee^{h\xi_i}<\infty$ for $0<h<H$, where $H>0$. For $0<\lambda<H$, put*

$$m(\lambda)=\sum_{i=1}^{n}E\xi_ie^{\lambda\xi_i}/Ee^{\lambda\xi_i},\quad \sigma^2(\lambda)=\sum_{i=1}^{n}\left(E\xi_i^2e^{\lambda\xi_i}/Ee^{\lambda\xi_i}-(E\xi_ie^{\lambda\xi_i}/Ee^{\lambda\xi_i})^2\right).$$

Then

$$P\left(\sum_{i=1}^{n}\xi_i\ge y\right)\ge\frac{3}{4}\left(\prod_{i=1}^{n}Ee^{\lambda\xi_i}\right)e^{-\lambda m(\lambda)-2\lambda\sigma(\lambda)}, \tag{7.35}$$

provided that

$$0<\lambda<H\quad \text{and}\quad m(\lambda)\ge y+2\sigma(\lambda). \tag{7.36}$$

Because the details in applying these lemmas to prove the propositions involve lengthy calculations, they are omitted here. Interested readers can find the detailed proofs at the Web site for the book given in the Preface.

7.3 Application to Self-Normalized LIL

It is known that the law of the iterated logarithm is usually a direct consequence of a moderate deviation result. We first show that condition (7.6) in Theorem 7.2 can be removed.

Theorem 7.17. *Let x_n be a sequence of real numbers such that $x_n \to \infty$ and $x_n = o(B_n)$. Assume*

$$B_n^{-2}\sum_{i=1}^n EX_i^2 I(|X_i| > \varepsilon B_n/x_n) \to 0 \qquad \text{for all } \varepsilon > 0. \tag{7.37}$$

Then

$$\log P(S_n/V_n \ge x_n) \sim -x_n^2/2. \tag{7.38}$$

As a direct consequence of Theorem 7.17, we have the following self-normalized law of the iterated logarithm for independent random variables:

Theorem 7.18. *If $B_n \to \infty$ and*

$$B_n^{-2}\sum_{i=1}^n EX_i^2 I\left(|X_i| > \varepsilon B_n/(\log\log B_n)^{1/2}\right) \to 0 \qquad \text{for all } \varepsilon > 0,$$

then

$$\limsup_{n\to\infty} \frac{S_n}{V_n(2\log\log B_n)^{1/2}} = 1 \ \ a.s. \tag{7.39}$$

Remark 7.19. Shao (1995) proved that if for every $\varepsilon > 0$,

$$B_n^{-2}\sum_{i=1}^n EX_i^2 I\left(|X_i| > \varepsilon B_n/(\log\log B_n)^{1/2}\right) \to 0 \qquad \text{as } n \to \infty \tag{7.40}$$

and

$$\sum_{n=1}^\infty P\left(|X_n| > \varepsilon B_n/(\log\log B_n)^{1/2}\right) < \infty \tag{7.41}$$

are satisfied, then .

$$\limsup_{n\to\infty} \frac{S_n}{B_n(2\log\log B_n)^{1/2}} = 1 \ \ a.s. \tag{7.42}$$

The following example shows that the self-normalized LIL (7.39) holds but the LIL (7.42) that normalizes by B_n instead fails. Let $X_1, X_2, \ldots$ be independent random variables satisfying

$$P(X_n = 0) = \frac{3}{4} - \frac{1}{n(\log\log n)^3} + \frac{1}{4\log\log n},$$
$$P(X_n = \pm 2) = \frac{1}{8} - \frac{1}{8\log\log n}, \quad P\left(X_n = \pm n^{1/2}\log\log n\right) = \frac{1}{2n(\log\log n)^3}.$$

Then $EX_n = 0$, $EX_n^2 = 1$, and $\{X_n^2, n \ge 1\}$ is uniformly integrable. Hence, by Theorem 7.18, (7.39) holds. On the other hand, note that for all $\varepsilon > 0$,

$$\sum_{n=1}^{\infty} P\left(|X_n| > \varepsilon B_n (\log\log B_n)^{1/2}\right) = \infty \qquad \text{with } B_n = \sqrt{n}.$$

Therefore, by the Borel–Cantelli lemma, (7.42) does not hold.

Proof (of Theorem 7.17). It suffices to show that for $0 < \varepsilon < 1/2$,

$$P(S_n/V_n \ge x_n) \le \exp\left(-(1-\varepsilon)x_n^2/2\right) \tag{7.43}$$

and

$$P(S_n/V_n \ge x_n) \ge \exp\left(-(1+\varepsilon)x_n^2/2\right) \tag{7.44}$$

for sufficiently large n. Let $\eta = \eta_\varepsilon > 0$ that will be specified later and define $\tau = \eta^2 B_n/x_n$. Set

$$\bar{X}_i = X_i I(|X_i| \le \tau), \qquad \bar{S}_n = \sum_{i=1}^{n} \bar{X}_i, \qquad \bar{V}_n^2 = \sum_{i=1}^{n} \bar{X}_i^2.$$

Observe that

$$\begin{aligned}
P(S_n/V_n \ge x_n) &\le P\left(\bar{S}_n/V_n \ge (1-\eta)x_n\right) \\
&\quad + P\left(\sum_{i=1}^{n} X_i I(|X_i| > \tau)/V_n \ge \eta x_n\right) \\
&\le P\left(\bar{S}_n/\bar{V}_n \ge (1-\eta)x_n\right) + P\left(\sum_{i=1}^{n} I(|X_i| > \tau) \ge (\eta x_n)^2\right) \\
&\le P\left(\bar{S}_n \ge (1-\eta)^{3/2} x_n B_n\right) + P\left(\bar{V}_n^2 \le (1-\eta)B_n^2\right) \\
&\quad + P\left(\sum_{i=1}^{n} I(|X_i| > \tau) \ge (\eta x_n)^2\right).
\end{aligned} \tag{7.45}$$

From (7.37), it follows that

$$\sum_{i=1}^{n} P(|X_i| > \tau) \le \tau^{-2} \sum_{i=1}^{n} EX_i^2 I(|X_i| > \tau) = o(x_n^2).$$

Therefore

$$\begin{aligned}
P\left(\sum_{i=1}^{n} I(|X_i| > \tau) \ge (\eta x_n)^2\right) &\le \left(\frac{3\sum_{i=1}^{n} P(|X_i| > \tau)}{(\eta x_n)^2}\right)^{\eta^2 x_n^2} \\
&= o(1)^{\eta^2 x_n^2} \le \exp(-2x_n^2)
\end{aligned} \tag{7.46}$$

for n sufficiently large. Note that

$$E\bar{V}_n^2 = B_n^2 - \sum_{i=1}^{n} EX_i^2 I(|X_i| > \tau) = (1 - o(1))B_n^2 \geq (1 - \eta/2)B_n^2$$

for sufficiently large n. Hence, by the Bernstein inequality (2.17),

$$\begin{aligned} P\left(\bar{V}_n^2 \leq (1-\eta)B_n^2\right) &\leq \exp\left(-\frac{(\eta B_n^2/2)^2}{2\sum_{i=1}^n EX_i^4 I(|X_i| \leq \tau)}\right) \\ &\leq \exp\left(-\frac{(\eta B_n^2/2)^2}{2\tau^2 B_n^2}\right) \\ &= \exp\left(-\frac{x_n^2}{8\eta^2}\right) \leq \exp(-2x_n^2), \end{aligned} \tag{7.47}$$

provided that $\eta < 1/16$ and that n is sufficiently large.

We now estimate $P(\bar{S}_n \geq (1-\eta)^{3/2} x_n B_n)$. Observe that

$$\begin{aligned} |E\bar{S}_n| &= \left|\sum_{i=1}^n EX_i I(|X_i| > \tau)\right| \\ &\leq \tau^{-1} \sum_{i=1}^n EX_i^2 I(|X_i| > \tau) = o(1) x_n B_n. \end{aligned}$$

It follows from the Bernstein inequality (2.17) that

$$\begin{aligned} P\left(\bar{S}_n \geq (1-\eta)^{3/2} x_n B_n\right) &\leq P\left(\bar{S}_n - E\bar{S}_n \geq (1-2\eta) x_n B_n\right) \\ &\leq \exp\left(-\frac{((1-2\eta)x_n B_n)^2}{2(B_n^2 + x_n B_n \tau)}\right) \\ &\leq \exp\left(-\frac{((1-2\eta)x_n B_n)^2}{2(1+\eta^2)B_n^2}\right) \\ &\leq \exp\left(-(1-\varepsilon)x_n^2/2\right), \end{aligned} \tag{7.48}$$

provided that $(1-2\eta)/(1+\eta^2) > 1-\varepsilon$. From (7.45)–(7.48), (7.43) follows.

To prove (7.44), let $0 < \varepsilon < 1/2$, $1/4 > \eta = \eta_\varepsilon > 0$,

$$G = \{1 \leq i \leq n : x_n^2 EX_i^2 > \eta^3 B_n^2\}, \qquad H = \{1 \leq i \leq n : x_n^2 EX_i^2 \leq \eta^3 B_n^2\}.$$

First we show that

$$\#(G) = o(x_n^2) \quad \text{and} \quad \sum_{i \in G} EX_i^2 = o(B_n^2). \tag{7.49}$$

Note that for $i \in G$,

$$\eta^3(B_n/x_n)^2 \le EX_i^2 = EX_i^2 I(|X_i| \le \eta^2 B_n/x_n) + EX_i^2 I(|X_i| > \eta^2 B_n/x_n)$$
$$\le \eta^4(B_n/x_n)^2 + EX_i^2 I(|X_i| > \eta^2 B_n/x_n).$$

Hence

$$EX_i^2 I(|X_i| > \eta^2 B_n/x_n) \ge \eta^4(B_n/x_n)^2$$

for $i \in G$, and by (7.37),

$$\eta^4(B_n/x_n)^2 \#(G) \le \sum_{i \in G} EX_i^2 I(|X_i| > \eta^2 B_n/x_n) = o(B_n^2),$$

which proves the first part of (7.49). For the second part of (7.49), we have

$$\begin{aligned}\sum_{i\in G} EX_i^2 &= \sum_{i\in G} EX_i^2 I\left(|X_i| \le \eta^2 B_n/x_n\right) + EX_i^2 I\left(|X_i| > \eta^2 B_n/x_n\right)\\ &\le \sum_{i\in G} (\eta^2 B_n/x_n)^2 + \sum_{i=1}^n EX_i^2 I\left(|X_i| > \eta^2 B_n/x_n\right)\\ &= o(x_n^2)(B_n/x_n)^2 + o(B_n^2) = o(B_n^2).\end{aligned}$$

Now we show that we only need to focus on $i \in H$ to prove (7.44). Let

$$S_H = \sum_{i\in H} X_i, \qquad S_G = \sum_{i\in G} X_i, \qquad V_H^2 = \sum_{i\in H} X_i^2, \qquad V_G^2 = \sum_{i\in G} X_i^2.$$

Noting that

$$|S_G/V_n| \le [\#(G)]^{1/2} = o(x_n),$$

we have

$$\begin{aligned}P(S_n/V_n \ge x_n) &= P(S_H/V_n \ge x_n - S_G/V_n) \\ &\ge P(S_H \ge (1+\eta)x_n V_n) \\ &\ge P\left(S_H \ge (1+\eta)x_n(V_H^2 + \eta B_n^2)^{1/2}, V_G^2 \le \eta B_n^2\right) \\ &= P\left(S_H \ge (1+\eta)x_n(V_H^2 + \eta B_n^2)^{1/2}\right) P(V_G^2 \le \eta B_n^2).\end{aligned} \tag{7.50}$$

From (7.49), it follows that

$$P(V_G^2 \le \eta B_n^2) \ge 1 - E(V_G^2)/(\eta B_n^2) \ge 1/2 \tag{7.51}$$

for n sufficiently large.

Let $\tau = \eta^2 B_n/x_n$ and let Y_i, $i \in H$, be independent random variables such that Y_i has the distribution function of X_i conditioned on $|X_i| \le \tau$. Put

$$\tilde{S}_H = \sum_{i\in H} Y_i, \qquad \tilde{V}_H^2 = \sum_{i\in H} Y_i^2.$$

Note that

$$\begin{aligned}\sum_{i\in H} EY_i^2 &= \sum_{i\in H} \frac{EX_i^2 I(|X_i|\le\tau)}{P(|X_i|\le\tau)}\\ &\le \sum_{i\in H} \frac{EX_i^2 I(|X_i|\le\tau)}{1-\eta} \le B_n^2/(1-\eta),\end{aligned}$$

$$\begin{aligned}\sum_{i\in H} EY_i^2 &\ge \sum_{i\in H} EX_i^2 I(|X_i|\le\tau)\\ &= \sum_{i\in H} EX_i^2 - \sum_{i\in H} EX_i^2 I(|X_i|>\tau)\\ &= B_n^2 - \sum_{i\in G} EX_i^2 - \sum_{i\in H} EX_i^2 I(|X_i|>\tau)\\ &= (1-o(1))B_n^2 \ge (1-\eta)B_n^2,\\ \sum_{i\in H} EY_i^4 &\le \tau^2 \sum_{i\in H} EY_i^2 \le \eta^4 B_n^4/x_n^2,\\ \sum_{i\in H} |EY_i| &\le 2\tau^{-1} \sum_{i\in H} EX_i^2 I(|X_i|>\tau) = o(x_n B_n).\end{aligned}$$

Moreover,

$$\begin{aligned}&P\left(S_H \ge (1+\eta)x_n(V_H^2+\eta B_n^2)^{1/2}\right)\\ &\ge P\left(S_H \ge (1+\eta)x_n(V_H^2+\eta B_n^2)^{1/2}, \max_{i\in H}|X_i|\le\tau\right) \qquad (7.52)\\ &= P\left(\max_{i\in H}|X_i|\le\tau\right) P\left(\tilde{S}_H \ge (1+\eta)x_n(\tilde{V}_H^2+\eta B_n^2)^{1/2}\right)\\ &= P\left(\max_{i\in H}|X_i|\le\tau\right) P\left(\tilde{S}_H \ge (1+\eta)x_n(\tilde{V}_H^2+\eta B_n^2)^{1/2}, \tilde{V}_H^2 \le (1+2\eta)B_n^2\right)\\ &\ge P\left(\max_{i\in H}|X_i|\le\tau\right) P\left(\tilde{S}_H \ge (1+\eta)(1+3\eta)x_n B_n\right) - P\left(\tilde{V}_H^2 > (1+2\eta)B_n^2\right).\end{aligned}$$

Similar to the proof of (7.47), we have

$$\begin{aligned}P\left(\tilde{V}_H^2 > (1+2\eta)B_n^2\right) &\le \exp\left(-\frac{((1+2\eta-1/(1-\eta))B_n^2)^2}{2\left(\sum_{i\in H} EY_i^4 + 2\tau^2 B_n^2\right)}\right) \qquad (7.53)\\ &\le \exp\left(-\frac{\eta^2 B_n^4}{16\eta^4 B_n^4/x_n^2}\right) \le \exp(-2x_n^2).\end{aligned}$$

Also note that

$$P\left(\max_{i\in H}|X_i|\le\tau\right) = \prod_{i\in H}(1-P(|X_i|>\tau)) \tag{7.54}$$
$$\ge \prod_{i\in H}(1-\tau^{-2}EX_i^2I(|X_i|>\tau))$$
$$\ge \exp\left(-2\sum_{i\in H}\tau^{-2}EX_i^2I(|X_i|>\tau)\right)$$
$$= \exp\left(-o(1)x_n^2\right).$$

Finally, by Kolmogorov's lower exponential bound (see Theorem 2.22),

$$P\left(\tilde{S}_H\ge(1+\eta)(1+3\eta)x_nB_n\right)\ge\exp\left(-\frac{(1+\varepsilon/2)(1+\eta)^2(1+3\eta)^2x_n^2B_n^2}{2(1-\eta)B_n^2}\right)$$
$$\ge \exp\left(-(1+\varepsilon)x_n^2/2\right) \tag{7.55}$$

for sufficiently large n, provided that η is chosen small enough. Combining the above inequalities yields (7.44). □

Proof (of Theorem 7.18). We follow the proof of Theorem 6.14. We first show that

$$\limsup_{n\to\infty}\frac{S_n}{V_n(2\log\log B_n)^{1/2}}\le 1\ \ a.s. \tag{7.56}$$

For $\theta>1$, let $m_k:=m_k(\theta)=\min\{n: B_n\ge\theta^k\}$. It follows from condition (7.40) that

$$B_{m_k}\sim\theta^k \qquad \text{as } k\to\infty. \tag{7.57}$$

Let $x_k=(2\log\log B_{m_k})^{1/2}$. Then, for $0<\varepsilon<1/2$,

$$P\left(\max_{m_k\le n\le m_{k+1}}\frac{S_n}{V_n}\ge(1+7\varepsilon)x_k\right)$$
$$\le P\left(\frac{S_{m_k}}{V_{m_k}}\ge(1+2\varepsilon)x_k\right)+P\left(\max_{m_k\le n\le m_{k+1}}\frac{S_n-S_{m_k}}{V_n}\ge 5\varepsilon x_k\right). \tag{7.58}$$

By Theorem 7.2,

$$P\left(\frac{S_{m_k}}{V_{m_k}}\ge(1+2\varepsilon)x_k\right)\le\exp\left(-(1+2\varepsilon)x_k^2/2\right)\le C\,k^{-1-\varepsilon} \tag{7.59}$$

for every sufficiently large k.

To bound the second term in the right-hand side of (7.58), let $\eta=(\theta^2-1)^{1/2}$ and define $z_k=\eta B_{m_k}/x_k$. Set $T_n=\sum_{i=m_k+1}^n X_iI(|X_i|\le z_k)$. Therefore

$$
\begin{aligned}
P&\left(\max_{m_k\le n\le m_{k+1}} \frac{S_n - S_{m_k}}{V_n} \ge 5\varepsilon x_k\right)\\
&\le P\left(\max_{m_k\le n\le m_{k+1}} T_n \ge 2\varepsilon x_k B_{m_k}\right)\\
&\quad + P(V_{m_k} \le B_{m_k}/2) + P\left(\sum_{i=1+m_k}^{m_{k+1}} I(|X_i| > z_k) \ge (\varepsilon x_k)^2\right). \qquad (7.60)
\end{aligned}
$$

Note that

$$\sum_{i=1+m_k}^{m_{k+1}} EX_i^2 I(|X_i| \le z_k) \sim (\theta^2 - 1)\theta^{2k}$$

and

$$
\begin{aligned}
\max_{m_k\le n\le m_{k+1}} |ET_n| &\le z_k^{-1} \sum_{i=1+m_k}^{m_{k+1}} EX_i^2 \sim z_k^{-1}(\theta^2-1)\theta^{2k}\\
&\sim (\theta^2-1)^{1/2} x_k B_{m_k} \le \varepsilon x_k B_{m_k}/2
\end{aligned}
$$

for $1 < \theta < 1 + \varepsilon^2/8$. By the Bernstein inequality (2.17), for all large k,

$$
\begin{aligned}
\log P\left(\max_{m_k\le n\le m_{k+1}} T_n \ge 2\varepsilon x_k B_{m_k}\right) &\le -\frac{(\varepsilon x_k B_{m_k})^2}{2\left((\theta^2-1)\theta^{2k} + \varepsilon x_k B_{m_k} z_k\right)}\\
&\sim -\frac{\varepsilon^2 x_k^2}{2\left((\theta^2-1) + \varepsilon(\theta^2-1)^{1/2}\right)}\\
&\le -x_k^2, \qquad (7.61)
\end{aligned}
$$

provided that $\theta(>1)$ is close enough to 1. By the Bernstein inequality again,

$$
\begin{aligned}
P(V_{m_k} \le B_{m_k}/2) &\le P\left(\sum_{i=1}^{m_k} X_i^2 I(|X_i| \le z_k) \le B_{m_k}^2/4\right) \qquad (7.62)\\
&\le \exp\left(-\frac{(3B_{m_k}^2/4)^2}{2\left\{\sum_{i=1}^{m_k} EX_i^4 I(|X_i| \le z_k) + B_{m_k}^2 z_k^2\right\}}\right)\\
&\le \exp\left(-\frac{B_{m_k}^4}{8B_{m_k}^2 z_k^2}\right)\\
&\le \exp(-x_k^2)
\end{aligned}
$$

for $\theta(>1)$ close to 1. Let

$$t = t_k := \log\left\{(\varepsilon x_k)^2 \Big/ \sum_{i=1}^{m_{k+1}} z_k^{-2} EX_i^2 I(|X_i| > z_k)\right\}.$$

By (7.40), $t \to \infty$. From the Markov inequality, it follows that

$$\begin{aligned}
&P\left(\sum_{i=1+m_k}^{m_{k+1}} I(|X_i| > z_k) \ge (\varepsilon x_k)^2\right) \\
&\le e^{-t(\varepsilon x_k)^2} \prod_{i=1+m_k}^{m_{k+1}} \left(1 + (e^t - 1)P(|X_i| > z_k)\right) \\
&\le \exp\left(-t(\varepsilon x_k)^2 + (e^t - 1)\sum_{i=1}^{m_{k+1}} z_k^{-2} EX_i^2 I(|X_i| > z_k)\right) \\
&\le \exp\left(-(\varepsilon x_k)^2 \log\left\{\frac{(\varepsilon x_k)^2}{3\sum_{i=1}^{m_{k+1}} z_k^{-2} EX_i^2 I(|X_i| > z_k)}\right\}\right) \\
&\le \exp(-x_k^2) \qquad (7.63)
\end{aligned}$$

for sufficiently large k. Combining the above inequalities yields (7.56) by the Borel–Cantelli lemma and the arbitrariness of ε.

Next, we prove that

$$\limsup_{n\to\infty} \frac{S_n}{V_n(2\log\log B_n)^{1/2}} \ge 1 \quad a.s. \qquad (7.64)$$

Let $n_k = \min\{m : B_m \ge e^{4k\log k}\}$. Then, $B_{n_k} \sim e^{4k\log k}$. Observe that

$$\begin{aligned}
&\limsup_{n\to\infty} \frac{S_n}{V_n(2\log\log B_n)^{1/2}} \qquad (7.65) \\
&\ge \limsup_{k\to\infty} \frac{S_{n_k}}{V_{n_k}(2\log\log B_{n_k})^{1/2}} \\
&\ge \limsup_{k\to\infty} \frac{S_{n_k} - S_{n_{k-1}}}{V_{n_k}(2\log\log B_{n_k})^{1/2}} + \liminf_{k\to\infty} \frac{S_{n_{k-1}}}{V_{n_k}(2\log\log B_{n_k})^{1/2}} \\
&= \limsup_{k\to\infty} \frac{(V_{n_k}^2 - V_{n_{k-1}}^2)^{1/2}}{V_{n_k}} \frac{S_{n_k} - S_{n_{k-1}}}{(V_{n_k}^2 - V_{n_{k-1}}^2)^{1/2}(2\log\log B_{n_k})^{1/2}} \\
&\quad + \liminf_{k\to\infty} \frac{V_{n_{k-1}}}{V_{n_k}} \frac{S_{n_{k-1}}}{V_{n_{k-1}}(2\log\log B_{n_k})^{1/2}}.
\end{aligned}$$

Since $(S_{n_k} - S_{n_{k-1}})/(V_{n_k}^2 - V_{n_{k-1}}^2)^{1/2}$, $k \ge 1$, are independent, it follows from Theorem 7.2 and the Borel–Cantelli lemma that

$$\limsup_{n\to\infty} \frac{S_{n_k} - S_{n_{k-1}}}{(V_{n_k}^2 - V_{n_{k-1}}^2)^{1/2}(2\log\log B_{n_k})^{1/2}} \ge 1 \quad a.s. \qquad (7.66)$$

Similar to (7.62) and by the Borel–Cantelli lemma, we have

$$\liminf_{k\to\infty} \frac{V_{n_k}}{B_{n_k}} \ge 1/2 \quad a.s.$$

Note that

$$P(V_{n_{k-1}} \geq B_{n_k}/k) \leq k^2 E V_{n_{k-1}}^2 / B_{n_k}^2 = k^2 B_{n_{k-1}}^2 / B_{n_k}^2 \leq k^{-2}.$$

Then, by the Borel–Cantelli lemma again,

$$\lim_{k\to\infty} \frac{V_{n_{k-1}}}{V_{n_k}} = 0 \;\; a.s. \tag{7.67}$$

From (7.65) and (7.66), (7.67), (7.64) follows. □

7.4 Cramér-Type Moderate Deviations for Two-Sample t-Statistics

Let $X_1,\ldots,X_{n_1}$ be a random sample from a population with mean μ_1 and variance σ_1^2, and $Y_1,\ldots,Y_{n_2}$ be a random sample from another population with mean μ_2 and variance σ_2^2. Assuming that the two random samples are independent, define the two-sample t-statistic

$$T = \frac{\bar{X} - \bar{Y} - (\mu_1 - \mu_2)}{\sqrt{s_1^2/n_1 + s_2^2/n_2}}, \tag{7.68}$$

where $\bar{X} = \sum_{i=1}^{n_1} X_i/n_1$, $\bar{Y} = \sum_{i=1}^{n_2} Y_i/n_2$,

$$s_1^2 = \frac{1}{n_1 - 1}\sum_{i=1}^{n_1}(X_i - \bar{X})^2, \qquad s_2^2 = \frac{1}{n_2 - 1}\sum_{i=1}^{n_2}(Y_i - \bar{Y})^2.$$

Two-sample t-statistics are commonly used for testing the difference between two population means or constructing confidence intervals for the difference. Cao (2007) has proved the following moderate deviation results, analogous to Theorems 6.1 and 7.4, for the two-sample t-statistic (7.68).

Theorem 7.20. *Let $n = n_1 + n_2$. Assume that $c_1 \leq n_1/n_2 \leq c_2$ for some $0 < c_1 \leq c_2 < \infty$ and all large n. Then for any $x := x(n_1,n_2)$ satisfying $x \to \infty$, $x = o(n^{1/2})$,*

$$\log P(T \geq x) \sim -x^2/2 \tag{7.69}$$

as $n \to \infty$. If, in addition $E|X_1|^3 < \infty$ and $E|Y_1|^3 < \infty$, then

$$\frac{P(T \geq x)}{1 - \Phi(x)} = 1 + O(1)(1+x)^3 n^{-1/2} d^3 \tag{7.70}$$

for $0 \leq x \leq n^{1/6}/d$, where $d^3 = (E|X_1 - \mu_1|^3 + E|Y_1 - \mu_2|^3)/(\sigma_1^2 + \sigma_2^2)^{3/2}$ and $O(1)$ depends only on c_1 and c_2. In particular,

$$\frac{P(T \geq x)}{1 - \Phi(x)} \to 1 \tag{7.71}$$

uniformly in $x \in [0, o(n^{1/6}))$.

Proof. Without loss of generality, assume $\mu_1 = \mu_2 = 0$. Let

$$V_{1,n_1}^2 = \sum_{i=1}^{n_1} X_i^2, \qquad V_{2,n_2}^2 = \sum_{i=1}^{n_2} Y_i^2,$$
$$s_1^{*2} = \frac{V_{1,n_1}^2}{n_1 - 1}, \qquad s_2^{*2} = \frac{V_{2,n_2}^2}{n_2 - 1},$$
$$T^* = \frac{\bar{X} - \bar{Y}}{\sqrt{s_1^{*2}/n_1 + s_2^{*2}/n_2}}.$$

Noting that

$$s_1^2 = s_1^{*2}\left(1 - n_1\bar{X}^2/V_{1,n_1}^2\right), \qquad s_2^2 = s_2^{*2}\left(1 - n_2\bar{Y}^2/V_{2,n_2}^2\right),$$

we have

$$P(T^* \ge x) \le P(T \ge x) \le P(T^* \ge x\sqrt{1-\varepsilon}) + P\left(\frac{n_1\bar{X}^2}{V_{1,n_1}^2} \ge \varepsilon\right) + P\left(\frac{n_2\bar{Y}_2}{V_{2,n_2}^2} \ge \varepsilon\right) \tag{7.72}$$

for any $0 < \varepsilon < 1$.

To prove (7.69), let $\varepsilon = (1+x)/\sqrt{n}$. It follows from Theorem 7.2 that

$$\log P(T^* \ge x) \sim -x^2/2 \tag{7.73}$$

and

$$\log P\left(T^* \ge x\sqrt{1-\varepsilon}\right) \sim -x^2/2. \tag{7.74}$$

By Theorem 6.1,

$$\log P\left(\frac{n_1\bar{X}^2}{V_{1,n_1}^2} \ge \varepsilon\right) \sim -\varepsilon n_1/2 \tag{7.75}$$

and

$$\log P\left(\frac{n_2\bar{Y}^2}{V_{2,n_2}^2} \ge \varepsilon\right) \sim -\varepsilon n_1/2. \tag{7.76}$$

Noting that $x^2 = o(\varepsilon \min(n_1, n_2))$, we obtain (7.69) by combining (7.72)–(7.76).

To prove (7.70), let $\varepsilon = (1+x)/\sqrt{n}$ again. By Theorem 7.4,

$$\frac{P(T^* \ge x)}{1 - \Phi(x)} = 1 + O(1)(1+x)^3 n^{-1/2} d^3 \tag{7.77}$$

and

$$\frac{P(T^* \ge x\sqrt{1-\varepsilon})}{1 - \Phi(x)} = 1 + O(1)(1+x)^3 n^{-1/2} d^3. \tag{7.78}$$

By Theorem 6.1,

$$P(n_1\bar{X}^2/V_{1,n_1}^2 \geq \varepsilon) = o\left((1+x)^3 n^{-1/2}(1-\Phi(x))\right), \tag{7.79}$$

$$P(n_2\bar{Y}^2/V_{2,n_2}^2 \geq \varepsilon) = o\left((1+x)^3 n^{-1/2}(1-\Phi(x))\right). \tag{7.80}$$

From (7.72) and (7.77)–(7.80), (7.70) follows. □

7.5 Supplementary Results and Problems

1. It has been shown by Chistyakov and Götze (2004b) that the bound in (7.3) is the best possible. Consult their paper for a detailed example.
2. Wang (2005) has proved the following result:
   Let $X, X_1, X_2, \ldots$ be i.i.d. random variables with $E(X) = 0$ and $E(X^4) < \infty$. Then
   $$\frac{P(S_n \geq xV_n)}{1-\Phi(x)} = \exp\left\{-\frac{x^3 EX^3}{3\sqrt{n}\sigma^3}\right\}\left(1+O\left(\frac{1+x}{\sqrt{n}}\right)\right)$$
   for $0 \leq x \leq O(n^{1/6})$.
   Compare his result and proof with Theorem 7.20. Can his assumption $E(X^4) < \infty$ be weakened to $E|X|^3 < \infty$?
3. Hu et al. (2008) have shown that if $X, X_1, X_2, \ldots$ are i.i.d. random variables with $E(X) = 0$ and $E(X^4) < \infty$, then
   $$\lim_{n\to\infty} \frac{P(\max_{1\leq k\leq n} S_k \geq xV_n)}{1-\Phi(x)} = 2$$
   uniformly in $x \in [0, o(n^{1/6}))$. It would be interesting to see if the finiteness of $E(X^4)$ can be weakened to that of $E|X|^3$.
4. Prove (7.78) and (7.79).

Chapter 8
Self-Normalized Empirical Processes and U-Statistics

Whereas previous chapters have considered limit theorems for self-normalized sums of independent random variables, we extend sums to more general structures in this chapter, namely, self-normalized empirical processes and U-statistics. In particular, we extend the methods and results of Chaps. 6 and 7 to self-normalized U-statistics.

8.1 Self-Normalized Empirical Processes

Let $X, X_1, \ldots, X_n$ be i.i.d random variables with values in a measurable space $(\mathscr{X}, \mathscr{C})$. Let $\mathscr{F}$ be a class of real-valued measurable functions on $(\mathscr{X}, \mathscr{C})$. Consider all functions f in $\mathscr{F}$ that are *centered* and *normalized*, i.e.,

$$Ef(X) = 0 \quad \text{and} \quad Ef^2(X) = 1.$$

Define the self-normalized empirical process by

$$W_n(f) = \frac{\sum_{i=1}^n f(X_i)}{\sqrt{\sum_{i=1}^n f^2(X_i)}}, \qquad f \in \mathscr{F}.$$

There is a comprehensive theory on the classical empirical process $\{\frac{1}{\sqrt{n}} \sum_{i=1}^n f(X_i), f \in \mathscr{F}\}$; see, e.g., Shorack and Wellner (1986) and van der Vaart and Wellner (1996). For the self-normalized empirical process, Bercu et al. (2002) have proved the following moderate and large deviation results and exponential bounds for

$$\mathscr{W}_n = \sup_{f \in \mathscr{F}} W_n(f).$$

Definition 8.1. $\mathscr{F}$ is said to have a *finite covering with brackets in* L^2 satisfying *concordance of signs* if for any $\delta > 0$, one can find a finite family $\mathscr{C}$ of pairs of measurable functions in L^2 such that, for any $f \in \mathscr{F}$, there exists (g,h) in $\mathscr{C}$ with

$$|g| \le |f| \le |h|, \quad gh \ge 0 \qquad \text{and} \qquad E\left[(h(X) - g(X)\right]^2 \le \delta.$$

V.H. de la Peña et al., *Self-Normalized Processes: Limit Theory and Statistical Applications*,
Probability and its Applications,

© Springer-Verlag Berlin Heidelberg 2009

Theorem 8.2. *Suppose that $\mathscr{F}$ is a countable class of centered and normalized functions. Assume that $\mathscr{F}$ has finite covering with brackets in L^2 satisfying concordance of signs, and that*

$$C_0 := \sup_{n\geq 1} E\left[\sup_{f\in\mathscr{F}} \max\left(\frac{1}{\sqrt{n}}\sum_{i=1}^{n} f(X_i), 0\right)\right] < \infty. \tag{8.1}$$

Then, for any sequence (x_n) such that $x_n \to \infty$ and $x_n = o(\sqrt{n})$,

$$\lim_{n\to\infty} \frac{1}{x_n^2} \log P(\mathscr{W}_n \geq x_n) = -\frac{1}{2}. \tag{8.2}$$

Theorem 8.3. *Let $t : \mathscr{X} \to \mathbb{R}$ be a measurable function such that $t(X)$ has a continuous distribution function, $Et(X) = 0$ and $e^{\psi(\theta)} = Ee^{\theta t(X)} < \infty$ for $2m \leq \theta \leq 2M$, where $m < 0 < M$. For $\theta \in [m,0)\cup(0,M]$ and $x \in \mathscr{X}$, define $f_\theta(x) = \exp\{\theta t(x) - \psi(\theta)\}$ and let $\mathscr{F} = \{f_\theta : \theta \in [m,0)\cup(0,M]\}$. Then for any $r \geq 0$,*

$$\lim_{n\to\infty} n^{-1} \log P\left(\mathscr{W}_n \geq r\sqrt{n}\right) = -\mathscr{I}(r), \tag{8.3}$$

where $\mathscr{I}(r) = \inf_{f\in\mathscr{F}} I_f(r)$ and

$$I_f(r) = -\log \sup_{a\geq 0} \inf_{t\geq 0} E \exp\left(t\left[af(X) - r\left(f^2(X) + a^2\right)/2\right]\right). \tag{8.4}$$

The proofs of Theorems 8.2 and 8.3 involve certain concentration inequalities and approximation arguments that are quite technical. Details can be found in Bercu et al. (2002).

8.2 Self-Normalized U-Statistics

Let $X, X_1, \ldots, X_n$ be i.i.d. random variables, and let $h(x_1,x_2)$ be a real-valued symmetric Borel measurable function such that $Eh(X_1,X_2) = \theta$. An unbiased estimate of θ is the U-statistic

$$U_n = \binom{n}{2}^{-1} \sum_{1\leq i<j\leq n} h(X_i,X_j), \tag{8.5}$$

typical examples of which include:

(1) Sample mean: $h(x_1,x_2) = \frac{1}{2}(x_1+x_2)$.
(2) Sample variance: $h(x_1,x_2) = \frac{1}{2}(x_1-x_2)^2$.
(3) Gini's mean difference: $h(x_1,x_2) = |x_1-x_2|$.
(4) One-sample Wilcoxon's statistic: $h(x_1,x_2) = I(x_1+x_2 \leq 0)$.

The function h in (8.5) is called the *kernel* of the U-statistic. For notational simplicity we only consider the case of U-statistics of order 2, i.e., h is a function of two variables. The kernel of a U-statistic of order m is a function of m variables, for which (8.5) is generalized to

$$U_n = \binom{n}{m}^{-1} \sum_{1<i_1<\cdots<i_m\leq n} h(X_{i_1},\ldots,X_{i_m}).$$

8.2.1 The Hoeffding Decomposition and Central Limit Theorem

Without loss of generality, assume $\theta = 0$. Let $g(x) = Eh(x,X)$. Hoeffding (1948) has shown that the U-statistic (8.5) has the following decomposition:

$$U_n = \frac{2}{n}\sum_{i=1}^{n} g(X_i) + \frac{2}{n}\Delta_n, \tag{8.6}$$

where

$$\Delta_n = \frac{1}{n-1}\sum_{1\leq i<j\leq n} \left\{h(X_i,X_j) - g(X_i) - g(X_j)\right\}. \tag{8.7}$$

By showing that Δ_n is usually negligible under some regularity conditions, he approximates a non-degenerate U-statistic (i.e., $g(X)$ is non-degenerate) by the sample mean of i.i.d. random variables. In particular, if $Eh^2(X_1,X_2) < \infty$ and $\sigma_1^2 = \mathrm{Var}(g(X_1)) > 0$, he thereby obtains the central limit theorem

$$\sup_x \left| P\left(\frac{\sqrt{n}}{2\sigma_1}U_n \leq x\right) - \Phi(x)\right| \to 0 \qquad \text{as } n \to \infty. \tag{8.8}$$

Likewise, if $E|h(X_1,X_2)|^{5/3} < \infty$, $E|g(X_1)|^3 < \infty$ and $\sigma_1^2 = \mathrm{Var}(g(X_1)) > 0$, then we have the Berry–Esseen bound

$$\sup_x \left| P\left(\frac{\sqrt{n}}{2\sigma_1}U_n \leq x\right) - \Phi(x)\right| = O(n^{-1/2}); \tag{8.9}$$

see Koroljuk and Borovskikh (1994), Alberink and Bentkus (2001, 2002), Wang and Weber (2006). There are also large deviation results for U-statistics; see Borovskich and Weber (2003a,b).

8.2.2 Self-Normalized U-Statistics and Berry–Esseen Bounds

Since σ_1 is typically unknown, what is used in statistical inference is the self-normalized U-statistic

$$T_n = \sqrt{n}(U_n - \theta)/R_n\,, \tag{8.10}$$

where R_n^2 is the *jackknife estimate* of σ_1^2 given by

$$R_n^2 = \frac{4(n-1)}{(n-2)^2} \sum_{i=1}^{n} (q_i - U_n)^2, \qquad \text{with} \quad q_i = \frac{1}{n-1} \sum_{\substack{j=1 \\ j \neq i}}^{n} h(X_i, X_j). \tag{8.11}$$

It is easy to see that the central limit theorem remains valid for the self-normalized U-statistic (8.10) provided that $E|h(X_1,X_2)|^2 < \infty$. The following Berry–Esseen bound is due to Wang et al. (2000):

Theorem 8.4. *Assume $\theta = 0$ and $E|h(X_1,X_2)|^3 < \infty$. Then*

$$\sup_x |P(T_n \leq x) - \Phi(x)| \leq A n^{-1/2} E|h(X_1,X_2)|^3 / \sigma_1^3, \tag{8.12}$$

where A is an absolute constant.

8.2.3 Moderate Deviations for Self-Normalized U-Statistics

In view of the moderate deviation results for self-normalized sums in Chaps. 6 and 7, it is natural to ask whether similar results hold for self-normalized U-statistics. Assuming that $0 < \sigma_1^2 = Eg^2(X_1) < \infty$, we describe here the approach of Lai et al. (2008) to establish results like (6.1) and (7.12) when the kernel satisfies

$$h^2(x_1,x_2) \leq c_0 \left[\sigma_1^2 + g^2(x_1) + g^2(x_2)\right] \tag{8.13}$$

for some $c_0 > 0$. This condition is satisfied by the typical examples (1)–(4) in the first paragraph of Sect. 8.2.

Theorem 8.5. *Assume that* (8.13) *holds for some $c_0 > 0$ and $0 < \sigma_1^2 = Eg^2(X_1) < \infty$. Then, for any sequence x_n with $x_n \to \infty$ and $x_n = o(n^{1/2})$,*

$$\log P(T_n \geq x_n) \sim -x_n^2/2. \tag{8.14}$$

If, in addition, $E|g(X_1)|^3 < \infty$, then

$$P(T_n \geq x) = (1 - \Phi(x))\,[1 + o(1)] \tag{8.15}$$

holds uniformly in $x \in [0, o(n^{1/6}))$.

Without loss of generality, assume $\theta = 0$. Write $S_n = \sum_{j=1}^{n} g(X_j)$ and $V_n^2 = \sum_{j=1}^{n} g^2(X_j)$. The following theorem shows that the self-normalized U-statistic T_n can be approximated by the self-normalized sum S_n/V_n under condition (8.13). As a result, (8.14) and (8.15) follow from (6.1) and (7.12).

Theorem 8.6. *Assume that $\theta = 0$, $0 < \sigma_1^2 = Eg^2(X_1) < \infty$ and the kernel $h(x_1,x_2)$ satisfies condition* (8.13). *Then there exists a constant $\eta > 0$ depending only on σ_1^2 and c_0 such that, for all $0 \leq \varepsilon_n < 1$, $0 \leq x \leq \sqrt{n}/3$ and n sufficiently large,*

$$P\{S_n/V_n \geq (1+\varepsilon_n)x\} - 5\sqrt{2}(n+2)e^{-\eta \min(n\varepsilon_n^2, \sqrt{n}\varepsilon_n x)} \leq P(T_n \geq x)$$
$$\leq P\{S_n/V_n \geq (1-\varepsilon_n)x\} + 5\sqrt{2}(n+2)e^{-\eta \min(n\varepsilon_n^2, \sqrt{n}\varepsilon_n x)}. \tag{8.16}$$

8.3 Proofs of Theorems 8.5 and 8.6

8.3.1 Main Ideas of the Proof

Assume that $\theta = 0$. Let

$$R_n^{*2} = \frac{4(n-1)}{(n-2)^2}\sum_{i=1}^n q_i^2, \qquad T_n^* = \frac{\sqrt{n}U_n}{R_n^*}. \tag{8.17}$$

Noting that $\sum_{i=1}^n (q_i - U_n)^2 = \sum_{i=1}^n q_i^2 - 2U_n \sum_{i=1}^n q_i + nU_n^2 = \sum_{i=1}^n q_i^2 - nU_n^2$, we have

$$T_n = \frac{T_n^*}{\left(1 - \frac{4(n-1)}{(n-2)^2}T_n^{*2}\right)^{1/2}}, \tag{8.18}$$

and therefore

$$\{T_n \geq x\} = \left\{T_n^* \geq \frac{x}{[1+4x^2(n-1)/(n-2)^2]^{1/2}}\right\}. \tag{8.19}$$

Thus, we only need to work on T_n^* instead of T_n. Without loss of generality, assume $\sigma_1^2 = 1$; otherwise, consider h/σ_1 in the place of h.

We next establish a relation between T_n^* and S_n/V_n. To do this, let

$$\psi(x_1, x_2) = h(x_1, x_2) - g(x_1) - g(x_2), \tag{8.20}$$

$$\Delta_n = \frac{1}{n-1}\sum_{1 \leq i \neq j \leq n} \psi(X_i, X_j), \qquad W_n^{(i)} = \sum_{\substack{j=1 \\ j \neq i}}^n \psi(X_i, X_j), \qquad \Lambda_n^2 = \sum_{i=1}^n \left(W_n^{(i)}\right)^2.$$

It is easy to see that

$$nU_n/2 = S_n + \Delta_n. \tag{8.21}$$

Also observe that $\sum_{j=1, j\neq i}^n h(X_i, X_j) = (n-2)g(X_i) + S_n + W_n^{(i)}$ and

$$\begin{aligned}
\frac{(n-1)(n-2)^2}{4}R_n^{*2} &= \sum_{i=1}^n \left(\sum_{\substack{j=1 \\ j \neq i}}^n h(X_i, X_j)\right)^2 \\
&= (n-2)^2 V_n^2 + \Lambda_n^2 + (3n-4)S_n^2 \\
&\quad + 2(n-2)\sum_{i=1}^n g(X_i)W_n^{(i)} + 2S_n \sum_{i=1}^n W_n^{(i)}.
\end{aligned}$$

Combining this with $|\sum_{i=1}^n g(X_i)W_n^{(i)}| \le V_n \Lambda_n$ and $|S_n \sum_{i=1}^n W_n^{(i)}| \le |S_n|\sqrt{n}\Lambda_n$, where $\Lambda_n^2 \le n \max_{1\le i\le n} |W_n^{(i)}|^2$, we can write

$$R_n^{*2} = \frac{4}{n-1} V_n^2 (1+\delta_n), \tag{8.22}$$

in which

$$\begin{aligned} |\delta_n| &\le \frac{1}{(n-2)^2}\left[\frac{\Lambda_n^2}{V_n^2} + \frac{3nS_n^2}{V_n^2} + \frac{2n\Lambda_n}{V_n} + 2\sqrt{n}\frac{|S_n|\Lambda_n}{V_n^2}\right] \\ &\le \frac{1}{(n-2)^2}\left(\frac{\Lambda_n^2}{V_n^2} + \frac{4n\Lambda_n}{V_n} + \frac{3nS_n^2}{V_n^2}\right). \end{aligned} \tag{8.23}$$

By (8.21)–(8.22) and (8.17),

$$T_n^* = \frac{S_n + \Delta_n}{d_n V_n (1+\delta_n)^{1/2}}, \qquad \text{where } d_n = \sqrt{n/(n-1)}. \tag{8.24}$$

We then make use of the following exponential bounds to conclude that Δ_n and δ_n are negligible.

Proposition 8.7. *There exist constants $\delta_0 > 0$ and $\delta_1 > 0$, depending only on σ_1^2 and c_0, such that for all $y > 0$,*

$$P(|\delta_n| \ge y) \le 4\sqrt{2}(n+2)\exp\left(-\delta_0 \min\{1, y, y^2\} n\right), \tag{8.25}$$

$$P(|\Delta_n| \ge yV_n) \le \sqrt{2}(n+2)\exp\left(-\delta_1 \min\{n, y\sqrt{n}\}\right). \tag{8.26}$$

8.3.2 Proof of Theorem 8.6

Lat $\tau_n' = \sqrt{\frac{n}{n-1}}\left[1 + \frac{4x^2(n-1)}{(n-2)^2}\right]^{-1/2}$. Since $x^2 \le n/9$ and $0 \le \varepsilon_n < 1$, it is easy to show that for $0 \le x \le \sqrt{n}/3$,

$$\tau_n := \left(1 - \frac{\varepsilon_n}{4}\right)^{1/2} \tau_n' \ge 1 - \varepsilon_n/2$$

when n is sufficiently large. Hence it follows from (8.19), (8.24) and Proposition 8.7 that

$$\begin{aligned} P(T_n \ge x) &\le P\{S_n/V_n \ge (1-\varepsilon_n)x\} + P\left\{|\Delta_n|/V_n \ge x(\varepsilon_n - 1) + x\tau_n'(1+\delta_n)^{1/2}\right\} \\ &\le P\{S_n/V_n \ge (1-\varepsilon_n)x\} + P\{|\Delta_n|/V_n \ge x(\varepsilon_n - 1) + x\tau_n\} + P\{|\delta_n| \ge \varepsilon_n/4\} \\ &\le P\{S_n/V_n \ge (1-\varepsilon_n)x\} + P\{|\Delta_n|/V_n \ge x\varepsilon_n/2\} + P\{|\delta_n| \ge \varepsilon_n/4\} \\ &\le P\{S_n/V_n \ge (1-\varepsilon_n)x\} + 5\sqrt{2}(n+2)e^{-\eta \min(n\varepsilon_n^2, \sqrt{n}\varepsilon_n x)}, \end{aligned}$$

where $\eta > 0$ is a constant depending only on σ_1^2 and c_0. This proves the upper bound of (8.16). The lower bound can be proved similarly.

8.3.3 *Proof of Theorem 8.5*

By the central limit theorem, the result (8.15) is obvious when $0 \le x \le 1$. In order to prove (8.15) for $x \in [1, o(n^{1/6}))$, we choose $\varepsilon_n = \max\{\varepsilon_n' x/\sqrt{n}, n^{-1/8}\}$, where $\varepsilon_n' \to \infty$ and $\varepsilon_n' x^3/\sqrt{n} \to 0$ for $x \in [1, o(n^{1/6}))$. Since

$$\min\left\{n\varepsilon_n^2, \sqrt{n}\varepsilon_n x\right\} \ge \sqrt{n}\varepsilon_n x \ge \max\left\{\varepsilon_n' x^2, n^{3/8} x\right\},$$

and since

$$\frac{1}{\sqrt{2\pi}}\left(\frac{1}{x} - \frac{1}{x^3}\right) e^{-x^2/2} \le 1 - \Phi(x) \le \frac{1}{\sqrt{2\pi}}\frac{1}{x} e^{-x^2/2} \qquad \text{for } x > 0,$$

we have uniformly in $x \in [1, o(n^{1/6}))$,

$$n e^{-\eta \min(n\varepsilon_n^2, \sqrt{n}\varepsilon_n x)} = o(1 - \Phi(x)). \tag{8.27}$$

Moreover, by (7.12),

$$\begin{aligned} P\{S_n/V_n \ge (1-\varepsilon_n)x\} &\le \{1 - \Phi[(1-\varepsilon_n)x]\}\left(1 + O(1)x^3/\sqrt{n}\right) \\ &\le [1 - \Phi(x)]\left\{1 + \frac{|\Phi[(1-\varepsilon_n)x] - \Phi(x)|}{1 - \Phi(x)}\right\}\left[1 + O(1)x^3/\sqrt{n}\right] \\ &= [1 - \Phi(x)]\,[1 + o(1)], \end{aligned} \tag{8.28}$$

where we have used the bound

$$|\Phi((1-\varepsilon_n)x) - \Phi(x)| \le \varepsilon_n x e^{-(1-\varepsilon_n)^2 x^2/2} = o(1 - \Phi(x))$$

uniformly in $[1, o(n^{1/6}))$, since $\varepsilon_n x^2 \le \varepsilon_n' x^3/\sqrt{n} = o(1)$.

Combining (8.27)–(8.28) with the upper bound in (8.16), we obtain $P(T_n \ge x) \le (1 - \Phi(x))(1 + o(1))$. Similarly we have $P(T_n \ge x) \ge (1 - \Phi(x))(1 + o(1))$. This proves (8.15). In a similar way, we can prove (8.14) by choosing $\varepsilon_n = \max\{n^{-1/8}, \varepsilon_n'\}$, where ε_n' are constants converging so slowly that $n\varepsilon_n'^2/x_n^2 \to \infty$.

8.3.4 *Proof of Proposition 8.7*

To prove (8.25), we make use of the exponential inequalities for sums of independent random variables in Chap. 2 to develop them further in the next two lemmas.

Lemma 8.8. *Let $\{\xi_i, i \geq 1\}$ be independent random variables with zero means and finite variances. Let $S_n = \sum_{i=1}^n \xi_i$, $V_n^2 = \sum_{i=1}^n \xi_i^2$, $B_n^2 = \sum_{i=1}^n E\xi_i^2$. Then for $x > 0$,*

$$P\left(|S_n| \geq x(V_n^2 + 5B_n^2)^{1/2}\right) \leq \sqrt{2}\exp(-x^2/8), \tag{8.29}$$

$$ES_n^2 I\left(|S_n| \geq x(V_n + 4B_n)\right) \leq 16B_n^2 e^{-x^2/4}. \tag{8.30}$$

Proof. Note that (8.29) is a variant of (2.14). It is easy to see that (8.30) holds when $0 < x < 3$. When $x > 3$, let $\{\eta_i, 1 \leq i \leq n\}$ be an independent copy of $\{\xi_i, 1 \leq i \leq n\}$. Set

$$S_n^* = \sum_{i=1}^n \eta_i, \qquad V_n^{*2} = \sum_{i=1}^n \eta_i^2.$$

Then

$$\begin{aligned} P\left(|S_n^*| \leq 2B_n,\ V_n^{*2} \leq 4B_n^2\right) &\geq 1 - P(|S_n^*| > 2B_n) - P(V_n^{*2} > 4B_n^2) \\ &\geq 1 - 1/4 - 1/4 = 1/2, \end{aligned}$$

by the Chebyshev inequality. Noting that

$$\begin{aligned} &\{|S_n| \geq x(4B_n + V_n), |S_n^*| \leq 2B_n, V_n^{*2} \leq 4B_n^2\} \\ &\subset \left\{|S_n - S_n^*| \geq x\left(4B_n + \left(\sum_{i=1}^n(\xi_i - \eta_i)^2\right)^{1/2} - V_n^*\right) - 2B_n, |S_n^*| \leq 2_n, V_n^{*2} \leq 4B_n^2\right\} \\ &\subset \left\{|S_n - S_n^*| \geq x\left(2B_n + \left(\sum_{i=1}^n(\xi_i - \eta_i)^2\right)^{1/2}\right) - 2B_n, |S_n^*| \leq 2B_n\right\} \\ &\subset \left\{|S_n - S_n^*| \geq x\left(\sum_{i=1}^n(\xi_i - \eta_i)^2\right)^{1/2},\ |S_n^*| \leq 2B_n\right\}, \end{aligned}$$

we have

$$\begin{aligned} &E\left\{S_n^2 I\left(|S_n| \geq x(V_n + 4B_n)\right)\right\} \\ &\quad = \frac{E\left\{S_n^2 I\left(|S_n| \geq x(V_n + 4B_n)\right) I(|S_n^*| \leq 2B_n, V_n^{*2} \leq 4B_n^2)\right\}}{P\left(|S_n^*| \leq 2B_n, V_n^{*2} \leq 4B_n^2\right)} \\ &\quad \leq 2E\left\{S_n^2 I\left(|S_n - S_n^*| \geq x\left(\sum_{i=1}^n(\xi_i - \eta_i)^2\right)^{1/2}, |S_n^*| \leq 2B_n\right)\right\} \\ &\quad \leq 4E\left\{(S_n - S_n^*)^2 I\left(|S_n - S_n^*| \geq x\left(\sum_{i=1}^n(\xi_i - \eta_i)^2\right)^{1/2}, |S_n^*| \leq 2B_n\right)\right\} \\ &\qquad +4E\left\{S_n^{*2} I\left(|S_n - S_n^*| \geq x\left(\sum_{i=1}^n(\xi_i - \eta_i)^2\right)^{1/2}, |S_n^*| \leq 2B_n\right)\right\} \\ &\quad \leq 4E\left\{(S_n - S_n^*)^2 I\left(|S_n - S_n^*| \geq x\left(\sum_{i=1}^n(\xi_i - \eta_i)^2\right)^{1/2}\right)\right\} \\ &\qquad +16B_n^2 P\left(|S_n - S_n^*| \geq x\left(\sum_{i=1}^n(\xi_i - \eta_i)^2\right)^{1/2}\right). \end{aligned} \tag{8.31}$$

Let $\{\varepsilon_i, 1 \leq i \leq n\}$ be a Rademacher sequence independent of $\{(\xi_i, \eta_i),\ 1 \leq i \leq n\}$. Noting that $\{\xi_i - \eta_i, 1 \leq i \leq n\}$ is a sequence of independent symmetric random variables, $\{\varepsilon_i(\xi_i - \eta_i), 1 \leq i \leq n\}$ and $\{\xi_i - \eta_i, 1 \leq i \leq n\}$ have the same joint distribution. By Theorem 2.14,

$$P\left(|\sum_{i=1}^n a_i\varepsilon_i| \geq x\left(\sum_{i=1}^n a_i^2\right)^{1/2}\right) \leq 2e^{-x^2/2} \tag{8.32}$$

for any real numbers a_i. Hence

$$\begin{aligned} E\left\{(\sum_{i=1}^n a_i\varepsilon_i)^2 I\left(|\sum_{i=1}^n \varepsilon_i| \geq x\left(\sum_{i=1}^n a_i^2\right)^{1/2}\right)\right\} &\leq (2+x^2)e^{-x^2/2}\sum_{i=1}^n a_i^2 \\ &\leq 1.2e^{-x^2/4}\sum_{i=1}^n a_i^2 \end{aligned} \tag{8.33}$$

for $x > 3$. By (8.32) and (8.33), for $x > 3$,

$$P\left(|S_n - S_n^*| \geq x\left(\sum_{i=1}^n (\xi_i - \eta_i)^2\right)^{1/2}\right) \leq 2e^{-x^2/2} \leq 0.22e^{-x^2/4}, \tag{8.34}$$

$$\begin{aligned} &E\left\{(S_n - S_n^*)^2 I\left(|S_n - S_n^*| \geq x\left(\sum_{i=1}^n (\xi_i - \eta_i)^2\right)^{1/2}\right) I(|S_n^*| \leq 2B_n)\right\} \\ &= E\left\{(\sum_{i=1}^n \varepsilon_i(\xi_i - \eta_i))^2 I\left(|\sum_{i=1}^n \varepsilon_i(\xi_i - \eta_i)| \geq x\left(\sum_{i=1}^n (\xi_i - \eta_i)^2\right)^{1/2}\right)\right\} \\ &\leq 1.2e^{-x^2/4} E\sum_{i=1}^n (\xi_i - \eta_i)^2 \\ &= 2.4B_n^2 e^{-x^2/4}. \end{aligned} \tag{8.35}$$

From (8.31), (8.34) and (8.35), (8.30) follows. □

Lemma 8.9. *With the same notations as in Sect. 8.3.1, assume $\sigma_1^2 = 1$. Then for all $y > 0$,*

$$P\left(|S_n| \geq y(V_n + \sqrt{5n})\right) \leq 2e^{-y^2/8} \tag{8.36}$$

and

$$P(V_n^2 \leq n/2) \leq e^{-\eta_0 n}, \tag{8.37}$$

where $\eta_0 = 1/(32a_0^2)$ and a_0 satisfies

$$E\left\{g(X_1)^2 I(|g(X_1)| \geq a_0)\right\} \leq 1/4. \tag{8.38}$$

Proof. Recall $Eg(X_1) = 0$ and $Eg^2(X_1) = 1$. Note that (8.36) is a special case of (8.29). To prove (8.37), let $Y_k = g(X_k)I(|g(X_k)| \leq a_0)$. Since $e^{-x} \leq 1 - x + x^2/2$ for $x > 0$, we have for $t = 1/(4a_0^2)$,

$$\begin{aligned} P(V_n^2 \leq n/2) &\leq P\left(\sum_{k=1}^n Y_k^2 \leq n/2\right) \\ &\leq e^{tn/2} E e^{-t\sum_{k=1}^n Y_k^2} = e^{tn/2}(Ee^{-tY_1^2})^n \\ &\leq e^{tn/2}\left(1 - tEY_1^2 + t^2 EY_1^4/2\right)^n \\ &\leq e^{tn/2}\left(1 - (3/4)t + t^2 a_0^2/2\right)^n \\ &\leq \exp\left(-(t/4 - t^2 a_0^2/2)n\right) = \exp\left(-n/(32a_0^2)\right). \end{aligned}$$

□

Lemma 8.10. *Assume* $\sigma_1^2 = 1$ *so that* $h^2(x_1,x_2) \le c_0\{1+g^2(x_1)+g^2(x_2)\}$. *Let* $a_0 = 2(c_0+4)$ *and define* $W_n^{(i)}$ *and* Λ_n^2 *by (8.20). Then for all* $y \ge 0$,

$$P\{\Lambda_n^2 \ge a_0 y^2 n(7V_n^2+11n)\} \le \sqrt{2}ne^{-y^2/8}. \tag{8.39}$$

Proof. Let $V_n^{(i)2} = \sum_{j=1,j\neq i}^n \psi^2(X_i,X_j)$ and $\tau^2(x) = E(\psi^2(X_1,X_j)|X_j = x)$. Conditional on X_i, $W_n^{(i)}$ is a sum of i.i.d. random variables with zero means. Hence it follows from (8.29) that

$$P\left\{|W_n^{(i)}| \ge y\left[V_n^{(i)2}+5(n-1)\tau^2(X_i)\right]^{1/2}\right\} \le \sqrt{2}e^{-y^2/8}. \tag{8.40}$$

Since $\psi^2(x_1,x_2) \le 2(c_0+4)[1+g^2(x_1)+g^2(x_2)]$,

$$V_n^{(i)2}+5(n-1)\tau^2(X_i) \le 2(c_0+4)\left[11n+6ng^2(X_i)+\textstyle\sum_{i=1}^n g^2(X_i)\right], \tag{8.41}$$

$$\sum_{i=1}^n\left(11n+6ng^2(X_i)+\sum_{i=1}^n g^2(X_i)\right) = n(7V_n^2+11n).$$

From (8.40)–(8.40), it follows that

$$\begin{aligned}P\{\Lambda_n^2 \ge a_0 y^2 n(7V_n^2+11n)\} &\le \textstyle\sum_{i=1}^n P\left\{|W_n^{(i)}| \ge y\left[V_n^{(i)2}+5(n-1)\tau^2(X_i)\right]^{1/2}\right\}\\ &\le \sqrt{2}ne^{-y^2/8}.\end{aligned}$$

□

Proof (of (8.25)*).* Without loss of generality, assume $\sigma_1^2 = 1$. By (8.36) and (8.37), for any $x > 0$,

$$\begin{aligned}P(|S_n| \ge 5xV_n) &\le P(V_n^2 \le n/2)+P\left\{|S_n| \ge x\left(V_n+\sqrt{5n}\right)\right\}\\ &\le 2e^{-x^2/8}+e^{-\eta_0 n}.\end{aligned}$$

By (8.39) and (8.37), for any $x > 0$,

$$\begin{aligned}P\left(\Lambda_n \ge \sqrt{7a_0+22}x\sqrt{n}V_n\right) &\le P(V_n^2 \le n/2)+P\{\Lambda_n^2 \ge a_0 x^2 n(7V_n^2+11n)\}\\ &\le \sqrt{2}ne^{-x^2/8}+e^{-\eta_0 n}.\end{aligned}$$

Moreover,

$$
\begin{aligned}
P(|\delta_n| \ge y) &\le 2P\left(|S_n| \ge \sqrt{y(n-2)}\,V_n/3\right) + 2P\left(\Lambda_n \ge y(n-2)\,V_n/4\right) \\
&\quad + P\left(\Lambda_n \ge \sqrt{y}\,(n-2)\,V_n/\sqrt{3}\right) \\
&\le 2\sqrt{2}(n+1)e^{-\delta_0' yn} + 2\sqrt{2}ne^{-\delta_0'' y^2 n} + 5e^{-\eta_0 n} \\
&\le 4\sqrt{2}(n+2)\exp\left(-\delta_0\, n \min\{1, y, y^2\}\right),
\end{aligned}
$$

where the constants δ_0, δ_0' and δ_0'' depend only on σ_1^2 and c_0, proving (8.25). □

We omit the proof of (8.26) because it is similar to that of (8.25) except that we use the following exponential inequality in place of (8.39):

$$
P\left\{\left|\sum_{1\le i<j\le n} \psi(X_i,X_j)\right| \ge a_1 y^2 \sqrt{n}\,(V_n^2 + 106n)^{1/2}\right\} \le \sqrt{2}\,(n+2)\,e^{-y^2/8}, \tag{8.42}
$$

where $a_1^2 = 46(c_0+4)$. The rest of this section is devoted to the proof of (8.42). Let $\mathscr{F}_n$ be the σ-field generated by $X_1,\dots,X_n$ and let

$$
Y_j = \sum_{i=1}^{j-1} \psi(X_i,X_j), \qquad T_{1n}^2 = \sum_{j=2}^{n} Y_j^2, \qquad T_{2n}^2 = \sum_{j=2}^{n} E(Y_j^2|\mathscr{F}_{j-1}). \tag{8.43}
$$

Note that $\sum_{1\le i<j\le n} \psi(X_i,X_j) = \sum_{j=2}^{n} Y_j$ and that $\{Y_j\}$ is a martingale difference sequence with respect to the filtration $\{\mathscr{F}_n\}$. The proof of (8.42) uses the following result on exponential inequalities for self-normalized martingales, which will be proved in Chap. 12 (Sect. 12.3.1) for more general self-normalized processes.

Lemma 8.11. *Let $\{\xi_i, \mathscr{F}_i, i \ge 1\}$ be a martingale difference sequence with $E\xi_i^2 < \infty$. Then for all $x > 0$,*

$$
P\left\{\frac{|\sum_{i=1}^{n} \xi_i|}{\left(\sum_{i=1}^{n}(\xi_i^2 + E(\xi_i^2|\mathscr{F}_{i-1}) + 2E\xi_i^2)\right)^{1/2}} \ge x\right\} \le \sqrt{2}\exp(-x^2/4).
$$

Proof (of (8.42)*).* Note that $EY_j^2 \le (j-1)Eh^2(X_1,X_2) \le 3(j-1)$, where the last inequality follows from (8.13) and $Eg^2(X_1) = 1$. With T_{1n}^2 and T_{2n}^2 defined in (8.43), we next show that

$$
P\{T_{1n}^2 \ge a_2 y^2 n\,(V_n^2 + n)\} \le \sqrt{2}ne^{-y^2/8}, \tag{8.44}
$$

$$
P\{T_{2n}^2 \ge a_3 y^2 n\,(V_n^2 + 50n)\} \le \sqrt{2}e^{-y^2/4}, \tag{8.45}
$$

where $a_2 = 14(c_0+4)$ and $a_3 = 16(c_0+4)$. Without loss of generality, assume $y \ge 1$; otherwise (8.44) and (8.45) are obvious. Write $V_j' = V_{\psi,j} + 4(j-1)^{1/2}\tau(X_j)$, where $V_{\psi,j}^2 = \sum_{i=1}^{j-1} \psi^2(X_i,X_j)$. To prove (8.45), note that

$$\begin{aligned}
&P\left\{T_{2n}^2 \ge 2y^2\left[4n\sum_{j=2}^n \tau^2(X_j)+64n^2E\tau^2(X_1)\right]\right\} \\
&\le P\left\{\sum_{j=2}^n E\left[Y_j^2 I(|Y_j| \le yV_j')|\mathscr{F}_{j-1}\right] \ge y^2\left[4n\sum_{j=2}^n \tau^2(X_j)+64n^2E\tau^2(X_1)\right]\right\} \\
&\quad +P\left\{\sum_{j=2}^n E\left[Y_j^2 I(|Y_j| > yV_j')|\mathscr{F}_{j-1}\right] \ge y^2\left[4n\sum_{j=2}^n \tau^2(X_j)+64n^2E\tau^2(X_1)\right]\right\} \\
&:= J_1 + J_2. \qquad (8.46)
\end{aligned}$$

Since $y \ge 1$,

$$\begin{aligned}
J_1 &\le P\left\{\sum_{j=2}^n E\left[V_j'^2|\mathscr{F}_{j-1}\right] \ge \left[4n\sum_{j=2}^n \tau^2(X_j)+64n^2E\tau^2(X_1)\right]\right\} \\
&= P\left\{\sum_{j=2}^n\sum_{i=1}^{j-1} 2\tau^2(X_i) + 32\sum_{j=2}^n (j-1)E\tau^2(X_1) \ge 4n\sum_{j=2}^n \tau^2(X_j)+64n^2E\tau^2(X_1)\right\} \\
&= 0, \qquad (8.47)
\end{aligned}$$

$$\begin{aligned}
J_2 &\le \frac{1}{64y^2n^2E\tau^2(X_1)}\sum_{j=2}^n E\left[Y_j^2 I(|Y_j| > yV_j')\right] \\
&= \frac{1}{64y^2n^2E\tau^2(X_1)}\sum_{j=2}^n E\left\{E\left[Y_j^2 I(|Y_j| > yV_j')|X_j\right]\right\} \\
&\le \frac{16}{64y^2n^2E\tau^2(X_1)}\sum_{j=2}^n E\left[j\tau^2(X_1)\right]e^{-y^2/4} \qquad \text{by (8.30)} \\
&\le e^{-y^2/4}. \qquad (8.48)
\end{aligned}$$

The inequality (8.45) follows from (8.46)–(8.48) and the bound

$$4n\sum_{j=2}^n \tau^2(X_j)+64n^2E\tau^2(X_1) \le 8(c_0+4)n(50n+V_n^2),$$

recalling that $\tau^2(x) \le 2(c_0+4)[2+g(x)]$. The proof of (8.44) is similar.

Since $\sum_{1\le i<j\le n}\psi(x_i,x_j) = \sum_{j=2}^n Y_j$ and $\{Y_j, \mathscr{F}_j,\ j\ge 2\}$ is a martingale difference sequence, it follows from Lemma 8.11 that

$$P\left\{\left|\sum_{j=2}^n Y_j\right| \ge y\left(\sum_{j=2}^n\left[Y_j^2+2EY_j^2+E(Y_j^2|\mathscr{F}_{j-1})\right]\right)^{1/2}\right\} \le \sqrt{2}e^{-y^2/4}. \quad (8.49)$$

Combining (8.49) with (8.44) and (8.45) yields (8.42). □

8.4 Supplementary Results and Problems

1. In Sect. 8.2.1, assuming that $Eh^2(X_1, X_2) < \infty$, prove that $E\Delta_n^2 = O(1)$ and that (8.8) holds.
2. In Sect. 8.2.2, under the assumption $Eh^4(X_1, X_2) < \infty$, prove that

$$\sup_x |P(T_n \le x) - \Phi(x)| = O(n^{-1/2}).$$

3. Give a kernel h and i.i.d. random variables X_1 and X_2 such that $E|h(X_1, X_2)|^3 < \infty$ but condition (8.13) is not satisfied. Does Theorem 8.5 still hold for this kernel?

Part II
Martingales and Dependent Random Vectors

Chapter 9
Martingale Inequalities and Related Tools

In this chapter we first review basic martingale theory and then introduce tangent sequences and decoupling inequalities which are used to derive exponential inequalities for martingales. These exponential inequalities will be used in Chap. 10 to show that a wide range of stochastic models satisfy certain "canonical assumptions," under which self-normalized processes can be treated by a general "pseudo-maximization" approach described in Chap. 11.

9.1 Basic Martingale Theory

Durrett (2005, Chap. 4) provides details of the basic results in martingale theory summarized in this section. Chow and Teicher (1988, Sect. 11.3) gives a comprehensive treatment of convex function inequalities for martingales that include the Burkholder–Davis–Gundy inequalities in Theorem 9.6 as a special case.

9.1.1 Conditional Expectations and Martingales

Let X be a random variable defined on the probability space $(\Omega, \mathscr{F}, P)$ such that $E|X| < \infty$. Let $\mathscr{G} \subset \mathscr{F}$ be a σ-field. A random variable Y is called a version of the *conditional expectation* of X given $\mathscr{G}$, denoted by $E(X \mid \mathscr{G})$, if it satisfies the following two properties:

(a) Y is $\mathscr{G}$-measurable.
(b) $\int_A X\, dP = \int_A Y\, dP$ for all $A \in \mathscr{G}$.

The conditional expectation is therefore the Radon–Nikodym derivative $d\nu/dP$, where $\nu(A) = \int_A X\, dP$ for $A \in \mathscr{G}$. Hence it is unique except for P-null sets. The special case $Y = I(B)$ gives the conditional probability $P(B \mid \mathscr{G})$ of B given $\mathscr{G}$. A *filtration* is a nondecreasing sequence (i.e., $\mathscr{F}_n \subset \mathscr{F}_{n+1}$) of σ-fields $\mathscr{F}_n \subset \mathscr{F}$.

V.H. de la Peña et al., *Self-Normalized Processes: Limit Theory and Statistical Applications*, 123
Probability and its Applications,
© Springer-Verlag Berlin Heidelberg 2009

The Borel–Cantelli lemma has a conditional counterpart involving $\sum_{i=1}^{\infty} P(A_i|\mathscr{F}_{i-1})$. Moreover, Freedman (1973) has provided exponential inequalities relating $\sum_{i=1}^{\tau} X_i$ and $\sum_{i=1}^{\tau} E(X_i|\mathscr{F}_{i-1})$ for stopping times τ and nonnegative, bounded random variables X_i that are adapted to a filtration $\{\mathscr{F}_i\}$ (i.e., X_i is $\mathscr{F}_i$-measurable). A random variable N taking values in $\{1,2,\dots\} \cup \{\infty\}$ is called a *stopping time* with respect to a filtration $\{\mathscr{F}_n\}$ if $\{N=n\} \in \mathscr{F}_n$ for all integers $n \geq 1$. These results are summarized in the following lemma, which will be applied in Chap. 13.

Lemma 9.1. *Let $\{\mathscr{F}_n, n \geq 0\}$ be a filtration.*

(a) Let A_n be a sequence of events with $A_n \in \mathscr{F}_n$. Then

$$\{A_n \; i.o.\} = \left\{ \sum_{n=1}^{\infty} P(A_n \mid \mathscr{F}_{n-1}) = \infty \right\}.$$

(b) Suppose X_n is $\mathscr{F}_n$-measurable and $0 \leq X_n \leq c$ for some non-random constant $c > 0$. Let $\mu_n = E(X_n|\mathscr{F}_{n-1})$ and let τ be a stopping time. Then for $0 \leq a \leq b$,

$$P\left\{ \sum_{i=1}^{\tau} X_i \leq a, \; \sum_{i=1}^{\tau} \mu_i \geq b \right\} \leq \left\{ \left(\frac{b}{a}\right)^a e^{a-b} \right\}^{1/c},$$

$$P\left\{ \sum_{i=1}^{\tau} \mu_i \leq a, \; \sum_{i=1}^{\tau} X_i \geq b \right\} \leq \left\{ \left(\frac{a}{b}\right)^b e^{b-a} \right\}^{1/c}.$$

An important property of conditional expectations is its "tower property": $E(X) = E[E(X|\mathscr{G})]$. A useful result on the conditional probability of the union of $A_1, \dots, A_m$ is Dvoretzky's lemma below; see Durrett (2005, pp. 413–414) for the proof and an application.

Lemma 9.2. *Let $\{\mathscr{F}_n, n \geq 0\}$ be a filtration and $A_n \in \mathscr{F}_n$. Then for any nonnegative random variable ζ that is $\mathscr{F}_0$-measurable,*

$$P\left(\bigcup_{k=1}^{m} A_k | \mathscr{F}_0 \right) \leq \zeta + P\left\{ \sum_{k=1}^{m} P(A_k \mid \mathscr{F}_{k-1}) > \zeta \,\middle|\, \mathscr{F}_0 \right\}.$$

Let M_n be a sequence of random variables adapted to a filtration $\{\mathscr{F}_n\}$ such that $E|M_n| < \infty$ for all n. If

$$E(M_n \mid \mathscr{F}_{n-1}) = M_{n-1} \;\; a.s. \qquad \text{for all } n \geq 1, \tag{9.1}$$

then $\{M_n, \mathscr{F}_n, n \geq 1\}$ is called a *martingale* and $d_n := M_n - M_{n-1}$ is called a *martingale difference sequence*. When the equality in (9.1) is replaced by $\geq$, $\{M_n, \mathscr{F}_n, n \geq 1\}$ is called a *submartingale*. It is called a *supermartingale* if the equality in (9.1) is replaced by $\leq$. By Jensen's inequality, if M_n is a martingale and $\varphi : \mathbb{R} \to \mathbb{R}$ is convex, then $\varphi(M_n)$ is a submartingale.

9.1.2 Martingale Convergence and Inequalities

Associated with a stopping time N is the σ-field

$$\mathscr{F}_N = \{A \in \mathscr{F} : A \cap \{N = n\}\} \in \mathscr{F}_n \qquad \text{for all } n \geq 1\}. \tag{9.2}$$

A sequence of random variables X_n is said to be *uniformly integrable* if

$$\sup_{n\geq 1} E\{|X_n| I(|X_n| \geq a)\} \to 0 \qquad \text{as } a \to \infty. \tag{9.3}$$

Uniform integrability, which has been used in Chapter 7, is an important tool in martingale theory. Two fundamental results in martingale theory are the *optional stopping theorem* and the *martingale convergence theorem.*

Theorem 9.3. *Let $\{X_n, \mathscr{F}_n, n \geq 1\}$ be a submartingale and $M \leq N$ be stopping times (with respect to $\{\mathscr{F}_n\}$). If $\{X_{N\wedge n}, n \geq 1\}$ is uniformly integrable, then $E(X_N \mid \mathscr{F}_M) \geq X_M$ a.s., and consequently, $EX_N \geq EX_M$.*

Theorem 9.4. *Let $\{X_n, \mathscr{F}_n, n \geq 1\}$ be a submartingale. If $\sup_{n\geq 1} E(X_n^+) < \infty$, then X_n converges a.s. to a limit X_∞ with $E|X_\infty| < \infty$.*

Before describing exponential inequalities in the next two sections, we review some classical martingale inequalities.

Theorem 9.5. *Let $\{X_n, \mathscr{F}_n, n \geq 1\}$ be a submartingale. Then for every $\lambda > 0$,*

$$\lambda P\left\{\max_{1\leq i\leq n} X_i > \lambda\right\} \leq E\left\{X_n I\left(\max_{1\leq i\leq n} X_i > \lambda\right)\right\} \leq EX_n^+,$$

$$\lambda P\left\{\min_{1\leq i\leq n} X_i < -\lambda\right\} \leq EX_n^+ - EX_1.$$

Moreover, for any $p > 1$,

$$E\left(\max_{1\leq i\leq n} X_i^+\right)^p \leq \left(\frac{p}{p-1}\right)^p E(X_n^+)^p.$$

Theorem 9.6. *Let $\{M_n = \sum_{i=1}^n d_i, \mathscr{F}_n, n \geq 1\}$ be a martingale. Then there exist finite positive constants a_p, b_p depending only on p such that*

$$a_p E\left(\sum_{i=1}^n d_i^2\right)^{p/2} \leq E\max_{j\leq n} |M_j|^p \leq b_p E\left(\sum_{i=1}^n d_i^2\right)^{p/2} \qquad \textit{for all } p \geq 1.$$

9.2 Tangent Sequences and Decoupling Inequalities

Decoupling inequalities are based on the idea of comparing sums of dependent random variables d_i to sums of conditionally independent (*decoupled*) random variables that have the same conditional distributions as d_i given the past history $\mathscr{F}_{i-1}$.

This section summarizes several key concepts and results in decoupling, a comprehensive treatment of which is given in de la Peña and Giné (1999).

Definition 9.7. Let $\{e_i\}$ and $\{d_i\}$ be two sequences of random variables adapted to the filtration $\{\mathscr{F}_i\}$. Then $\{e_i\}$ and $\{d_i\}$ are *tangent* with respect to $\{\mathscr{F}_i\}$ if for all i,

$$\mathscr{L}(d_i|\mathscr{F}_{i-1}) = \mathscr{L}(e_i|\mathscr{F}_{i-1}),$$

where $\mathscr{L}(d_i|\mathscr{F}_{i-1})$ denotes the conditional probability law of d_i given $\mathscr{F}_{i-1}$.

9.2.1 Construction of Decoupled Tangent Sequences

Definition 9.8.

(a) A sequence $\{e_i\}$ of random variables adapted to the filtration $\{\mathscr{F}_i\}$ is said to satisfy the CI (*conditional independence*) condition if there exists a σ-field $\mathscr{G}$ contained in $\mathscr{F}$ such that $e_1, e_2, \ldots$ are conditionally independent given $\mathscr{G}$ and $\mathscr{L}(e_i|\mathscr{F}_{i-1}) = \mathscr{L}(e_i|\mathscr{G})$ for all i.
(b) A sequence $\{e_i\}$ which satisfies the CI condition and which is also tangent to $\{d_i\}$ is called a *decoupled tangent* sequence with respect to $\{d_i\}$.

Proposition 9.9. *For any sequence of random variables $\{d_i\}$ adapted to a filtration $\{\mathscr{F}_i\}$, there exists a decoupled sequence $\{e_i\}$ (on a possibly enlarged probability space) which is tangent to $\{d_i\}$ and conditionally independent of some σ-field $\mathscr{G}$; we can take $\mathscr{G}$ to be the σ-field generated by $\{d_i\}$.*

The *decoupled* sequence of Proposition 9.9 can be defined recursively as follows: Let e_1 be an independent copy of d_1. At the ith stage, given $\{d_1, \ldots, d_{i-1}\}$, e_i is independent of $e_1, \ldots, e_{i-1}, d_i$ and sampled according to the probability law $\mathscr{L}(d_i|\mathscr{F}_{i-1})$. In this case, if we let $\tilde{\mathscr{F}}_i = \sigma(\mathscr{F}_i, e_1, \ldots, e_i)$, then $\{e_i\}$ is $\{\tilde{\mathscr{F}}_i\}$-tangent to $\{d_i\}$ and satisfies the CI condition with respect to $\mathscr{G} = \sigma(\{d_i\})$. Therefore, any sequence of random variables $\{d_i\}$ has a *decoupled version* $\{e_i\}$. The following diagram illustrates the construction:

$$\begin{array}{ccccccccccccc} d_1 & \to & d_2 & \to & d_3 & \to & d_4 & \to & \cdots & \to & d_{i-1} & \to & d_i \\ & \searrow & & \searrow & & \searrow & & \searrow & & & & \searrow & \\ & e_1 & & e_2 & & e_3 & & e_4 & \ldots & & e_{i-1} & & e_i \end{array}$$

9.2.2 Exponential Decoupling Inequalities

The following theorem provides decoupling inequalities for the moment generating functions of two tangent sequences, one of which satisfies the CI condition. Its proof uses a simple lemma.

Theorem 9.10. *Let d_i be random variables adapted to the filtration $\{\mathcal{F}_i\}$.*

(a) On a possibly enlarged probability space, there exists $\{\tilde{\mathcal{F}}_i\}$, a σ-field $\mathcal{G}$ contained in $\mathcal{F}$ and sequence $\{e_i\}$ satisfying the CI condition given $\mathcal{G}$ and $\{\tilde{\mathcal{F}}_i\}$-tangent to $\{d_i\}$ such that for all $\mathcal{G}$-measurable random variables $g \geq 0$ and all real λ,

$$Eg\exp\left\{\lambda\sum_{i=1}^{n}d_i\right\} \leq \sqrt{Eg^2\exp\left\{2\lambda\sum_{i=1}^{n}e_i\right\}}.$$

(b) Let $\{e_i\}$ be any $\{\mathcal{F}_i\}$-tangent sequence to $\{d_i\}$ and satisfying the CI condition given $\mathcal{G} \subseteq \mathcal{F}$. Then, for all $\mathcal{G}$-measurable functions $g \geq 0$ and all real λ,

$$Eg\exp\left\{\lambda\sum_{i=1}^{n}d_i\right\} \leq \sqrt{Eg^2\exp\left\{2\lambda\sum_{i=1}^{n}e_i\right\}}.$$

Lemma 9.11. *Let X, Y be two nonnegative random variables such that $X = 0$ when $Y = 0$, and $E(X/Y) \leq K$ for some constant K. Then*

$$E\sqrt{X} \leq \sqrt{KEY}. \tag{9.4}$$

Proof. $E\sqrt{X} = E(\sqrt{\frac{X}{Y}} \times \sqrt{Y}) \leq \sqrt{E\frac{X}{Y}} \times \sqrt{EY} \leq \sqrt{KEY}$, by the Cauchy–Schwarz inequality. □

Proof (of Theorem 9.10). First assume that the d_i's are nonnegative. It follows from Proposition 9.9 that one can find a sequence $\{e_i\}$ which is tangent to $\{d_i\}$ and conditionally independent given some σ-field $\mathcal{G}$. Let $\mathcal{F}_i$ be the σ-field generated by $\{d_1,\ldots,d_i,e_1,\ldots,e_i\}$. We can use induction and the tower property of conditional expectations to show that

$$E\left(\frac{\prod_{i=1}^{n}d_i}{\prod_{i=1}^{n}E(d_i|\mathcal{F}_{i-1})}\right) = 1, \tag{9.5}$$

noting that if (9.5) is valid for $n-1$, then

$$\begin{aligned} E\left(\frac{\prod_{i=1}^{n}d_i}{\prod_{i=1}^{n}E(d_i|\mathcal{F}_{i-1})}\right) &= E\left[\frac{\prod_{i=1}^{n-1}d_i}{\prod_{i=1}^{n-1}E(d_i|\mathcal{F}_{i-1})} \times \frac{E(d_n|\mathcal{F}_{n-1})}{E(d_n|\mathcal{F}_{n-1})}\right] \\ &= E\left(\frac{\prod_{i=1}^{n-1}d_i}{\prod_{i=1}^{n-1}E(d_i|\mathcal{F}_{i-1})}\right) = 1. \end{aligned}$$

Since $\{e_i\}$ is tangent to $\{d_i\}$ and conditionally independent given $\mathcal{G}$ and since g is $\mathcal{G}$-measurable,

$$g\prod_{i=1}^{n}E(d_i|\mathscr{F}_{i-1}) = g\prod_{i=1}^{n}E(e_i|\mathscr{F}_{i-1}) = g\prod_{i=1}^{n}E(e_i|\mathscr{G})$$
$$= gE\left(\prod_{i=1}^{n}e_i|\mathscr{G}\right) = E\left(g\prod_{i=1}^{n}e_i|\mathscr{G}\right).$$

From Lemma 9.11 with $K = 1, X = g\prod_{i=1}^{n}d_i$ and $Y = g\prod_{i=1}^{n}E(d_i|\mathscr{F}_{i-1})$, it follows that

$$E\sqrt{g\prod_{i=1}^{n}d_i} \le \sqrt{E\left[E(g\prod_{i=1}^{n}e_i|\mathscr{G})\right]} = \sqrt{E(g\prod_{i=1}^{n}e_i)}.$$

To complete the proof, replace g, d_i, e_i in the preceding argument by $g^2, \exp(2\lambda d_i)$ and $\exp(2\lambda e_i)$. $\square$

9.3 Exponential Inequalities for Martingales

9.3.1 Exponential Inequalities via Decoupling

In this section we summarize the decoupling methods and results of de la Peña (1999) on exponential inequalities for the tail probability of the ratio of a martingale to its conditional variance.

Theorem 9.12. *Let $\{d_i, \mathscr{F}_i, i \ge 1\}$ be a martingale difference sequence with $E(d_j^2|\mathscr{F}_{j-1}) = \sigma_j^2$, $V_n^2 = \sum_{j=1}^{n}\sigma_j^2$. Assume that $E(|d_j|^k|\mathscr{F}_{j-1}) \le (k!/2)\sigma_j^2 c^{k-2}$ a.e. or $P\{|d_j| \le c|\mathscr{F}_{j-1}\} = 1$ for all $k > 2$ and some $c > 0$. Then for all $x > 0$ and $y > 0$,*

$$P\left\{\sum_{i=1}^{n}d_i > x,\ V_n^2 \le y \text{ for some } n\right\} \le \exp\left\{-\frac{x^2}{y(1+\sqrt{1+2cx/y})}\right\} \le \exp\left\{-\frac{x^2}{2(y+cx)}\right\}. \tag{9.6}$$

In the special case of independent random variables, (9.6) with $y = V_n^2$ is the classical Bernstein inequality (2.17). If the L_∞-norm $\|\sum_{i=1}^{n}\sigma_i^2\|_\infty$ is finite a.s., we can also set $y = \|\sum_{i=1}^{n}\sigma_i^2\|_\infty$ in (9.6) and obtain Bernstein's inequality for martingales.

Theorem 9.13. *With the same notations and assumptions as in Theorem 9.12, let $M_n = \sum_{i=1}^{n}d_i$. Then for all $\mathscr{F}_\infty$-measurable sets A and $x > 0$,*

$$P\left\{\frac{M_n}{V_n^2} > x, A\right\} \le E\left(\exp\left\{-\left(\frac{x^2}{1+cx+\sqrt{2cx+1}}\right)V_n^2\right\}\Bigg|\frac{M_n}{V_n^2} > x, A\right) \le E\left(\exp\left\{-\frac{x^2V_n^2}{2(1+cx)}\right\}\Bigg|\frac{M_n}{V_n^2} > x, A\right), \tag{9.7}$$

$$P\left\{\frac{M_n}{V_n^2} > x, A\right\} \le \sqrt{E\exp\left\{-\left(\frac{x^2}{1+\sqrt{2cx+1}+cx}\right)V_n^2\right\}I(A)}, \tag{9.8}$$

$$\begin{aligned} P\left\{\frac{M_n}{V_n^2} > x,\ \frac{1}{V_n^2} \le y \ \textit{for some } n\right\} &\le \exp\left\{-\frac{1}{y}\left(\frac{x^2}{1+\sqrt{2cx+1}+cx}\right)\right\} \\ &\le \exp\left\{-\frac{x^2}{2y(1+cx)}\right\}. \end{aligned} \tag{9.9}$$

Theorem 9.14. *Let $\{d_i, \mathscr{F}_i, i \ge 1\}$ be a martingale difference sequence with $|d_j| \le c$ for some $c > 0$, $E(d_j^2|\mathscr{F}_{j-1}) = \sigma_j^2$ and $V_n^2 = \sum_{i=1}^n \sigma_i^2$ or $V_n^2 = \|\sum_{i=1}^n \sigma_i^2\|_\infty$. Then for all $x > 0$ and $y > 0$,*

$$P\left\{\sum_{i=1}^n d_i > x, V_n^2 \le y \ \textit{for some } n\right\} \le \exp\left\{-\frac{x}{2c}\ \textit{arc sinh}\ \left(\frac{xc}{2y}\right)\right\}. \tag{9.10}$$

Moreover, for every $\mathscr{F}_\infty$-measurable set A, $\beta > 0$, $\alpha \ge 0$ and $x \ge 0$,

$$\begin{aligned} P\left\{\frac{\sum_{i=1}^n d_i}{\alpha+\beta V_n^2} > x, A\right\} &\le \exp\left\{-\frac{\alpha x}{c}\ \textit{arc sinh}\ \frac{c\beta x}{2}\right\} \\ &\times E\left[\exp\left\{-\left(\frac{\beta x}{2c}\ \textit{arc sinh}\ \left(\frac{\beta x c}{2}\right)\right)V_n^2\right\}\middle| M_n > (\alpha+\beta)V_n^2 x, A\right], \end{aligned} \tag{9.11}$$

$$\begin{aligned} &P\left\{\frac{\sum_{i=1}^n d_i}{\alpha+\beta V_n^2} > x, A\right\} \\ &\le \sqrt{\exp\left\{-\frac{\alpha x}{c}\ \textit{arc sinh}\ \frac{c\beta x}{2}\right\}E\exp\left\{-\left(\frac{\beta x}{2c}\ \textit{arc sinh}\ \left(\frac{\beta x c}{2}\right)\right)V_n^2\right\}I(A)}, \end{aligned} \tag{9.12}$$

$$\begin{aligned} &P\left\{\frac{\sum_{i=1}^n d_i}{\alpha+\beta V_n^2} > x,\ \frac{1}{V_n^2} \le y \ \textit{for some } n\right\} \\ &\le \exp\left\{-\frac{\alpha x}{c}\ \textit{arc sinh}\frac{c\beta x}{2}\right\}\exp\left\{-\frac{\beta x}{2cy}\ \textit{arc sinh}\ \left(\frac{\beta x c}{2}\right)\right\}. \end{aligned}$$

In what follows we provide the proofs of Theorems 9.12 and 9.13; the proof of Theorem 9.14 is similar and is therefore omitted. Let $M_n = \sum_{i=1}^n d_i$ and $\mathscr{G}$ be the σ-field generated by $\{d_i\}$. We will use the following variant of the Bennett–Hoeffding inequality for sums of independent random variables (Theorem 2.17).

Lemma 9.15. *Let $\{X_i\}$ be a sequence of independent random variables with $EX_i = 0$, $B_n^2 = \sum_{i=1}^n EX_i^2 > 0$, and such that there exists $c > 0$ for which*

$$E|X_j|^k \le \frac{k!}{2}c^{k-2}EX_j^2 \qquad \textit{for } k > 2,\ 1 \le j \le n,$$

Then

$$E\exp\left\{r\sum_{i=1}^{n}X_i\right\}\le\exp\left\{\frac{B_n^2r^2}{2(1-cr)}\right\}\qquad \textit{for } 0<r<\frac{1}{c}. \tag{9.13}$$

Proof (of Theorem 9.12). Let $\tau=\inf\{n: M_n>x \text{ and } V_n^2\le y\}$, with $\inf\emptyset=\infty$. Let $A=\{M_n>x \text{ and } V_n^2\le y \text{ for some } n\}$. Note that $P(A)=P(\tau<\infty, M_\tau>x, A)$. Applying Markov's inequality first, followed by Fatou's lemma (valid since $\tau<\infty$ on A), we obtain

$$\begin{aligned}
P(A)&\le P\left\{\sum_{i=1}^{\tau}d_i>x,A\right\}\\
&\le\inf_{\lambda>0}E\left[\exp\left\{\frac{\lambda}{2}\left(\sum_{i=1}^{\tau}d_i-x\right)\right\}I(M_\tau>x,A)\right]\\
&=\inf_{\lambda>0}E\left[\lim_{n\to\infty}\exp\left\{\frac{\lambda}{2}\left(\sum_{i=1}^{\tau\wedge n}d_i-x\right)\right\}I(M_{\tau\wedge n}>x,A)\right]\\
&\le\inf_{\lambda>0}\liminf_{n\to\infty}E\exp\left\{\frac{\lambda}{2}\left(\sum_{i=1}^{\tau\wedge n}d_i-x\right)\right\}I(M_{\tau\wedge n}>x,A)\\
&\le\inf_{\lambda>0}\liminf_{n\to\infty}\sqrt{E\exp\left\{\lambda\left(\sum_{i=1}^{\tau\wedge n}e_i-x\right)\right\}I(M_{\tau\wedge n}>x,A)}, \text{ by Theorem 9.10(b),}\\
&=\inf_{\lambda>0}\liminf_{n\to\infty}\sqrt{E\left[I(M_{\tau\wedge n}>x,A)e^{-\lambda x}E\left(\exp\left\{\lambda\sum_{i=1}^{\tau\wedge n}e_i\right\}\middle|\mathscr{G}\right)\right]},
\end{aligned}$$

recalling that the random variables outside the conditional expectation are $\mathscr{G}$ measurable. Observe that since $\{d_i\}$ and $\{e_i\}$ are tangent and $\{e_i\}$ is conditionally independent given $\mathscr{G}$, the moment assumptions on the distribution of d_i translate to conditions on the e_i's and therefore we can apply (9.13) to obtain

$$E\left(\exp\left\{\lambda\sum_{i=1}^{\tau\wedge n}e_i\right\}\middle|\mathscr{G}\right)\le\exp\left\{h(\lambda)V_{\tau\wedge n}^2\right\}, \tag{9.14}$$

where $h(\lambda)=\frac{\lambda^2}{2(1-\lambda c)}$. Replacing this in the above bound one obtains

$$P\left\{\sum_{i=1}^{\tau}d_i>x,A\right\}\le\inf_{\lambda>0}\liminf_{n\to\infty}\sqrt{E\left[\exp\left\{-\left(\lambda x-h(\lambda)V_{\tau\wedge n}^2\right)\right\}I(M_{\tau\wedge n}>x,A)\right]}.$$

Since the variable inside the expectation is dominated by

$$\exp\left\{-\left(\lambda x-h(\lambda)V_\tau^2\right)\right\}I(M_{\tau\wedge n}>x,A),$$

and since $V_\tau\le y$ on A, application of the dominated convergence theorem yields

$$P\left\{\sum_{i=1}^{\tau} d_i > x, A\right\} \le \inf_{\lambda>0} \sqrt{E\exp\{-(\lambda x - h(\lambda)V_\tau^2)\} I(M_\tau > x, A)}.$$

Dividing both sides by $\sqrt{P\{M_\tau > x, A\}}$ gives

$$P\left\{\sum_{i=1}^{\tau} d_i > x, A\right\} \le \inf_{\lambda>0} E\left[\exp\{-(\lambda x - h(\lambda)V_\tau^2)\} \,\middle|\, A \cap \{M_\tau > x\}\right].$$

Then, since $M_\tau > x$ and $V_\tau^2 \le y$ on A, we have

$$P\left\{\sum_{i=1}^{n} d_i > x, V_n^2 \le y \text{ for some } n\right\} \le \inf_{\lambda>0} \exp\{-(\lambda x - h(\lambda)y)\},$$

from which (9.6) follows by minimizing $\exp\{-(\lambda x - h(\lambda)y)\}$ over $\lambda > 0$. □

Proof (of Theorem 9.13). Application of Markov's inequality similar to that in the proof of Theorem 9.12 yields

$$\begin{aligned}
P\left\{\sum_{i=1}^{n} d_i > V_n^2 x, A\right\} &\le \inf_{\lambda>0} E\left[\exp\left\{\frac{\lambda}{2}\left(\sum_{i=1}^{n} d_i - V_n^2 x\right)\right\} I(M_n > V_n^2 x, A)\right] \\
&\le \inf_{\lambda>0} \sqrt{E\left[\exp\left\{\lambda\left(\sum_{i=1}^{n} e_i - V_n^2 x\right)\right\} I(M_n > V_n^2 x, A)\right]} \\
&= \inf_{\lambda>0} \sqrt{E\left[I(M_n > V_n^2 x, A)\exp\{-\lambda V_n^2 x\} E\left(\exp\left\{\lambda \sum_{i=1}^{n} e_i\right\} \,\middle|\, \mathscr{G}\right)\right]}.
\end{aligned}$$

Since $\{d_i\}$ and $\{e_i\}$ are tangent and $\{e_i\}$ is conditionally independent given $\mathscr{G}$, the moment assumptions on the distribution of d_i translate to conditions on the e_i's and therefore we can apply (9.13) to show that

$$E\left(\exp\left\{\lambda \sum_{i=1}^{n} e_i\right\} \,\middle|\, \mathscr{G}\right) \le \exp\left\{\frac{\lambda^2}{2(1-\lambda c)} V_n^2\right\}, \tag{9.15}$$

which can be combined with the preceding bound to yield

$$\begin{aligned}
P\left\{\sum_{i=1}^{n} d_i > V_n^2 x, A\right\} &\le \inf_{\lambda>0} \sqrt{E\exp\left\{-\left(\lambda x - \frac{\lambda^2}{2(1-\lambda c)}\right) V_n^2\right\} I(M_n > V_n^2 x, A)} \\
&\le \sqrt{E\exp\left\{-\frac{x^2}{1+\sqrt{2cx+1}+cx} V_n^2\right\} I(M_n > V_n^2 x, A)}.
\end{aligned}$$

Dividing both sides by $\sqrt{P\{M_n > V_n^2 x, A\}}$ gives (9.7), while (9.9) is obtained by adapting the stopping time argument used in the proof of Theorem 9.12. □

9.3.2 Conditionally Symmetric Random Variables

Let $\{d_i\}$ be a sequence of variables adapted to a filtration $\{\mathscr{F}_i\}$. Then we say that the d_i's are *conditionally symmetric* if $\mathscr{L}(d_i \mid \mathscr{F}_{i-1}) = \mathscr{L}(-d_i \mid \mathscr{F}_{i-1})$. Note that any sequence of real-valued random variables X_i can be "symmetrized" to produce an exponential supermartingale by introducing random variables X_i' such that

$$\mathscr{L}(X_n'|X_1, X_1', \ldots, X_{n-1}, X_{n-1}', X_n) = \mathscr{L}(X_n|X_1, \ldots, X_{n-1})$$

and setting $d_n = X_n - X_n'$; see Sect. 6.1 of de la Peña and Giné (1999).

Theorem 9.16. *Let $\{d_i\}$ be a sequence of conditionally symmetric random variables with respect to the filtration $\{\mathscr{F}_n\}$. Then for all $x > 0$, $y > 0$,*

$$P\left(\sum_{i=1}^{n} d_i \geq x, \sum_{i=1}^{n} d_i^2 \leq y \text{ for some } n\right) \leq \exp\left\{-\frac{x^2}{2y}\right\}. \tag{9.16}$$

Moreover, for all sets $A \in \mathscr{F}_\infty$ and all $\beta > 0$, $x \geq 0$, $y > 0$ and $\alpha \in \mathbb{R}$,

$$P\left(\frac{\sum_{i=1}^n d_i}{\alpha + \beta \sum_{i=1}^n d_i^2} \geq x, A\right) \leq E\left[\exp\left\{-x^2\left(\frac{\beta^2}{2}\sum_{i=1}^{n} d_i^2 + \alpha\beta\right)\right\} \middle| \frac{\sum_{i=1}^n d_i}{\alpha + \beta \sum_{i=1}^n d_i^2} \geq x, A\right], \tag{9.17}$$

$$P\left(\frac{\sum_{i=1}^n d_i}{\alpha + \beta \sum_{i=1}^n d_i^2} \geq x, A\right) \leq \sqrt{E \exp\left\{-x^2\left(\frac{\beta^2}{2}\sum_{i=1}^{n} d_i^2 + \alpha\beta\right)\right\}}, \tag{9.18}$$

$$P\left(\frac{\sum_{i=1}^n d_i}{\alpha + \beta \sum_{i=1}^n d_i^2} \geq x, \frac{1}{\sum_{i=1}^n d_i^2} \leq y \text{ for some } n\right) \leq \exp\left\{-x^2\left(\frac{\beta^2}{2y} + \alpha\beta\right)\right\}. \tag{9.19}$$

Lemma 9.17. *Let $\{d_i\}$ be a sequence of conditionally symmetric random variables with respect to the filtration $\{\mathscr{F}_n\}$. Then for all $\lambda > 0$,*

$$\left\{\frac{\exp\{\sum_{i=1}^n \lambda d_i\}}{\exp\{(\lambda^2/2)\sum_{i=1}^n d_i^2\}}, \mathscr{F}_n, n \geq 1\right\} \tag{9.20}$$

is a supermartingale.

Proof. Let $\mathscr{H}_0$ be the trivial σ-field $\{\Omega, \emptyset\}$ and for $n \geq 1$, let $\mathscr{H}_n$ be the σ-field generated by $(d_1, \ldots, d_{n-1}, |d_n|)$. Similarly, let $\mathscr{F}_0$ be the trivial σ-field and for $n \geq 2$, let $\mathscr{F}_{n-1}$ be the σ-field generated by $(d_1, \ldots, d_{n-1})$. Then the conditional symmetry of

$\{d_i\}$ implies that the conditional distribution of d_n given $\mathscr{H}_n$ and that of $-d_n$ given $\mathscr{H}_n$ are the same. Hence

$$E\left[\exp\{\lambda d_n\}\middle|\mathscr{H}_n\right] = E\left[\exp\{-\lambda d_n\}\middle|\mathscr{H}_n\right], \tag{9.21}$$

which can be shown by noting that for all $H_n \in \mathscr{H}_n$, $F_{n-1} \in \mathscr{F}_{n-1}$ and $\lambda > 0$,

$$\int_{(d_1,\dots,d_{n-1},|d_n|)\in H_n} \exp(\lambda d_n)\,dP = \int_{(d_1,\dots,d_{n-1},|d_n|)\in H_n} \exp(-\lambda d_n)\,dP.$$

Making use of (9.21) and the fact that $\{\exp(\lambda d_n) + \exp(-\lambda d_n)\}/2$ is measurable with respect to $\mathscr{H}_n$, we obtain

$$\begin{aligned} E\left[\exp(\lambda d_n) \mid \mathscr{H}_n\right] &= E\left[\frac{\exp(\lambda d_n)+\exp(-\lambda d_n)}{2}\middle|\mathscr{H}_n\right] \\ &= \frac{\exp(\lambda d_n)+\exp(-\lambda d_n)}{2} \le \exp\left(\frac{\lambda^2 d_n^2}{2}\right). \end{aligned}$$

Hence $E\{\exp(\lambda d_n - \frac{1}{2}\lambda^2 d_n^2)|\mathscr{H}_n\} \le 1$, and an induction argument can then be used to complete the proof of Lemma 9.17. □

Proof (of Theorem 9.16). We only consider the case $\alpha = 0$, $\beta = 1$ because the general case follows similarly. For all $A \in \mathscr{F}_\infty$ and $\lambda > 0$,

$$\begin{aligned} &P\left(\frac{\sum_{i=1}^n d_i}{\sum_{i=1}^n d_i^2} \ge x, A\right) \\ &\le E\exp\left\{\frac{\lambda}{2}\sum_{i=1}^n d_i - \frac{\lambda x}{2}\sum_{i=1}^n d_i^2\right\} I\left(\frac{\sum_{i=1}^n d_i}{\sum_{i=1}^n d_i^2} \ge x, A\right) \\ &= E\frac{\exp\{(\lambda/2)\sum_{i=1}^n d_i\}}{\exp\left\{\frac{\lambda^2}{4}\sum_{i=1}^n d_i^2\right\}} \exp\left\{\frac{\lambda^2}{4}\sum_{i=1}^n d_i^2 - \frac{\lambda x}{2}\sum_{i=1}^n d_i^2\right\} I\left(\frac{\sum_{i=1}^n d_i}{\sum_{i=1}^n d_i^2} \ge x, A\right) \\ &\le \sqrt{E\exp\left\{\frac{\lambda^2}{2}\sum_{i=1}^n d_i^2 - \lambda x\sum_{i=1}^n d_i^2\right\} I\left(\frac{\sum_{i=1}^n d_i}{\sum_{i=1}^n d_i^2} \ge x, A\right)} \\ &\le \sqrt{E\exp\left\{-\frac{x^2}{2}\sum_{i=1}^n d_i^2\right\} I\left(\frac{\sum_{i=1}^n d_i}{\sum_{i=1}^n d_i^2} \ge x, A\right)}, \end{aligned}$$

where the last inequality follows by minimizing over λ and the one that precedes it follows from the Cauchy–Schwarz inequality and Lemma 9.17. Dividing both sides by $\sqrt{P(\sum_{i=1}^n d_i/\sum_{i=1}^n d_i^2 \ge x, A)}$ yields the desired conclusion. □

9.3.3 Exponential Supermartingales and Associated Inequalities

Theorem 9.18 (Stout, 1973). *Let $\{d_n\}$ be a sequence of random variables adapted to a filtration $\{\mathscr{F}_n\}$ and such that $E(d_n|\mathscr{F}_{n-1}) \le 0$ and $d_n \le c$ a.s. for all n and some constant $c > 0$. For $\lambda > 0$ with $\lambda c \le 1$, let*

$$T_n = \exp\left(\lambda \sum_{i=1}^{n} d_i\right) \exp\left[-\frac{\lambda^2}{2}\left(1+\frac{\lambda c}{2}\right)\sum_{i=1}^{n} E(d_i^2|\mathscr{F}_{i-1})\right] \tag{9.22}$$

for $n \ge 1$, with $T_0 = 1$. Then $\{T_n, \mathscr{F}_n, n \ge 1\}$ is a nonnegative supermartingale with mean ≤ 1 and

$$P\left(\sup_{n\ge 0} T_n > \alpha\right) \le 1/\alpha \tag{9.23}$$

for all $\alpha \ge 1$.

Proof. Since $d_i \le c$ *a.s.* and $0 < \lambda < c^{-1}$,

$$e^{\lambda d_i} \le 1 + \lambda d_i + \frac{1}{2}(\lambda d_i)^2\left(1+\frac{\lambda c}{2}\right) \quad a.s.$$

Combining this with the assumption that $E(d_i|\mathscr{F}_{i-1}) \le 0$ *a.s.* yields

$$\begin{aligned} E(e^{\lambda d_i}|\mathscr{F}_{i-1}) &\le \frac{\lambda^2}{2} E(d_i^2|\mathscr{F}_{i-1})\left[1+\frac{\lambda c}{2}\right] \\ &\le \exp\left[\frac{\lambda^2}{2}\left(1+\frac{\lambda c}{2}\right)E(d_i^2|\mathscr{F}_{i-1})\right] \quad a.s., \end{aligned}$$

since $1+x \le e^x$. From this, (9.22) follows. To prove (9.23), let $\alpha > 0$ be fixed and let $\tau = \inf\{n \ge 0 : T_n > \alpha\}$, $\inf\emptyset = \infty$. Then the sequence $\{T_{n\wedge\tau}, n \ge 0\}$ is also a nonnegative supermartingale with $T_0 = 1$, and therefore for all $n \ge 1$,

$$1 \ge ET_{\tau\wedge n} \ge \alpha P(\tau \le n).$$

Letting $n \longrightarrow \infty$ completes the proof since $1 \ge \alpha P(\tau < \infty) = \alpha P(\sup_{n\ge 0} T_n > \alpha)$. □

While Theorem 9.18 can be regarded as a supermartingale "relative" of the Bennett–Hoeffding inequality, Lemma 9.15 (which is a variant of the Bennett–Hoeffding inequality) is likewise related to the following supermartingale.

Theorem 9.19. *Let $\{d_n\}$ be a sequence of random variables adapted to a filtration $\{\mathscr{F}_n\}$ such that $E(d_n|\mathscr{F}_{n-1}) = 0$ and $\sigma_n^2 = E(d_n^2|\mathscr{F}_{n-1}) < \infty$. Assume that there exists a positive constant M such that $E(|d_n|^k|\mathscr{F}_{n-1}) \le (k!/2)\sigma_n^2 M^{k-2}$ a.s. or $P(|d_n| \le M|\mathscr{F}_{n-1}) = 1$ a.s. for all $n \ge 1$, $k > 2$. Let $A_n = \sum_{i=1}^n d_i$, $V_n^2 = \sum_{i=1}^n E(d_i^2|\mathscr{F}_{i-1})$, $A_0 = V_0 = 0$. Then $\{\exp(\lambda A_n - \frac{1}{2(1-M\lambda)}\lambda^2 V_n^2), \mathscr{F}_n, n \ge 0\}$ is a supermartingale for every $0 \le \lambda \le 1/M$.*

The martingale $(M_n, \mathscr{F}_n, n \geq 1)$ is said to be *square-integrable* if $EM_n^2 < \infty$ for all n. A stochastic sequence $\{M_n\}$ adapted to a filtration $\{\mathscr{F}_n\}$ is said to be a *locally square-integrable martingale* if there are stopping times τ_m with respect to $\{\mathscr{F}_n\}$ such that $\lim_{m\to\infty} \tau_m = \infty$ *a.s.* and $\{M_{\tau_m \wedge n}, \mathscr{F}_n, n \geq 1\}$ is a square-integrable martingale for every $m \geq 1$. Azuma (1967) proved the following extension of the Bennett–Hoeffding inequality for locally square-integrable martingales.

Theorem 9.20 (Azuma, 1967). *Let $\{M_n = \sum_{i=1}^n d_i, \mathscr{F}_n, n \geq 1\}$ be a locally square-integrable martingale such that there exist nonrandom constants $a_i < b_i$ for which $a_i \leq d_i \leq b_i$ for all $i \geq 1$. Then for all $x \geq 0$,*

$$P(|M_n| \geq x) \leq 2\exp\left(-\frac{2x^2}{\sum_{i=1}^n (b_i - a_i)^2}\right).$$

Whereas the exponential inequalities in Theorems 9.12–9.14 and 9.18 involve conditional variances $\sum_{i=1}^n E(d_i^2|\mathscr{F}_{i-1})$, those in Theorem 9.16 and Lemma 9.17 involve the squared function $\sum_{i=1}^n d_i^2$ of the (local) martingale. Bercu and Touati (2008) have derived the following analogs of Theorems 9.12 and 9.14 by using $\sum_{i=1}^n E(d_i^2|\mathscr{F}_{i-1}) + \sum_{i=1}^n d_i^2$ for normalization to dispense with the boundedness assumptions of d_i in Theorems 9.12 and 9.14.

Theorem 9.21 (Bercu and Touati, 2008). *Let $\{M_n = \sum_{i=1}^n d_i\}$ be a locally square-integrable martingale adapted to the filtration $\{\mathscr{F}_n\}$ with $M_0 = 0$. Let $\langle M \rangle_n = \sum_{k=1}^n E(d_k^2|\mathscr{F}_{k-1})$ and $[M]_n = \sum_{k=1}^n d_k^2$. Then for all $\lambda \in \mathbb{R}$,*

$$\left\{\exp\left(\lambda M_n - \frac{\lambda^2}{2}(\langle M_n \rangle + [M]_n)\right), \mathscr{F}_n,\ n \geq 1\right\} \tag{9.24}$$

is a supermartingale with mean ≤ 1. Moreover, for all $x > 0$, $y > 0$,

$$P(|M_n| \geq x,\ [M]_n + \langle M \rangle_n \leq y) \leq 2\exp\left(-\frac{x^2}{2y}\right),$$

and for all $a \geq 0$, $b > 0$,

$$P\left(\frac{|M_n|}{a + b\langle M \rangle_n} \geq x,\ \langle M \rangle_n \geq [M]_n + y\right) \leq 2\exp\left[-x^2\left(ab + \frac{b^2 y^2}{2}\right)\right].$$

9.4 Supplementary Results and Problems

1. Let $X_1, X_2, \ldots$ be independent random variables and let T be a stopping time adapted to the filtration $\{\mathscr{F}_n\}$, where $\mathscr{F}_n$ is the σ-field generated by $X_1, \ldots, X_n$. Let $\{\tilde{X}_i,\ i \geq 1\}$ be an independent copy of $\{X_i,\ i \geq 1\}$:

   (a) Show that, on a possibly enlarged probability space, $\{X_n I(T \geq n),\ n \geq 1\}$ is tangent to $\{\tilde{X}_n I(T \geq n),\ n \geq 1\}$ with respect to the filtration $\{\mathscr{G}_n\}$, where $\mathscr{G}_n$ is the σ-field generated by $X_1, \ldots, X_n, \tilde{X}_1, \ldots, \tilde{X}_n$.

(b) Show that $\sum_{i=1}^{T\wedge n} \tilde{X}_i$ is a sum of conditionally independent random variables but $\sum_{i=1}^{T\wedge n} X_i$ is not.

2. Prove Lemma 9.15.
3. The exponential inequalities (9.7)–(9.9), (9.11), (9.12) and (9.17)–(9.19) are related to the strong law of large numbers since they involve a martingale divided by its conditional variance or quadratic variation, as noted by de la Peña et al. (2007) who have extended this approach to obtain exponential bounds for the ratio of two processes in a more general setting involving what they call the "canonical assumptions." These canonical assumptions are described in Chap. 10 where we introduce a general framework for self-normalization, in which we use the *square root* of the conditional variance or quadratic variation for normalization, instead of the strong-law-type normalization in the exponential inequalities in this chapter. In particular, if A and $B > 0$ are two random variables satisfying the canonical assumption that

$$E\exp(\lambda A - \lambda^2 B^2/2) \leq 1 \qquad \text{for all } \lambda \in \mathbb{R}, \tag{9.25}$$

de la Peña et al. (2007) have shown that for all $x \geq 0$ and $y > 0$,

$$P(A/B^2 > x,\ 1/B^2 \leq y) \leq e^{-x^2/(2y)}, \tag{9.26}$$

$$P(|A|/B > x,\ y \leq B \leq ay) \leq 4\sqrt{e}x(1+\log a)e^{-x^2/2} \quad \text{for all } a \geq 1. \tag{9.27}$$

Compare (9.26) with (9.9) and discuss their connection.
Hint: See Sect. 10.2.

Chapter 10
A General Framework for Self-Normalization

In this chapter we provide a general framework for the probability theory of self-normalized processes. We begin by describing another method to prove the large deviation result (3.8) for self-normalized sums of i.i.d. random variables. This approach leads to an exponential family of supermartingales associated with self-normalization in Sect. 10.1. The general framework involves these supermartingales, or weaker variants thereof, called "canonical assumptions" in Sect. 10.2, which also provides a list of lemmas showing a wide range of stochastic models that satisfy these canonical assumptions. Whereas Sect. 9.3 gives exponential inequalities for discrete-time martingales that are related to the canonical assumptions, Sect. 10.3 gives continuous-time analogs of these results.

10.1 An Exponential Family of Supermartingales Associated with Self-Normalization

10.1.1 The I.I.D. Case and Another Derivation of (3.8)

A key idea underlying the proof of Theorem 3.1 in Chap. 3 is the representation (3.13) so that $S_n \geq x n^{1/2} V_n \Leftrightarrow \sup_{b \geq 0} \sum_{i=1}^n \{bX_i - x(X_i^2 + b^2)/2\} \geq 0$. For each fixed b, letting $Y_i = bX_i - x(X_i^2 + b^2)/2$, the Cramér–Chernoff large deviation theory yields that the rate of decay of $n^{-1} \log P\{\sum_{i=1}^n Y_i \geq 0\}$ is $\inf_{t>0} \log E e^{tY}$. A technical argument, which involves partitioning $\{b \geq 0\}$ into a finite union of disjoint sets and truncation of X_i, is used in Sect. 3.2.2 to prove Theorem 3.1.

An alternative method to prove Theorem 3.1 is to use the finiteness of the moment generating function $e^{\psi(\theta,\rho)} = E\exp\{\theta X - \rho(\theta X)^2\}$ for all $\theta \in \mathbb{R}$ and $\rho > 0$ (without any moment condition on X), which yields the large-deviations rate function

$$\phi(\mu_1, \mu_2) = \sup_{\theta \in \mathbb{R}, \rho > 0} \{\theta \mu_1 - \rho \theta^2 \mu_2 - \psi(\theta, \rho)\}, \qquad \mu_1 \in \mathbb{R},\ \mu_2 \geq \mu_1^2. \quad (10.1)$$

V.H. de la Peña et al., *Self-Normalized Processes: Limit Theory and Statistical Applications*, Probability and its Applications,

© Springer-Verlag Berlin Heidelberg 2009

Since $S_n/(\sqrt{n}V_n) = g(n^{-1}\sum_{i=1}^n X_i, n^{-1}\sum_{i=1}^n X_i^2)$, where $g(\mu_1,\mu_2) = \mu_1/\sqrt{\mu_2}$, we can express $P\{S_n \ge x\sqrt{n}V_n\}$ as $P\{g(\hat{\mu}_1,\hat{\mu}_2) \ge x\}$, where $\hat{\mu}_1 = n^{-1}\sum_{i=1}^n X_i$, $\hat{\mu}_2 = \sum_{i=1}^n X_i^2$. A standard method to analyze large deviation probabilities via the moment generating function $e^{\psi(\theta,\rho)}$ is to introduce the family of measures $P_{\theta,\rho}$ under which the X_i are i.i.d. with density function $f_{\theta,\rho}(x) = \exp\{\theta x - \rho(\theta x)^2 - \psi(\theta,\rho)\}$ with respect to the measure P that corresponds to the case $\theta = 0$. We can use the change of measures as in Sect. 3.1 to show that $P\{g(\hat{\mu}_1,\hat{\mu}_2) \ge b\} = e^{(\kappa+o(1))n}$, where $\kappa = -\inf\{\phi(\mu_1,\mu_2) : g(\mu_1,\mu_2) \ge x\}$. As shown in Chan and Lai (2000, p. 1648), who use this approach to obtain a Bahadur–Ranga Rao-type refined large deviation approximation, the right-hand side of (3.8) is equal to e^{κ}.

10.1.2 A Representation of Self-Normalized Processes and Associated Exponential Supermartingales

Let $P_{\theta,\rho}^{(n)}$ denote the restriction of $P_{\theta,\rho}$ to the σ-field $\mathscr{F}_n$ generated by $X_1,\ldots,X_n$. The change of measures mentioned in the preceding paragraph involves the likelihood ratio (or Radon–Nikodym derivative)

$$\frac{dP_{\theta,\rho}^{(n)}}{dP^{(n)}} = \exp\left\{\theta S_n - \rho\theta^2 n V_n^2 - n\psi(\theta,\rho)\right\}, \qquad (10.2)$$

which is a martingale with mean 1 under P. When $EX = 0$ and $EX^2 < \infty$, Taylor's theorem yields

$$\psi(\theta,\rho) = \log\left(E\exp\left\{\theta X - \rho(\theta X)^2\right\}\right) = \left\{\left(\frac{1}{2} - \rho + o(1)\right)\theta^2 EX^2\right\}$$

as $\theta \to 0$. Let $\gamma > 0$, $A_n = S_n$ and $B_n^2 = (1+\gamma)\sum_{i=1}^n X_i^2$. It then follows that ρ and ε can be chosen sufficiently small so that for $|\lambda| < \varepsilon$,

$$\left\{\exp(\lambda A_n - \lambda^2 B_n^2/2), \mathscr{F}_n, n \ge 1\right\} \text{ is a supermartingale with mean } \le 1. \qquad (10.3)$$

The assumption (10.3) and its variants provide a general framework to analyze self-normalized processes of the form A_n/B_n with $B_n > 0$. A key observation is

$$\frac{A_n^2}{2B_n^2} = \max_{\lambda}\left(\lambda A_n - \frac{\lambda^2 B_n^2}{2}\right). \qquad (10.4)$$

Although maximizing the supermartingale in (10.3) over λ would not yield a supermartingale and the maximum may also occur outside the range $|\lambda| < \varepsilon$, integrating the supermartingale with respect to the measure $f(\lambda)d\lambda$ still preserves the supermartingale property. Laplace's method for asymptotic evaluation of integrals (see Sect. 11.1) still has the effect of maximizing $(\lambda A_n - \lambda^2 B_n^2/2)$, and the maximum is still $\frac{1}{2}(A_n/B_n)^2$ if the maximizer A_n/B_n^2 lies inside $(-\varepsilon,\varepsilon)$. This "pseudo-maximization" approach, which will be described in the next chapter, was introduced by de la Peña et al. (2000, 2004) to study self-normalized processes in the general framework of (10.3) or some even weaker variants, which replace the moment generating function $e^{\psi(\theta,\rho)}$ in the i.i.d. case and which they call the *canonical assumptions*.

10.2 Canonical Assumptions and Related Stochastic Models

A continuous-time analog of (10.3) simply replaces $n\,(\geq 1)$ by $t\,(\geq 0)$, with $A_0 = 0$. A canonical assumption that includes both cases, therefore, is

$$\{\exp(\lambda A_t - \lambda^2 B_t^2/2),\ \mathscr{F}_t,\ t \in T\} \text{ is a supermartingale with mean } \leq 1, \tag{10.5}$$

where T is either $\{1,2,\dots\}$ or $[0,\infty)$. A weaker canonical assumption considers a pair of random variables A, B, with $B > 0$ such that

$$E\exp(\lambda A - \lambda^2 B^2/2) \leq 1, \tag{10.6}$$

either (a) for all real λ, or (b) for all $\lambda \geq 0$, or (c) for all $0 \leq \lambda < \lambda_0$. Lemma 9.17 and Theorems 9.18, 9.19 and 9.21, in Chap. 9 and the following lemmas provide a wide range of stochastic models that satisfy (10.3), (10.5) or (10.6). In particular, Lemma 10.2 follows from Proposition 3.5.12 of Karatzas and Shreve (1991). Lemma 10.3 is taken from Proposition 4.2.1 of Barlow et al. (1986); see Sect. 10.3 for an introduction to continuous-time martingales.

Lemma 10.1. *Let W_t be a standard Brownian motion. Assume that T is a stopping time such that $T < \infty$ a.s. Then*

$$E\exp\{\lambda W_T - \lambda^2 T/2\} \leq 1,$$

for all $\lambda \in \mathbb{R}$.

Lemma 10.2. *Let M_t be a continuous local martingale, with $M_0 = 0$. Then $\exp\{\lambda M_t - \lambda^2 \langle M\rangle_t/2\}$ is a supermartingale for all $\lambda \in \mathbb{R}$, and therefore*

$$E\exp\{\lambda M_t - \lambda^2 \langle M\rangle_t/2\} \leq 1.$$

Lemma 10.3. *Let $\{M_t, \mathscr{F}_t, t \geq 0\}$ be a locally square-integrable right-continuous martingale, with $M_0 = 0$. Let $\{V_t\}$ be an increasing process, which is adapted, purely discontinuous and locally integrable; let $V^{(p)}$ be its dual predictable projection. Set $X_t = M_t + V_t$, $C_t = \sum_{s\leq t}((\Delta X_s)^+)^2$, $D_t = \{\sum_{s\leq t}((\Delta X_s)^-)^2\}_t^{(p)}$, $H_t = \langle M^c\rangle_t + C_t + D_t$. Then for all $\lambda \in \mathbb{R}$, $\exp\{\lambda(X_t - V_t^{(p)}) - \frac{1}{2}\lambda^2 H_t\}$ is a supermartingale and hence*

$$E\exp\{\lambda(X_t - V_t^{(p)}) - \lambda^2 H_t/2\} \leq 1.$$

In Sect. 10.3, we give two additional lemmas (Lemmas 10.6 and 10.7) on continuous-time martingales that satisfy the canonical assumption (10.5). In Sect. 13.2, we derive the following two lemmas that are associated with the theory of self-normalized LIL.

Lemma 10.4. *Let $\{d_n\}$ be a sequence of random variables adapted to a filtration $\{\mathscr{F}_n\}$ such that $E(d_n|\mathscr{F}_{n-1}) \le 0$ and $d_n \ge -M$ a. s. for all n and some non-random positive constant M. Let $A_n = \sum_{i=1}^n d_i$, $B_n^2 = 2C_\gamma \sum_{i=1}^n d_i^2$, $A_0 = B_0 = 0$ where $C_\gamma = -\{\gamma + \log(1-\gamma)\}/\gamma^2$. Then $\{\exp(\lambda A_n - \frac{1}{2}\lambda^2 B_n^2), \mathscr{F}_n, n \ge 0\}$ is a supermartingale for every $0 \le \lambda \le \gamma M^{-1}$.*

Lemma 10.5. *Let $\{\mathscr{F}_n\}$ be a filtration and Y_n be $\mathscr{F}_n$-measurable random variables. Let $0 \le \gamma_n < 1$ and $0 < \lambda_n \le 1/C_{\gamma_n}$ be $\mathscr{F}_{n-1}$-measurable random variables, with C_γ given in Lemma 10.4. Let $\mu_n = E\{Y_n I(-\gamma_n \le Y_n < \lambda_n)|\mathscr{F}_{n-1}\}$. Then $\exp\{\sum_{i=1}^n (Y_i - \mu_i - \lambda_i^{-1} Y_i^2)\}$ is a supermartingale whose expectation is ≤ 1.*

10.3 Continuous-Time Martingale Theory

As shown in Chap. 1 of Karatzas and Shreve (1991) or Chap. II of Revuz and Yor (1999), the basic martingale theory in Sect. 9.1 can be readily extended to continuous-time martingales/submartingales/supermartingales if the sample paths are a.s. right-continuous. In particular, such processes have left-hand limits and are therefore *cadlag* (continu à droit, limité à gauche).

Here we summarize some of the main results that are related to the lemmas in Sect. 10.2 and the inequalities in Chap. 9. Comprehensive treatments of these and other results can be found in the monographs by Elliott (1982), Karatzas and Shreve (1991) and Revuz and Yor (1999). A filtration $(\mathscr{F}_t)$ is said to be *right-continuous* if $\mathscr{F}_t = \mathscr{F}_{t+} := \bigcap_{\varepsilon>0} \mathscr{F}_{t+\varepsilon}$. It is said to be *complete* if $\mathscr{F}_0$ contains all the P-null sets (that have zero probability) in $\mathscr{F}$. In what follows we shall assume that the process $\{X_t,\ t \ge 0\}$ is right-continuous and adapted to a filtration $\{\mathscr{F}_t\}$ that is right-continuous and complete. The σ-field generated on $\Omega \times [0,\infty)$ by the space of adapted processes which are left-continuous on $(0,\infty)$ is called the *predictable σ-field.* A process $\{X_t\}$ is *predictable* if the map $(\omega,t) \mapsto X_t(\omega)$ from $\Omega \times [0,\infty)$ to $\mathbb{R}$ is measurable with respect to the predictable σ-field. An extended random variable T taking values in $[0,\infty]$ is called (a) a *stopping time* (with respect to the filtration $\{\mathscr{F}_t\}$) if $\{T \le t\} \in \mathscr{F}_t$ for all $t \ge 0$, (b) an *optional time* if $\{T < t\} \in \mathscr{F}_t$ for all $t > 0$, and (c) a *predictable time* if there exists an increasing sequence of stopping times T_n such that $T_n < T$ on $\{T > 0\}$ and $\lim_{n\to\infty} T_n = T$ *a.s.*

A stochastic process $X = \{X_t,\ t \ge 0\}$ is called *measurable* if the map $(\omega,t) \mapsto X_t(\omega)$ from $\Omega \times [0,\infty)$ to $\mathbb{R}$ is measurable. For a measurable process, there exists a predictable process $Y = \{Y_t,\ t \ge 0\}$ such that

$$E\{X_T I(T < \infty)|\mathscr{F}_{T-}\} = Y_T I(T < \infty) \ a.s., \tag{10.7}$$

for every predictable time T. The process Y is essentially unique and is called the *predictable projection* of X, denoted by X^{Π}. Suppose $A = \{A_t,\ t \geq 0\}$ is a nondecreasing, measurable process such that $E(A_\infty) < \infty$. Then there exists an essentially unique nondecreasing, predictable process $A^{(p)}$ such that for all bounded, measurable processes $\{X_t,\ t \geq 0\}$,

$$E(X_t^{\Pi} A_t) = E(X_t A_t^{(p)}). \tag{10.8}$$

The process $A^{(p)}$ is called the *dual predictable projection* of A; see Elliott (1982, p. 72).

10.3.1 Doob–Meyer Decomposition and Locally Square-Integrable Martingales

Let $\mathscr{T}_a$ be the class of stopping times such that $P(T \leq a) = 1$ for all $T \in \mathscr{T}_a$. A right-continuous process $\{X_t,\ t \geq 0\}$ adapted to a filtration $\{\mathscr{F}_t\}$ is said to be of class DL if $\{X_T, T \in \mathscr{T}_a\}$ is uniformly integrable for every $a > 0$. If $\{X_t, \mathscr{F}_t,\ t \geq 0\}$ is a nonnegative right-continuous submartingale, then it is of class DL. The *Doob–Meyer decomposition* says that if a right-continuous submartingale $\{X_t, \mathscr{F}_t,\ t \geq 0\}$ is of class DL, then it admits the decomposition

$$X_t = M_t + A_t, \tag{10.9}$$

in which $\{M_t, \mathscr{F}_t,\ t \geq 0\}$ is a right-continuous martingale with $M_0 = 0$ and A_t is predictable, nondecreasing and right-continuous. Moreover, the decomposition is essentially unique in the sense that if $X_t = M_t' + A_t'$ is another decomposition, then $P\{M_t = M_t', A_t = A_t' \text{ for all } t\} = 1$. The process A_t in the Doob-Meyer decomposition is called the *compensator* of the submartingale $\{X_t, \mathscr{F}_t,\ t \geq 0\}$.

Suppose $\{M_t, \mathscr{F}_t,\ t \geq 0\}$ is a right-continuous martingale that is square integrable, i.e., $EM_t^2 < \infty$ for all t. Since M_t^2 is a right-continuous, nonnegative submartingale (by Jensen's inequality), it has the Doob–Meyer decomposition whose compensator is called the *predictable variation* process and denoted by $\langle M \rangle_t$, i.e., $M_t^2 - \langle M \rangle_t$ is a martingale. If $\{N_t, \mathscr{F}_t,\ t \geq 0\}$ is another right-continuous square integrable martingale, then $(M_t + N_t)^2 - \langle M+N \rangle_t$ and $(M_t - N_t)^2 - \langle M-N \rangle_t$ are martingales, and the *predictable covariation* process $\langle M,N \rangle_t$ is defined by

$$\langle M,N \rangle_t = \frac{1}{4} \{\langle M+N \rangle_t - \langle M-N \rangle_t\}, \qquad t \geq 0. \tag{10.10}$$

Let $\mathscr{M}_2$ denote the linear space of all right-continuous, square-integrable martingales M with $M_0 = 0$. Two processes X and Y on $(\Omega, \mathscr{F}, P)$ are *indistinguishable* if $P(X_t = Y_t \text{ for all } t \geq 0) = 1$. Define a norm on $\mathscr{M}_2$ by

$$\|M\| = \sum_{n=1}^{\infty} \frac{\min\left(\sqrt{EM_n^2}, 1\right)}{2^n}. \tag{10.11}$$

This induces a metric $\rho(M,M') = \|M - M'\|$ on $\mathcal{M}_2$ if indistinguishable processes are treated as the same process. A subspace $\mathcal{H}$ of $\mathcal{M}_2$ is said to be *stable* if it is a closed set in this metric space and has two additional "closure" properties:

(a) $M \in \mathcal{H} \Rightarrow M^T \in \mathcal{H}$ for all stopping times T, where $M_t^T = M_{T\wedge t}$.
(b) $M \in \mathcal{H}$ and $A \in \mathcal{F}_0 \Rightarrow MI(A) \in \mathcal{H}$.

Two martingales M,N belonging to $\mathcal{M}_2$ are said to be *orthogonal* if $\langle M,N\rangle_t = 0$ *a.s.* for all $t \geq 0$, or equivalently, if $\{M_t N_t, \mathcal{F}_t, t \geq 0\}$ is a martingale. If $\mathcal{H}$ is a stable subspace of $\mathcal{M}_2$, then so is

$$\mathcal{H}^\perp := \{N \in \mathcal{M}_2 : N \text{ is orthogonal to } M \text{ for all } M \in \mathcal{H}\}. \tag{10.12}$$

Moreover, every $X \in \mathcal{M}_2$ has a unique (up to indistinguishability) decomposition

$$X = M + N, \qquad \text{with } M \in \mathcal{H} \text{ and } N \in \mathcal{H}^\perp. \tag{10.13}$$

Besides the dual predictable projection defined by (10.8), Lemma 10.3 also involves the "continuous part" M^c of $M \in \mathcal{M}_2$ (or a somewhat more general M which is the a.s. limit of elements of $\mathcal{M}_2$), which is related to the decomposition (10.13) with $\mathcal{H} = \mathcal{M}_2^c$, where

$$\mathcal{M}_2^c = \{M \in \mathcal{M}_2 : M \text{ has continuous sample paths}\}. \tag{10.14}$$

It can be shown that $\mathcal{M}_2^c$ is a stable subspace of $\mathcal{M}_2$, and therefore (10.13) means that every $M \in \mathcal{M}_2$ has an essentially unique decomposition

$$M = M^c + M^d, \qquad \text{with } M^c \in \mathcal{M}_2^c \text{ and } M^d \in (\mathcal{M}_2^c)^\perp. \tag{10.15}$$

While M^c is called the continuous part of M, M^d is called its "purely discontinuous" part. Note that M^c and M^d are orthogonal martingales.

For $M \in \mathcal{M}_2$ and $t > 0$, let Π be a partition $0 = t_0 < t_1 < \cdots < t_k = t$ of $[0,t]$. Then as $\|\Pi\| := \max_{1\leq i\leq k} |t_i - t_{i-1}| \to 0$, $\sum_{i=1}^k (M_{t_i} - M_{t_{i-1}})^2$ converges in probability to a limit, which we denote by $[M]_t$. The random variable $[M]_t$ is called the *quadratic variation* process of M. For $M \in \mathcal{M}_2^c$, $[M] = \langle M\rangle$. More generally, for $M \in \mathcal{M}_2$,

$$[M]_t = \langle M^c\rangle_t + \sum_{0<s\leq t} (\triangle M_s)^2, \tag{10.16}$$

where $\triangle M_s = M_s - M_{s-}$, noting that $M_{s-} = \lim_{u\uparrow s} M_u$ exists since M is cadlag. The *quadratic covariation* of M and N, which both belong to $\mathcal{M}_2$, is defined by $[M,N]_t = \{[M+N] - [M-N]\}/4$, and (10.16) can be generalized to

$$[M,N]_t = \langle M^c, N^c\rangle_t + \sum_{0<s\leq t} (\triangle M_s)(\triangle N_s). \tag{10.17}$$

We can relax the integrability assumptions above by using *localization*. If there exists a sequence of stopping times T_n such that $\{M_{T_n \wedge t}, \mathscr{F}_t,\ t \geq 0\}$ is a martingale (or a square-integrable martingale, or bounded), then $\{M_t, \mathscr{F}_t,\ t \geq 0\}$ is called a *local martingale* (or *locally square-integrable martingale*, or *locally bounded*). By a limiting argument, we can again define $\langle M \rangle_t$, $\langle M,N \rangle_t$, $[M]_t$, $[M,N]_t$, M^c and M^d for locally square integrable martingales. Moreover, a continuous local martingale M_t can be expressed as time-changed Brownian motion:

$$M_t = W_{\langle M \rangle_t}, \tag{10.18}$$

which provides the background for Lemma 10.2; see Karatzas and Shreve (1991, Sect. 3.4B). If V is an adapted process with finite variation on bounded intervals, then its dual predictable projection process $V^{(p)}$ (see (10.8)) is the essentially unique predictable process having finite variation on bounded intervals and such that $V - V^{(p)}$ is a local martingale; see Elliott (1982, p. 121).

10.3.2 Inequalities and Stochastic Integrals

Let $\mathscr{M}_2^{\text{loc}}$ denote the class of right-continuous, locally square-integrable martingales M with $M_0 = 0$. Let M,N belong to $\mathscr{M}_2^{\text{loc}}$ and H,K be two measurable processes. For $0 \leq s \leq t$, let $\langle M,N \rangle_{s,t} = \langle M,N \rangle_t - \langle M,N \rangle_s$ and note that

$$\langle M + rN \rangle_{s,t} = \langle M,M \rangle_{s,t} + 2r\langle M,N \rangle_{s,t} + r^2 \langle N,N \rangle_{s,t}$$

is a nonnegative quadratic function of r. Therefore

$$|\langle M,N \rangle_{s,t}| \leq \{\langle M,M \rangle_{s,t}\}^{1/2} \{\langle N,N \rangle_{s,t}\}^{1/2}. \tag{10.19}$$

Hence, approximating the Lebesgue–Stieltjes integral below by a sum, we obtain from (10.19) the *Kunita–Watanabe inequality*

$$\int_0^t |H_s||K_s| d\overline{\langle M,N \rangle}_s \leq \left(\int_0^t H_s^2 d\langle M \rangle_s \right)^{1/2} \left(\int_0^t K_s^2 d\langle N \rangle_s \right)^{1/2}, \tag{10.20}$$

where we use the notation $\bar{\zeta}_t$ to denote the total variation of a process ζ on $[0,t]$.

A continuous-time analog of Theorem 9.12 is the following: If M is a continuous local martingale with $M_0 = 0$, then for $x > 0$ and $y > 0$,

$$P\{M_t \geq x \text{ and } \langle M \rangle_t \leq y \text{ for some } t \geq 0\} \leq \exp(-x^2/2y); \tag{10.21}$$

see Revuz and Yor (1999, p. 145) whose Sect. IV.4 provides an extension of the Burkholder–Davis–Gundy inequalities (Theorem 9.6) to continuous local martingales. Barlow et al. (1986) have derived convex function inequalities, which generalize the Burkholder–Davis–Gundy inequalities, for right-continuous, locally square-integrable martingales. In connection with Lemma 10.3, they have also proved the following analog of (10.21) for right-continuous, locally square-integrable martingales:

$$P\{M_T \geq x, H_T \leq y\} \leq \exp(-x^2/2y) \qquad \text{for all stopping times } T, \tag{10.22}$$

where $H_t = \langle M^c \rangle_t + \sum_{s \leq t}((\triangle M_s)^+)^2 + \{\sum_{s \leq t}((\triangle M_s)^-)^2\}_t^{(p)}$, following the notation of Lemma 10.3. They note that in the case $\triangle M \leq c$ *a.s.* for some constant $c \geq 0$, (10.22) follows from the sharper inequality

$$P\{M_T \geq x, \langle M \rangle_T \leq y\} \leq \exp\left\{-\frac{x^2}{2y}\psi\left(\frac{cx}{y}\right)\right\}, \tag{10.23}$$

where $\psi(0) = 1$ and $\psi(\lambda) = (2/\lambda^2)\int_0^\lambda \log(1+t)dt$ for $\lambda > 0$. They derive (10.23) from Lemma 10.6 below, which is a continuous-time analog of Theorem 9.18. They also derive (10.22) in the case $\triangle M \geq 0$ *a.s.* from another exponential supermartingale, given in Lemma 10.7 below.

Lemma 10.6. *Let $\{M_t, \mathscr{F}_t, t \geq 0\}$ be a locally square-integrable, right-continuous martingale such that $M_0 = 0$ a.s. and $\triangle M_t \leq c$ a.s. for all $t \geq 0$ and some $c > 0$. Then for $\lambda > 0$,*

$$\{\exp(\lambda M_t - \varphi_c(\lambda)\langle M \rangle_t), \mathscr{F}_t, t \geq 0\} \quad \textit{is a supermartingale}, \tag{10.24}$$

where $\varphi_c(\lambda) = c^{-2}(e^{\lambda c} - 1 - \lambda c)$.

Let $\{X_t, t \geq 0\}$ be a cadlag process of locally bounded variation. Then $\prod_{s \leq t}(1 + \triangle X_s)$ is well defined as a limit and is a cadlag process of locally bounded variation; see Problem 10.1. For a locally square-integrable, right-continuous martingale M, the Volterra equation

$$Z_t = 1 + \int_0^t Z_{s-}\, dM_s \tag{10.25}$$

has a unique cadlag solution

$$Z_t = \mathscr{E}(M)_t := \exp\left\{M_t - \frac{1}{2}\langle M^c \rangle_t\right\} \prod_{s \leq t}(1 + \triangle M_s)e^{-\triangle M_s},$$

in which $\mathscr{E}(M)$ is called the *Doléans exponential* of M.

Lemma 10.7. *Let $\{M_t, \mathscr{F}_t, t \geq 0\}$ be a locally square-integrable, right-continuous martingale M with $M_0 = 0$. Then $\{\mathscr{E}(M)_t, \mathscr{F}_t, t \geq 0\}$ is a martingale. If $\triangle M_t \geq 0$ for all $t \geq 0$, then $\exp(M_t - \frac{1}{2}[M]_t) \leq \mathscr{E}(M)_t$ and $\{\exp(M_t - \frac{1}{2}[M]_t), \mathscr{F}_t, t \geq 0\}$ is a supermartingale.*

We now define the stochastic integral $\int_0^t X_s dY_s$ with *integrand* $X = \{X_s, 0 \leq s \leq t\}$ and *integrator* $Y = \{Y_s, 0 \leq s \leq t\}$. If Y has bounded variation on $[0,t]$, then the integral can be taken as an ordinary pathwise Lebesgue–Stieltjes integral over $[0,t]$. If Y is a right-continuous, square-integrable martingale and X is a predictable process such that $\int_0^t X_s^2\, d\langle Y\rangle_s < \infty$ *a.s.*, then $\int_0^t X_s\, dY_s$ can be defined by the limit (in probability) of integrals (which reduce to sums) whose integrands are step functions and converge to X in an L_2-sense. More generally, one can define $\int_0^t X_s\, dY_s$ when X is a predictable, locally bounded process and Y is a semimartingale, which is defined below. Moreover, the process $\{\int_0^t X_s dY_s,\ t \geq 0\}$ is also a semimartingale.

Definition 10.8. A process Y which is adapted to the filtration $\{\mathscr{F}_t, t \geq 0\}$ is called a *semimartingale* if it has a decomposition of the form $Y_t = Y_0 + M_t + V_t$, where M is a locally square-integrable, right-continuous martingale, V_t is an adapted cadlag process with finite variation on bounded intervals, and $M_0 = V_0 = 0$.

Theorem 10.9 (Ito's formula for semimartingales). *Let $X(t) = (X_1(t), \ldots, X_m(t))$, $t \geq 0$, be a vector-valued process whose components X_i are semimartingales, which can be decomposed as $X_i(0) + M_i(t) + V_i(t)$. Let $f : [0,\infty) \times \mathbb{R}^m \to \mathbb{R}$ be of class $C^{1,2}$ (i.e., $f(t,x)$ is twice continuously differentiable in x and continuously differentiable in t). Then $\{f(t,X(t)), t \geq 0\}$ is a semimartingale and*

$$\begin{aligned} f(t,X(t)) &= f(0,X(0)) + \sum_{i=1}^m \int_{0+}^t \frac{\partial}{\partial x_i} f(s,X(s-))\, dX_i(s) \\ &\quad + \frac{1}{2}\sum_{i=1}^m \sum_{j=1}^m \int_{0+}^t \frac{\partial^2}{\partial x_i \partial x_j} f(s,X(s-))\, d\langle M_i^c, M_j^c\rangle_s \\ &\quad + \sum_{0<s\leq t} \left\{ f(s,X(s)) - f(s,X(s-)) - \sum_{i=1}^m (\partial/\partial x_i) f(s,X(s-))\, \triangle X_i(s) \right\}. \end{aligned}$$

An important corollary of Theorem 10.9 is the *product rule* for semimartingales X and Y: XY is a semimartingale and

$$d(X_t Y_t) = X_{t-} dY_t + Y_{t-} dX_t + d[X,Y]_t. \tag{10.26}$$

In particular, for a locally square-integrable, right-continuous martingale M, setting $X = Y = M$ in (10.26) yields

$$[M]_t = M_t^2 - 2\int_0^t M_{s-}\, dM_s. \tag{10.27}$$

By (10.27), $M^2 - [M]$ is a martingale but $[M]$ is not predictable. Since $M^2 - \langle M\rangle$ is a martingale and $\langle M\rangle$ is predictable, it follows that $\langle M\rangle_t$ is the compensator of the nondecreasing process $[M]_t$.

Barlow et al. (1986, Sect. 2) give a general theory of inqualities of the form $P\{X_T \geq x, Y_T \leq y\}$ for stopping time T and two nonnegative adapted processes X and Y. Earlier results along this line are Lenglart's (1977) inequalities. Let X be a right-continuous adapted process and Y a nondecreasing predictable process with $Y_0 = 0$ such that

$$E|X_T| \leq EY_T \quad \text{for every bounded stopping time } T. \tag{10.28}$$

Then for any $\varepsilon > 0$, $\delta > 0$ and stopping time τ,

$$P\left\{\sup_{t\leq\tau}|X_t| \geq \varepsilon, |Y_\tau| < \delta\right\} \leq \varepsilon^{-1}E(\delta \wedge Y_\tau), \tag{10.29}$$

$$P\left\{\sup_{t\leq\tau}|X_t| \geq \varepsilon\right\} \leq (\delta/\varepsilon) + P\{Y_\tau \geq \delta\}. \tag{10.30}$$

Note that (10.30) follows from (10.29) since

$$P\left\{\sup_{t\leq\tau}|X_t| \geq \varepsilon\right\} \leq P\left\{\sup_{t\leq\tau}|X_t| \geq \varepsilon, Y_\tau < \delta\right\} + P\{Y_\tau \geq \delta\}.$$

In particular, since (10.28) holds with $X = M^2$ and $Y = \langle M\rangle$ for a locally square-integrable right-continuous martingale M, it follows from (10.30) that for any $\eta > 0$, $\delta > 0$ and stopping time τ,

$$P\left\{\sup_{t\leq\tau}|M_t| \geq \eta\right\} \leq (\delta/\eta^2) + P\{\langle M\rangle_\tau \geq \delta\}.$$

10.4 Supplementary Results and Problems

1. Let $X = \{X_s, 0 \leq s \leq t\}$ be a cadlag process with bounded variation. Show that $\prod_{i=1}^{k}(1 + \triangle X_{t_i})$ is absolutely convergent as $\|\mathscr{P}\| \to 0$, where $\mathscr{P}$ is a partition $t_0 = 0 < t_1 < \cdots < t_k = t$ of $[0,t]$. The limit is called the *product-integral* of X and is denoted by $\prod_{0\leq s\leq t}(1 + \triangle X_s)$. We can clearly extend the definition of the product-integral $Y_t := \prod_{-\infty<s\leq t}(1 + \triangle X_s)$ to cadlag processes $X = \{X_t, t \in \mathbb{R}\}$ such that X has bounded variation on $(-\infty, a]$ for all a. Show that the product-integral Y is cadlag and has bounded variation on $(-\infty, a]$ for all a.
2. Let $X = \{X_t, -\infty < t < \infty\}$ be a cadlag process that has bounded variation on $(-\infty, a]$ for all a. Show that $Y_t := \prod_{-\infty<s\leq t}(1 + \triangle X_s)$ satisfies the Volterra equation

$$Y_t = 1 + \int_{-\infty}^{t} Y_{s-}dX_s. \tag{10.31}$$

By applying this result to $X_t = S(t)$, where $S(t) = P(T > t)$ is the survival function of a random variable T, show that

$$S(t) = \prod_{s\leq t}(1 - \triangle A(s)) \qquad \text{if } A(t) < \infty, \tag{10.32}$$

where A is the *cumulative hazard function* of T defined by

$$A(t) = \int_{-\infty}^{t} \frac{dF(s)}{1 - F(s-)} = -\int_{-\infty}^{t} \frac{dS(u)}{S(u-)}, \tag{10.33}$$

in which $F = 1 - S$ is the distribution function of T.

3. Let $X_1,\dots,X_n$ be i.i.d. random variables with common distribution function F. Let $C_1,\dots,C_n$ be independent random variables that are also independent of $\{X_1,\dots,X_n\}$. Let $\tilde{X}_i=\min(X_i,C_i)$, $\delta_i=I(X_i\le C_i)$ for $i=1,\dots,n$. Let $\mathscr{F}_t$ be the σ-field generated by

$$I(\tilde{X}_i\le t),\quad \delta_i I(\tilde{X}_i\le t),\quad \tilde{X}_i I(\tilde{X}_i\le t),\qquad i=1,\dots,n. \tag{10.34}$$

Define

$$R_n(t)=\sum_{i=1}^n I(\tilde{X}_i\ge t),\qquad N_n(t)=\sum_{i=1}^n I(\tilde{X}_i\le t,\,\delta_i=1). \tag{10.35}$$

Show that $\{N_n(t),\,\mathscr{F}_t,\,-\infty<t<\infty\}$ is a right-continuous, nonnegative submartingale with compensator $\int_{-\infty}^t R_n(s)dA(s)$. This result enables one to apply continuous-time martingale theory to analyze statistical methods for the analysis of censored data $(\tilde{X}_i,\delta_i)$, $1\le i\le n$, that are observed instead of $X_1,\dots,X_n$.

4. Because the X_i in the preceding problem are not fully observable, one cannot use the empirical distribution of the X_i to estimate F. In view of the preceding problem,

$$M_n(t):=N_n(t)-\int_{-\infty}^t R_n(s)dA(s) \tag{10.36}$$

is a martingale with respect to the filtration $\{\mathscr{F}_t\}$. By (10.36), $dM_n(t)=dN_n(t)-R_n(t)dA(t)$, which suggests the following estimate of $A(t)$ based on $\{(\tilde{X}_i,\delta_i),\,1\le i\le n\}$:

$$\hat{A}_n(t)=\int_{-\infty}^t \frac{I(R_n(s)>0)}{R_n(s)}dN_n(t)=\sum_{s\le t}\frac{\triangle N_n(s)}{R_n(s)}, \tag{10.37}$$

using the convention $0/0=0$ and noting that $\int_{-\infty}^t\{I(R_n(s)>0)/R_n(s)\}dM_n(s)$ is a martingale since $R_n(s)$ is left-continuous and therefore predictable. This and (10.32) suggest the following estimator of S:

$$\hat{S}_n(t)=\prod_{s\le t}(1-\triangle\hat{A}_n(s))=\prod_{s\le t}\left(1-\frac{\triangle N_n(s)}{R_n(s)}\right). \tag{10.38}$$

The estimator $\hat{A}_n$ is called the *Nelson–Aalen estimator* of the cumulative hazard function A, and $\hat{S}_n$ (or $1-\hat{S}_n$) is called the *Kaplan–Meier estimator* of the survival function S (or distribution function F). Show that for any τ such that $F(\tau)<1$ and $\liminf_{n\to\infty} n^{-1}\sum_{i=1}^n P(C_i\ge\tau)>0$,

$$P\left\{\lim_{n\to\infty}\sup_{t\le\tau}|\hat{A}_n(t)-A(t)|=0\right\}=P\left\{\lim_{n\to\infty}\sup_{t\le\tau}|\hat{S}_n(t)-S(t)|=0\right\}=1, \tag{10.39}$$

thereby establishing the strong uniform consisting of $\hat{A}_n$ and $\hat{S}_n$ on $(-\infty,\tau]$.

Hint: Use the Borel–Cantelli lemma and exponential bounds for martingales and sums of independent random variables.

5. Rebolledo (1980) has proved the following functional central limit theorem for continuous-time locally square-integrable, right-continuous martingales $\{M_n, \mathscr{F}_t, t \in T\}$, where T is an interval (possibly infinite):

   Suppose that there exists a nonrandom function V such that $\langle M_n \rangle(t) \xrightarrow{P} V(t)$ for every $t \in T$ and that $\langle M_n^{(\varepsilon)} \rangle(t) \xrightarrow{P} 0$ for every $t \in T$ and $\varepsilon > 0$, where $M_n^{(\varepsilon)}$ is the subset of the purely discontinuous part of M_n that consists of jumps larger in absolute value than ε. Then M_n converges weakly to $W(V(\cdot))$ in $D(T)$, where W is Brownian motion and $D(T)$ denotes the space of cadlag functions on T with the Skorohod metric.

   Making use of Rebolledo's central limit theorem, show that $\sqrt{n}(\hat{A}_n - A)$ and $\sqrt{n}(\hat{S}_n - S)$ converge weakly in $D((-\infty, \tau])$ as $n \to \infty$, for any τ such that $F(\tau) < 1$ and $\liminf_{n\to\infty} n^{-1} \sum_{i=1}^{n} P(C_i \geq \tau) > 0$. See Problem 2.7 and Sect. 15.3.1 for related weak convergence results and functional central limit theorems.

Chapter 11
Pseudo-Maximization via Method of Mixtures

In this chapter we describe the method of mixtures to perform pseudo-maximization that generates self-normalized processes via (10.4). Section 11.1 describes a prototypical example and Laplace's method for asymptotic evaluation of integrals. Section 11.2 reviews the method of mixtures used by Robbins and Siegmund (1970) to evaluate boundary crossing probabilities for Brownian motion, and generalizes the method to analyze boundary crossing probabilities for self-normalized processes. In Sect. 11.3 we describe a class of mixing density functions that are particularly useful for developing L_p and exponential inequalities for self-normalized processes, details of which are given in the next chapter.

11.1 Pseudo-Maximization and Laplace's Method

We begin with a review of Laplace's method for asymptotic evaluation of the integral $\int_{-\infty}^{\infty} f(\theta)e^{ag(\theta)}d\theta$ as $a \to \infty$, where f and g are continuous functions on $\mathbb{R}$ such that g has unique maximum at θ^* and is twice continuously differentiable in some neighborhood of θ^*, $\limsup_{|\theta|\to\infty} g(\theta) < \min\{g(\theta^*), 0\}$ and $\limsup_{|\theta|\to\infty} |f(\theta)|e^{Ag(\theta)} < \infty$ for some $A > 0$. Since $g'(\theta^*) = 0$, $g''(\theta^*) < 0$ and

$$e^{ag(\theta)} = e^{ag(\theta^*)} \exp\left\{a\left[g''(\theta^*) + o(1)\right](\theta - \theta^*)^2/2\right\} \qquad \text{as } \theta \to \theta^*, \tag{11.1}$$

and since the assumptions on f and g imply that for every $\varepsilon > 0$, there exists $\eta_\varepsilon > 0$ such that as $a \to \infty$,

$$\left(\int_{-\infty}^{\theta^*-\varepsilon} + \int_{\theta^*+\varepsilon}^{\infty}\right) f(\theta)e^{ag(\theta)}d\theta = O\left(\exp\left(a\left[g(\theta^*) - \eta_\varepsilon\right]\right)\right),$$

V.H. de la Peña et al., *Self-Normalized Processes: Limit Theory and Statistical Applications*, 149
Probability and its Applications,
© Springer-Verlag Berlin Heidelberg 2009

it follows that

$$\int_{-\infty}^{\infty} f(\theta)e^{ag(\theta)}d\theta \sim f(\theta^*)e^{ag(\theta^*)}\left(-ag''(\theta^*)\right)^{-1/2}\int_{-\infty}^{\infty} e^{-t^2/2}dt$$

$$= \sqrt{\frac{2\pi}{a|g''(\theta^*)|}} f(\theta^*)e^{ag(\theta^*)} \tag{11.2}$$

as $a \to \infty$, using the change of variables $t = \left(-af''(\theta^*)\right)^{1/2}(\theta - \theta^*)$.

Laplace's asymptotic formula (11.2) relates the integral $\int_{-\infty}^{\infty} f(\theta)e^{ag(\theta)}d\theta$ to the maximum of $e^{ag(\theta)}$ over θ. This is the essence of the *pseudo-maximization* approach that we use to analyze a self-normalized process (which can be represented as a maximum by (10.4)) via an integral with respect to a probability measure. Note that there is much flexibility in choosing the probability measure (or the mixing density function f), and that the function $g(\theta) := \theta A_n - \theta^2 B_n^2/2$ in (10.4) is a random function whose maximizer θ^* is a random variable A_n/B_n. Whereas f is assumed to be continuous, and therefore bounded on finite intervals, in the preceding paragraph which follows the conventional exposition of Laplace's method, allowing f to approach ∞ in regions where A_n/B_n^2 tends to be concentrated can tailor f to the analysis of $Eh(A_n/B_n)$ for given h (e.g., $h(x) = |x|^p$ or $h(x) = e^{\lambda x}$). In Sect. 11.2 we describe a class of mixing density functions f with this property, which will be used in Chap. 12 for the analysis of $Eh(A_n/B_n)$. Similar density functions were introduced by Robbins and Siegmund (1970) who used the method of mixtures to analyze boundary crossing probabilities for Brownian motion. Section 11.3 reviews their method and results and extends them to derive boundary crossing probabilities for self-normalized processes under the canonical assumption (10.3) or (10.6).

11.2 A Class of Mixing Densities

Let $L : (0,\infty) \to [0,\infty)$ be a nondecreasing function such that

$$L(cy) \le 3cL(y) \qquad \text{for all } c \ge 1 \text{ and } y > 0, \tag{11.3}$$

$$L(y^2) \le 3L(y) \qquad \text{for all } y \ge 1, \tag{11.4}$$

$$\int_1^{\infty} \frac{dx}{xL(x)} = \frac{1}{2}. \tag{11.5}$$

An example is the function

$$L(y) = \beta\{\log(y+\alpha)\}\{\log\log(y+\alpha)\}\{\log\log\log(y+\alpha)\}^{1+\delta}, \tag{11.6}$$

where $\delta > 0$, α is sufficiently large so that (11.3) and (11.4) hold and β is a normalizing constant to ensure (11.5). By a change of variables,

$$\int_0^1 \left(\lambda L\left(\frac{1}{\lambda}\right)\right)^{-1} d\lambda = \int_1^{\infty} (\lambda L(\lambda))^{-1} d\lambda,$$

so condition (11.5) ensures that

$$f(\lambda) = \frac{1}{\lambda L\left(\max\left(\lambda, \frac{1}{\lambda}\right)\right)}, \qquad \lambda > 0, \tag{11.7}$$

is a probability density on $(0,\infty)$. Therefore the canonical assumption (10.6) holding for all $\lambda \geq 0$ implies that

$$1 \geq E \int_0^\infty \exp\left\{Ax - (B^2x^2/2)\right\} f(x)\, dx. \tag{11.8}$$

Lemma 11.1. *Let $\gamma \geq 1$. Then $yL(y/B \vee B/y) \leq 3\gamma\{L(\gamma) \vee L(B \vee B^{-1})\}$ for any $0 < y \leq \gamma$ and $B > 0$. Consequently, for any $A \geq B > 0$ and any $-\frac{A}{B} < x \leq 0$,*

$$\left(x + \frac{A}{B}\right) L\left(\frac{x+\frac{A}{B}}{B} \vee \frac{B}{x+\frac{A}{B}}\right) \leq 3\frac{A}{B}\left\{L\left(\frac{A}{B}\right) \vee L\left(B \vee \frac{1}{B}\right)\right\}. \tag{11.9}$$

Proof. First consider the case $y \leq 1$. From (11.3) and the fact L is nondecreasing, it follows that

$$yL\left(\frac{y}{B} \vee \frac{B}{y}\right) \leq yL\left(\frac{1}{y}\left(\frac{1}{B} \vee B\right)\right) \leq 3L\left(B \vee \frac{1}{B}\right).$$

For the remaining case $1 < y \leq \gamma$, since L is nondecreasing, we have

$$\begin{aligned} yL\left(\frac{y}{B} \vee \frac{B}{y}\right) &\leq \gamma L\left(\gamma\left(\frac{1}{B} \vee B\right)\right) \\ &\leq \gamma\left\{L(\gamma^2) \vee L\left(\left(B \vee \frac{1}{B}\right)^2\right)\right\} \\ &\leq 3\gamma\left\{L(\gamma) \vee L\left(B \vee \frac{1}{B}\right)\right\}, \end{aligned}$$

where the last inequality follows from (11.4). □

Lemma 11.2. *Let $A, B > 0$ be two random variables satisfying the canonical assumption (10.6) for all $\lambda \geq 0$. Define*

$$g(x) = x^{-1} \exp(x^2/2) I(x \geq 1). \tag{11.10}$$

Then

$$E \frac{g\left(\frac{A}{B}\right)}{L\left(\frac{A}{B}\right) \vee L\left(B \vee \frac{1}{B}\right)} \leq \frac{3}{\int_0^1 \exp\{-x^2/2\}\, dx}.$$

Proof. From (11.7) and (11.8), it follows that

$$
\begin{aligned}
1 &\geq E\int_0^\infty \frac{\exp\{Ax-(B^2x^2/2)\}}{xL\left(x\vee\frac{1}{x}\right)}dx \\
&= E\int_0^\infty \frac{\exp\left\{\frac{Ay}{B}-\frac{y^2}{2}\right\}}{yL\left(\frac{y}{B}\vee\frac{B}{y}\right)}dy \quad \text{(letting } y=Bx\text{)} \\
&\geq E\left[e^{A^2/2B^2}\int_{-\frac{A}{B}}^\infty \frac{\exp\{-(x^2/2)\}}{\left(x+\frac{A}{B}\right)L\left(\frac{x+\frac{A}{B}}{B}\vee\frac{B}{x+\frac{A}{B}}\right)}I\left(\frac{A}{B}\geq 1\right)dx\right] \left(\text{letting } x=y-\frac{A}{B}\right) \\
&\geq E\left[e^{A^2/2B^2}\int_{-1}^0 \frac{\exp\{-(x^2/2)\}dx}{\frac{3A}{B}\left(L\left(\frac{A}{B}\right)\vee L\left(B\vee\frac{1}{B}\right)\right)}I\left(\frac{A}{B}\geq 1\right)\right] \quad \text{by (11.9)} \\
&= \left\{\frac{1}{3}\int_0^1 e^{-x^2/2}dx\right\}E\frac{g\left(\frac{A}{B}\right)}{L\left(\frac{A}{B}\right)\vee L\left(B\vee\frac{1}{B}\right)}. \qquad \square
\end{aligned}
$$

The class of mixing densities (11.3)–(11.5) was introduced by de la Peña et al. (2000, 2004) who made use of the properties in Lemmas 11.1, 11.2 and the following easy lemma to prove moment and exponential bounds for self-normalized processes satisfying the canonical assumption (10.5) or (10.6). Details will be given in Sect. 12.2.2.

Lemma 11.3. *Let* $r>0$, $0<\delta<1$, $g_r(x)=x^{-1}\exp(rx^2/2)I(x\geq 1)$. *If*

$$g_r^{1-\delta}(x)\leq L\left(B\vee\frac{1}{B}\right),$$

then

$$x\leq\sqrt{\frac{2}{r(1-\delta)}\log^+ L\left(B\vee\frac{1}{B}\right)}.$$

11.3 Application of Method of Mixtures to Boundary Crossing Probabilities

11.3.1 The Robbins–Siegmund Boundaries for Brownian Motion

Let $W_t, t\geq 0$, be standard Brownian motion. Then $\exp(\theta W_t-\theta^2t/2), t\geq 0$, is a continuous martingale with mean 1, and therefore (10.5) holds with $A_t=W_t, B_t=t$ and $\varepsilon=\infty$. Hence $f(W_t,t), t\geq 0$, is also a continuous martingale, where

$$f(x,t) = \int_0^\infty \exp(\theta x - \theta^2 t/2)dF(\theta), \qquad x \in \mathbb{R},\ t \geq 0, \tag{11.11}$$

and F is any measure on $(0,\infty)$ which is finite on bounded intervals. Robbins and Siegmund (1970, Sect. 3) make use of this and the following lemma to evaluate a class of boundary crossing probabilities for Brownian motion.

Lemma 11.4. *Let $\varepsilon > 0$ and let $\{Z_t, \mathscr{F}_t, t \geq a\}$ be a nonnegative martingale with continuous sample paths on $\{Z_a < \varepsilon\}$ and such that*

$$Z_t I\left(\sup_{s>a} Z_s < \varepsilon\right) \xrightarrow{P} 0 \qquad \text{as } t \to 0. \tag{11.12}$$

Then

$$P\left\{\sup_{t>a} Z_t \geq \varepsilon \mid \mathscr{F}_a\right\} = Z_a/\varepsilon \ \ a.s. \qquad \text{on } \{Z_a < \varepsilon\}. \tag{11.13}$$

Consequently, $P\{\sup_{t\geq a} Z_t \geq \varepsilon\} = P\{Z_a \geq \varepsilon\} + \varepsilon^{-1} E\{Z_a I(Z_a < \varepsilon)\}$.

Proof. Let $T = \inf\{t \geq a : Z_t \geq \varepsilon\}$ ($\inf\emptyset = \infty$). Then $\{Z_{T\wedge t}, \mathscr{F}_t, t \geq 0\}$ is a nonnegative martingale by the optional stopping theorem. Therefore for $A \in \mathscr{F}_a$ and $t \geq a$,

$$\begin{aligned} E\{Z_a I(A, Z_a < \varepsilon)\} &= E\{Z_{T\wedge t} I(A, Z_a < \varepsilon)\} \\ &= \varepsilon P(A \cap \{T \leq t, Z_a < \varepsilon\}) \\ &\quad + E\{Z_t I(A, T > t, Z_a < \varepsilon)\}. \end{aligned} \tag{11.14}$$

By (11.12), $Z_t I(T > t) \xrightarrow{P} 0$ as $t \to \infty$, and therefore letting $t \to \infty$ in (11.14) yields

$$\begin{aligned} E\{Z_a I(A, Z_a < \varepsilon)\} &= \varepsilon P(A \cap \{T < \infty, Z_a < \varepsilon\}) \\ &= \varepsilon \int_{A\cap\{Z_a<\varepsilon\}} P(T < \infty \mid \mathscr{F}_a) dP, \end{aligned}$$

proving (11.13). □

Corollary 11.5. *Define f by (11.11). Then for any $b \in \mathbb{R}, h \geq 0$ and $a > 0$,*

$$\begin{aligned} &P\{f(W_t + b, t + h) \geq \varepsilon \text{ for some } t \geq a\} \\ &\quad = P\{f(W_a + b, a + h) \geq \varepsilon\} \\ &\quad + \frac{1}{\varepsilon}\int_0^\infty \exp(b\theta - \frac{h}{2}\theta^2)\Phi\left(\frac{\beta_F(a+h,\varepsilon) - b}{\sqrt{a}} - \sqrt{a}\theta\right) dF(\theta), \end{aligned}$$

where Φ is the standard normal distribution function and

$$\beta_F(t,\varepsilon) = \inf\{x : f(x,t) \geq \varepsilon\}. \tag{11.15}$$

Proof. Without loss of generality we can assume that $f(x,a+h) < \infty$ for some x because otherwise the result is trivial. Then for $t \geq a$, the equation $f(x,t) = \varepsilon$ has a unique solution $\beta_F(t,\varepsilon)$, and $\beta_F(t,\varepsilon)$ is continuous and increasing for $t \geq a$, We next show that $f(b+W_t,\ h+t) \xrightarrow{P} 0$ as $t \to \infty$. Let ϕ be the standard normal density function. For any $c > 0$,

$$f(b+c\sqrt{t},h+t) = (\phi(c))^{-1} \int_0^\infty \phi(c-\theta\sqrt{t})\exp(b\theta - h\theta^2/2)dF(\theta) \to 0 \tag{11.16}$$

as $c \to \infty$, by the dominated convergence theorem. From (11.16), it follows that for any $\varepsilon > 0$ and $c > 0$, the following inequality holds for all sufficiently large t:

$$P\{f(b+W_t,h+t) \geq \varepsilon\} \leq P\{W_t \geq c\sqrt{t}\} = 1 - \Phi(c), \tag{11.17}$$

which can be made arbitrarily small by choosing c sufficiently large. Hence $Z_t \xrightarrow{P} 0$, where $Z_t = f(b+W_t,h+t)$. With this choice of Z_t, the desired conclusion follows from Lemma 11.4. □

Robbins and Siegmund (1970) make use of Corollary 11.5 to obtain boundary crossing probabilities for boundaries of the form β_F, noting that

$$\{f(W_t+b,t+h) \geq \varepsilon\} = \{W_t \geq \beta_F(t+h,\varepsilon) - b\}. \tag{11.18}$$

They also make use of Lemma 11.4 to prove the following two variants of Corollary 11.5: (a) If $f(b,h) < \varepsilon$ for some $b \in \mathbb{R}$ and $h \geq 0$, then

$$P\{W_t \geq \beta_F(t+h,\varepsilon) - b \ \text{ for some } t \geq 0\} = f(b,h)/\varepsilon. \tag{11.19}$$

(b) If F is a probability measure on $(0,\infty)$ and $\varepsilon > 1$, then

$$P\{W_t \geq \beta_F(t,\varepsilon) \text{ for some } t \geq 0\} = P\{f(W_t,t) \geq \varepsilon \text{ for some } t \geq 0\} = \varepsilon^{-1}. \tag{11.20}$$

11.3.2 Extensions to General Self-Normalized Processes

Replacing (W_t,t) by (A_t,B_t^2) in the preceding argument, de la Peña et al. (2004, pp. 1920–1921) have derived boundary crossing probabilities for the self-normalized process A_t/B_t under the canonical assumption (10.5), or (10.3) for discrete-time processes. In fact, letting

$$\Phi_r(x) = x^r/r \qquad \text{for } x \geq 0,\ \ 1 < r \leq 2, \tag{11.21}$$

they generalize the canonical assumption (10.5) to

$$\{\exp(\theta A_t - \Phi_r(\theta B_t)), \mathscr{F}_t, t \in T\} \text{ is a supermartingale with mean 1 for } 0 < \theta < \theta_0, \tag{11.22}$$

where $B_t > 0$, $T = \{0,1,2,\dots\}$ or $T = [0,\infty)$. In the case $T = [0,\infty)$, they also assume that A_t and B_t are right-continuous.

Let F be any finite measure on $(0,\lambda_0)$,with $F(0,\lambda_0) > 0$ and define the function

$$\psi(u,v) = \int_0^{\lambda_0} \exp(\lambda u - \lambda^r v/r) dF(\lambda). \tag{11.23}$$

Given any $c > 0$ and $v > 0$, the equation $\psi(u,v) = c$ has a unique solution $u = \beta_F(v,c)$. For the case $r = 2$, the function $v \mapsto \beta_F(v,c)$ is called a *Robbins–Siegmund boundary* in Lai (1976a), in which such boundaries are shown to have the following properties:

(a) $\beta_F(v,c)$ is a concave function of v.
(b) $\lim_{v\to\infty} \beta_F(v,c)/v = b_F/2$, where $b_F = \sup\{b > 0 : F(0,b) = 0\}$ $(\sup \emptyset = 0)$.
(c) If $dF(\lambda) = f(\lambda)d\lambda$ for $0 < \lambda < \lambda_0$, $\inf_{0<\lambda<\lambda_0} f(\lambda) > 0$ while $\sup_{0<\lambda<\lambda_0} f(\lambda) < \infty$, then $\beta_F(v,c) \sim (v \log v)^{1/2}$ as $v \to \infty$.
(d) If $dF(\lambda) = f(\lambda)d\lambda$ for $0 < \lambda < e^{-2}$, and $= 0$ elsewhere, where

$$f(\lambda) = 1/\left\{\lambda(\log \lambda^{-1})(\log\log \lambda^{-1})^{1+\delta}\right\} \tag{11.24}$$

for some $\delta > 0$, then as $v \to \infty$,

$$\beta_F(v,c) = \left\{2v\left[\log_2 v + \left(\frac{3}{2} + \delta\right)\log_3 v + \log\left(\frac{c}{2\sqrt{\pi}}\right) + o(1)\right]\right\}^{1/2}, \tag{11.25}$$

where, as in Robbins and Siegmund (1970), we write $\log_k v = \log(\log_{k-1} v)$ for $k \ge 2$, $\log_1 v = \log v$. For general $1 < r \le 2$, (a) still holds, (b) holds with $b_F/2$ replaced by b_F^{r-1}/r, and (c) can be generalized to $\beta_F(v,c) \sim v^{1/r}\{(\log v)/(r-1)\}^{(r-1)/r}$ as $v \to \infty$. Moreover, if f is given by (11.24), then

$$\beta_F(v,c) \sim v^{1/r}\{r(\log\log v)/(r-1)\}^{(r-1)/r} \qquad \text{as } v \to \infty. \tag{11.26}$$

Note that $\psi(A_t, B_t^r)$ is right-continuous when $T = [0,\infty)$ by the assumption on (A_t, B_t) in this case. It follows from (11.22) that $\{\psi(A_t, B_t^r), t \ge 0\}$ is a nonnegative supermartingale with mean $\le F(0,\lambda_0)$ and therefore,

$$\begin{aligned} P\{A_t \ge \beta_F(B_t^r, c) \text{ for some } t \in T\} \\ = P\{\psi(A_t, B_t^r) \ge c \text{ for some } t \in T\} \le F(0,\lambda_0)/c, \end{aligned} \tag{11.27}$$

for every $c > 0$. In particular, by choosing c in (11.27) arbitrarily large, we obtain from (11.26) and (11.27) the following:

Corollary 11.6. *Let $1 < r \le 2, \Phi_r(x) = x^r/r$ for $x \ge 0$ and suppose that* (11.22) *holds for the process $(A_t, B_t), t \in T$, and that A_t and B_t are right-continuous in the case $T = [0, \infty)$. Then*

$$\limsup_{t\to\infty} \frac{A_t}{B_t(\log\log B_t)^{(r-1)/r}} \le \left\{\frac{r}{r-1}\right\}^{(r-1)/r} \quad a.s. \quad on \left\{\lim_{t\to\infty} B_t = \infty\right\}.$$

We conclude this section with a discussion of the properties (c) and (d) of the Robbins–Siegmund boundaries, relating them to the results and methods of Sects. 11.1 and 11.2. Suppose $dF(\lambda) = f(\lambda)$ for $0 \le \lambda \le \lambda_0$ and consider the case $r = 2$ in (11.23). First assume that f is continuous and positive on $[0, \lambda_0]$, then since $\lambda u - \lambda^2 v/2$ is maximized at $\lambda = u/v$, Laplace's asymptotic formula (11.2) yields (as $v \to \infty$)

$$\psi(u,v) \sim \sqrt{\frac{2\pi}{v}} f\left(\frac{u}{v}\right) \exp\left(\frac{u^2}{2v}\right) \qquad \text{if } \varepsilon \le \frac{u}{v} \le \lambda_0 - \varepsilon, \tag{11.28}$$

for every $\varepsilon > 0$. From (11.28), it follows that for given $c > 0$,

$$\psi(u,v) \ge c \Longleftrightarrow u \ge \left\{2v\left[\log\left(\sqrt{\frac{v}{2\pi}}\right) + \log\left(\frac{c}{f(u/v)}\right) + o(1)\right]\right\}^{1/2}$$

as $v \to \infty$, proving property (c). If $\sup_{0<\lambda<\lambda_0} f(\lambda) < \infty$ and $\inf_{0<\lambda<\lambda_0} f(\lambda) > 0$, then we can use this assumption to bound the integral in (11.2) above and below to prove property (c) without assuming f to be continuous.

When f can become infinite as in (11.24), we have to modify Laplace's method accordingly by bounding f above and below and combining with certain bounds on $\exp(u\lambda - v\lambda^2/2)$, as in the proof of Lemma 11.2, to derive the asymptotic behavior of $\beta_F(v,c)$ via that of $\psi(u,v)$. This is the basic idea behind Robbins and Siegmund (1970, Sect. 5) derivation of (11.25). First we can rewrite the equation $\psi(u,v) = c$ as

$$\begin{aligned} c &= \int_0^{e^{-2}} \exp\left(\frac{u}{\sqrt{v}}\lambda\sqrt{v} - \frac{1}{2}\lambda^2 v\right) f(\lambda)d\lambda \\ &= \frac{1}{\phi(u/\sqrt{v})} \int_0^{e^{-2}} \phi\left(\lambda\sqrt{v} - \frac{u}{\sqrt{v}}\right) f(\lambda)d\lambda, \end{aligned} \tag{11.29}$$

from which it follows that

$$u/\sqrt{v} \to \infty \quad \text{and} \quad u/\sqrt{v} = o(\sqrt{v}) \qquad \text{as } v \to \infty. \tag{11.30}$$

Let $\gamma > 1$. Since f is decreasing on $(0, \lambda)$ for sufficiently small $\lambda > 0$, (11.29) and (11.30) imply that for all v sufficiently large,

$$c \geq \frac{1}{\phi(u/\sqrt{v})} \int_0^{\gamma u/v} \phi\left(\lambda\sqrt{v} - \frac{u}{\sqrt{v}}\right) f(\lambda)d\lambda$$
$$\geq \frac{f(\gamma u/v)}{\phi(u/\sqrt{v})} \int_0^{\gamma u/v} \phi\left(\lambda\sqrt{v} - \frac{u}{\sqrt{v}}\right) d\lambda$$
$$= \frac{\Phi((\gamma-1)u/\sqrt{v}) - \Phi(-u/\sqrt{v})}{\sqrt{v}\phi(u/\sqrt{v})} f(\gamma u/v).$$

Letting $v \to \infty$ and then $\gamma \to 1$, it follows that

$$c \geq \frac{1+o(1)}{(u/\sqrt{v})\phi(u/\sqrt{v})\{\log(v/u)\}\{\log\log(v/u)\}^{1+\delta}}. \tag{11.31}$$

To obtain a reverse inequality, take $0 < a < b < 1$ and split the integral in (11.29) as $\int_0^{au/v} + \int_{au/v}^{bu/v} + \int_{bu/v}^{e^{-2}}$. The first integral can be bounded by

$$\int_0^{au/v} \phi\left(\lambda\sqrt{v} - \frac{u}{\sqrt{v}}\right) f(\lambda)d\lambda \leq \phi\left((a-1)\frac{u}{\sqrt{v}}\right) \int_0^{au/v} f(\lambda)d\lambda$$
$$= \frac{\phi((a-1)u/\sqrt{v})}{\delta(\log\log(v/au))^{\delta}}.$$

An upper bound for $\int_{au/v}^{bu/v}$ can be obtained by bounding $f(\lambda)$ in this range by $f(au/v)$, and a similar upper bound can be obtained for $\int_{bu/v}^{e^{-2}}$. It follows by combining these upper bounds that as $v \to \infty$,

$$c \leq \frac{b^{-1}+o(1)}{(u/\sqrt{v})\phi(u/\sqrt{v})\{\log(v/u)\}\{\log\log(v/u)\}^{1+\delta}}. \tag{11.32}$$

Letting $b \to 1$ in (11.32) and combining the result with (11.31), (11.30) follows. This idea that makes use of the properties of f has been used in Sect. 11.2 for f of the more general form (11.7).

11.4 Supplementary Results and Problems

1. Let P_θ be a probability measure under which $X_1, X_2, \ldots$ are i.i.d. random variables with density function $g_\theta(x) = e^{\theta x - \psi(\theta)}$ with respect to some probability measure m on $\mathbb{R}$ such that $\int_{-\infty}^{\infty} e^{\theta x} dm(x) < \infty$ for some $\theta \neq 0$; this is the *exponential family* with natural parameter $\theta \in \Theta := \{\theta \in \mathbb{R} : \int e^{\theta x} dm(x) < \infty\}$. Let $S_n = X_1 + \cdots + X_n$, and let $\mathscr{F}_n$ be the σ-field generated by $X_1, \ldots, X_n$:
   (a) Show that Θ is an interval (possibly infinite) containing 0 and that X_1 has distribution m under P_0. Denote P_0 by P for simplicity.

(b) Let F be a probability measure on Θ and let

$$f(S_n,n) = \int_{\Theta} e^{\theta S_n - n\psi(\theta)} dF(\theta). \tag{11.33}$$

Show that $P\{f(S_n,n) \geq c \text{ for some } n \geq 1\} \leq c^{-1}$ for every $c > 1$.

(c) The Bernoulli distribution $P(X_1 = 1) = p = 1 - P(X_1 = 0)$ can be embedded in an exponential family with $\theta = \log(p/(1-p))$. Suppose the probability distribution F on θ has density function $f(\theta)d\theta = p^{a-1}(1-p)^{b-1}dp/B(a,b)$, $0 < p < 1$, where $B(\cdot,\cdot)$ is the beta function and $a > 0$, $b > 0$. Let $0 < p_0 < 1$ and let P_0 correspond to the case $p = p_0$. Show that in this case $f(S_n,n)$ has an explicit formula and

$$f(S_n,n) \geq c \Longleftrightarrow p_0 \in I_n(S_n,c), \tag{11.34}$$

where I_n is an interval. Hence conclude from (b) that

$$P_p\{p \in I_n(S_n,c) \text{ for every } n \geq 1\} \geq 1 - c^{-1}. \tag{11.35}$$

Thus, the random sequence $\{I_n(S_n,c),\ n \geq 1\}$ simultaneously covers the true parameter p with probability no smaller than $1 - c^{-1}$. This is called a $(1-c^{-1})$-level *confidence sequence.*

2. For the confidence sequence $I_n = I_n(S_n,c)$ defined by (11.34), show that the width of I_n converges a.s. to 0. Hence, for $p \neq p_0$, $P_{p_0}\{p \notin I_n \text{ for all large } n\} = 1$. Robbins (1970) has made use of this property to construct sequential tests of $H_0 : p = p_0$ with Type I error probability no larger than c^{-1} and with power 1.
3. Show that for given p_0, it is possible to choose F such that the associated confidence sequence I_n (depending on S_n, c and F) has width of the order $n^{-1/2}\{2p_0(1-p_0)\log\log n\}^{1/2}$ *a.s.*
4. Robbins and Siegmund (1970) have proved the following limit theorem for boundary crossing probabilities of the random walk $S_n = X_1 + \cdots + X_n$, where $X_1, X_2, \ldots$ are i.i.d. random variables having mean 0 and variance 1, relating them to those of the Brownian motion W_t:

(a) *Let $h > 0$ and let $\beta : [h,\infty) \to \mathbb{R}$ be a continuous function such that $t^{-1/2}\beta(t)$ is ultimately nondecreasing as $t \to \infty$ and*

$$\int_h^{\infty} \frac{\beta(t)}{t^{3/2}} \exp\left(-\frac{\beta^2(t)}{2t}\right) dt < \infty, \tag{11.36}$$

which is related to the upper-lower class test in Problem 2.9. Then

$$\begin{aligned} &\lim_{m\to\infty} P\{S_n \geq \sqrt{m}\beta(n/m) \text{ for some } n \geq hm\} \\ &= P\{W_t \geq \beta(t) \text{ for some } t \geq h\}. \end{aligned} \tag{11.37}$$

(b) *Assume furthermore that β is defined and continuous on $(0,h)$, that $\beta(t)/\sqrt{t}$ is nonincreasing for t sufficiently small and that*

$$\int_0^1 \frac{\beta(t)}{t^{3/2}} \exp\left(-\frac{\beta^2(t)}{2t}\right) dt < \infty. \tag{11.38}$$

Then (11.37) *continues to hold with $n \geq hm$ replaced by $n \geq 1$ and $t \geq h$ by $t > 0$.*

Apply this limit theorem to the Robbins–Siegmund boundaries β_F considered in (c) and (d) of Sect. 11.3.2, and explain how you can use the result to construct an approximate 95% confidence sequence for the mean of a distribution with known variance 1 based on successive observations $X_1, X_2, \ldots$ drawn independently from the distribution.

Chapter 12
Moment and Exponential Inequalities for Self-Normalized Processes

Inspired by three continuous-time martingale inequalities that are described in Sect. 12.1, de la Peña et al. (2000, 2004) have developed moment and exponential inequalities for general self-normalized process by making use of the method of mixtures described in Chap. 11. In Sect. 12.2 we present these moment and exponential inequalities under the canonical assumption (10.5) or (10.6) and explain how the method of mixtures can be used to derive them. Their applications are given in Sect. 12.3.

12.1 Inequalities of Caballero, Fernandez and Nualart, Graversen and Peskir, and Kikuchi

Caballero et al. (1998) provide an estimate for the L_p-norm of a continuous martingale divided by its quadratic variation, where $1 \leq p < q$. This norm is bounded by a universal constant times the L_q-norm of the inverse of the square root of the quadratic variation. The following theorem generalizes this result to random variables satisfying the canonical assumption (10.6) for all λ. Let $\|X\|_p = (E|X|^p)^{1/p}$.

Theorem 12.1. *Let $B > 0$ and A be two random variables satisfying* (10.6) *for all $\lambda \in \mathbb{R}$. Then for $1 \leq p < q$, there exists a universal constant $C = C(p,q)$ such that*

$$\left\| \frac{A}{B^2} \right\|_p \leq C \|B^{-1}\|_q.$$

Proof. Note that

$$E\left[\left|\frac{A}{B^2}\right|^p\right] = \int_0^\infty p x^{p-1} P\left[|A| > xB^2\right] dx,$$

$$P\left\{|A| > xB^2\right\} \leq P\left\{A > xB^2\right\} + P\left\{-A > xB^2\right\}.$$

V.H. de la Peña et al., *Self-Normalized Processes: Limit Theory and Statistical Applications*, 161
Probability and its Applications,
© Springer-Verlag Berlin Heidelberg 2009

Take $\alpha > 1$ and $\beta > 1$ such that $1/\beta + 1/\alpha = 1$. Choose $\lambda > 0$ and $\theta > 0$ such that $(\lambda\alpha)^2/2 = \theta\alpha$, that is, $\theta = \lambda^2\alpha/2$. Then

$$\begin{aligned} P(A > xB^2) &\le P(e^{\lambda A - \theta B^2} > e^{(\lambda x - \theta)B^2}) \\ &\le E(e^{\lambda A - \theta B^2} e^{-(\lambda x - \theta)B^2}) \\ &\le \left(E\left[e^{\lambda\alpha A - \theta\alpha B^2}\right]\right)^{1/\alpha} \left(E\left[e^{-\beta(\lambda x - \theta)B^2}\right]\right)^{1/\beta}. \end{aligned}$$

Since $\theta = \lambda^2\alpha/2$, it then follows from (10.6) that

$$P(A > xB^2) \le \left(E\left[e^{-\beta(\lambda x - \lambda^2\alpha/2)B^2}\right]\right)^{1/\beta}.$$

The optimal λ is given by $\lambda = x/\alpha$, which yields

$$P(A > xB^2) \le \left(E\left[e^{-(\beta x^2/2\alpha)B^2}\right]\right)^{1/\beta} = \left(E\left[e^{-(\beta-1)(x^2/2)B^2}\right]\right)^{1/\beta}$$

since $1/\alpha = (\beta - 1)/\beta$. Therefore, for any $\varepsilon > 0$ and $\delta > 1$,

$$\begin{aligned} E\left[\left|\frac{A}{B^2}\right|^p\right] &\le 2\int_0^\infty px^{p-1} \left(E\left[e^{-(\beta-1)(x^2/2)B^2}\right]\right)^{1/\beta} dx \\ &\le 2\varepsilon^p + 2\int_\varepsilon^\infty px^{p-1} \left(E\left[e^{-(\beta-1)(x^2/2)B^2}\right]\right)^{1/\beta} dx \\ &= 2\varepsilon^p + 2p\int_\varepsilon^\infty x^{-\delta} \left(E\left[x^{(\delta+p-1)\beta} e^{-(\beta-1)(x^2/2)B^2}\right]\right)^{1/\beta} dx \\ &\le 2\varepsilon^p + 2p\left(\int_\varepsilon^\infty x^{-\delta}\,dx\right) \left(E\left[\sup_{x\in\mathbb{R}}\left(x^{(\delta+p-1)\beta} e^{-(\beta-1)(x^2/2)B^2}\right)\right]\right)^{1/\beta}. \end{aligned}$$

Let $\psi(x) = x^{(\delta+p-1)\beta} e^{-(\beta-1)(x^2/2)B^2}$. Then

$$\psi'(x) = \left[\beta(\delta+p-1)x^{(\delta+p-1)\beta-1} - x^{\beta(\delta+p-1)+1}(\beta-1)B^2\right] e^{-(\beta-1)(x^2/2)B^2}.$$

This function ψ is minimized when

$$x_0 = \sqrt{\frac{\beta(\delta+p-1)}{\beta-1}}\, B^{-1}.$$

Therefore,

$$\psi(x_0) = \left(\frac{\beta(\delta+p-1)}{\beta-1}\right)^{\frac{\delta+p-1}{2}\beta} (B^2)^{-(\delta+p-1)\beta/2} e^{-(\delta+p-1)\beta/2}.$$

Hence,

$$
\begin{aligned}
E\left[\left|\frac{A}{B^2}\right|^p\right] &\le 2\varepsilon^p + \frac{2p}{\delta-1}\varepsilon^{1-\delta}e^{-(\delta+p-1)/2}\left(\frac{\beta(\delta+p-1)}{\beta-1}\right)^{(\delta+p-1)/2} \\
&\quad \times \left(E\left[\left(B^2\right)^{-(\delta+p-1)\beta/2}\right]\right)^{1/\beta} \\
&= 2\varepsilon^p + \varepsilon^{1-\delta}K\,,
\end{aligned}
$$

where

$$
K = \frac{2p}{\delta-1}e^{-(\delta+p-1)/2}\left(\frac{\beta(\delta+p-1)}{\beta-1}\right)^{(\delta+p-1)/2}\left(E\left[(B^2)^{-(\delta+p-1)\beta/2}\right]\right)^{1/\beta}.
$$

Next we optimize over ε. Set $Q(\varepsilon) = 2\varepsilon^p + \varepsilon^{1-\delta}K$. Then

$$
Q'(\varepsilon) = 2p\varepsilon^{p-1} + (1-\delta)e^{-\delta}K,
$$

and the unique solution of $Q'(\varepsilon) = 0$ is given by

$$
\varepsilon_0 = \left(\frac{\delta-1}{2p}\right)^{1/(p+\delta-1)}K^{1/(p+\delta-1)},
$$

for which

$$
\begin{aligned}
Q(\varepsilon_0) &= 2\left(\frac{\delta-1}{2p}\right)^{p/(p+\delta-1)}K^{p/(p+\delta-1)} \\
&\quad + \left(\frac{\delta-1}{2p}\right)^{(1-\delta)/(p+\delta-1)}K^{(1-\delta)/(p+\delta-1)+1} \\
&= K^{p/(p+\delta-1)}\left(2\left(\frac{\delta-1}{2p}\right)^{p/(p+\delta-1)} + \left(\frac{\delta-1}{2p}\right)^{(1-\delta)/(p+\delta-1)}\right) \\
&= \left(\frac{2p}{\delta-1}\right)^{p/(p+\delta-1)}e^{-p/2}\left(\frac{\beta(\delta+p-1)}{\beta-1}\right)^{p/2} \\
&\quad \times \left(2\left(\frac{\delta-1}{2p}\right)^{p/(p+\delta-1)} + \left(\frac{\delta-1}{2p}\right)^{(1-\delta)/(p+\delta-1)}\right) \\
&\quad \times \left(E\left[\left(B^2\right)^{-(\delta+p-1)\beta/2}\right]\right)^{p/[\beta(\delta+p-1)]}.
\end{aligned}
$$

Therefore,

$$
\left\|\frac{A}{B^2}\right\|_p \le 2^{1/p}e^{-1/2}\sqrt{\frac{\beta(\delta+p-1)}{\beta-1}}\left(1+\frac{p}{\delta-1}\right)^{1/p}\|B^{-1}\|_{(\delta+p-1)\beta}.
$$

We now choose $\beta > 1$ and $\delta > 1$ by the equations $q-p = (\beta-1)(p+1)$, $\delta = 2-1/\beta$. Then $q = (\delta-1+p)\beta$. Hence,

$$\left\|\frac{A}{B^2}\right\|_p \le C_{p,q}\|B^{-1}\|_q, \qquad \text{where } C_{p,q} = \frac{2^{1/p}}{\sqrt{e}}\left((p+1)\left(1+\frac{p}{q-p}\right)\right)^{1/2+1/p}. \qquad \square$$

For continuous local martingales M_t, Revuz and Yor (1999, p.168) give a closely related inequality for $p > q > 0$:

$$E\left\{\sup_{s\ge 0}|M_s|^p/\langle M\rangle_\infty^{q/2}\right\} \le C_{pq}E\left(\sup_{s\ge 0}|M_S|\right)^{p-q}, \tag{12.1}$$

where C_{pq} is a universal constant depending on p,q. In Sect. 12.2 we consider what is arguably the more important case $p = q$ under the canonical assumption (10.5), which we also use to extend the following two inequalities to a more general setting.

Theorem 12.2 (Graversen and Peskir, 2000). *Let $M = (M_t)_{t\ge 0}$ be a continuous local martingale with predictable variation process $(\langle M\rangle_t)_{t\ge 0}$. Then there exist universal constants $D_1 > 0$ and $D_2 > 0$ such that*

$$\begin{aligned} D_1 E\sqrt{\log(1+\log(1+\langle M\rangle_\tau))} &\le E\left(\max_{0\le t\le\tau}\frac{|M_t|}{\sqrt{1+\langle M\rangle_t}}\right) \\ &\le D_2 E\sqrt{\log(1+\log(1+\langle M\rangle_\tau))} \end{aligned} \tag{12.2}$$

for all stopping times τ.

Theorem 12.3 (Kikuchi, 1991). *Let $(M_t)_{t\ge 0}$ be a continuous local martingale such that $\tilde{M}_\infty := \sup_{t\ge 0}|M_t| < \infty$ a.s. Then for every $p > 0$ and $0 < \alpha < \frac{1}{2}$, there exists an absolute constant $C_{\alpha,p}$ such that*

$$E\left[\tilde{M}_\infty^p \exp\left(\alpha\tilde{M}_\infty^2/\langle M\rangle_\infty\right)\right] \le C_{\alpha,p}E(\tilde{M}_\infty^p).$$

12.2 Moment Bounds via the Method of Mixtures

In this section we show how the method of mixtures to perform pseudo-maximization in Chap. 11 can be applied to develop moment bounds that generalize Theorems 12.1–12.3 to more general functions h than the special cases $h(x) = |x|^p$ and $h(x) = \exp(\alpha x^2)$ considered in these theorems and to more general processes. We use the canonical assumption (10.5), or (10.6), or a variant thereof. While Sect. 11.1 has related pseudo-maximization to Laplace's method when the method of mixtures uses a bounded continuous density function, we use in Sect. 12.2.1 a special class of these mixing densities, namely, Gaussian density functions, for which integration can be performed exactly in an explicit form, without resorting to Laplace's approximation. In Sect. 12.2.2 we use the class of mixing densities in Sect. 11.2 and modify Laplace's method by bounding the integrand, using arguments similar to those in the last paragraph of Sect. 11.3.

12.2.1 Gaussian Mixing Densities

We begin with a simple application of the method of mixtures to derive exponential and L_p-bounds for $A/\sqrt{B^2+(EB)^2}$ when (10.6) holds for all $\lambda \in \mathbb{R}$.

Theorem 12.4. *Let $B \geq 0$ and A be two random variables satisfying* (10.6) *for all $\lambda \in \mathbb{R}$. Then for any $y > 0$,*

$$E\frac{y}{\sqrt{B^2+y^2}}\exp\left\{\frac{A^2}{2(B^2+y^2)}\right\} \leq 1. \tag{12.3}$$

Consequently, if $EB > 0$, then $E\exp(A^2/[4(B^2+(EB)^2)]) \leq \sqrt{2}$ and

$$E\exp\left(x|A|\Big/\sqrt{B^2+(EB)^2}\right) \leq \sqrt{2}\exp\left(x^2\right) \qquad \textit{for all } x > 0. \tag{12.4}$$

Moreover, for all $p > 0$,

$$E\left(|A|\Big/\sqrt{B^2+(EB)^2}\right)^p \leq 2^{p-1/2}p\Gamma(p/2). \tag{12.5}$$

Proof. Multiplying both sides of (10.6) by $(2\pi)^{-1/2}y\exp(-\lambda^2y^2/2)$ and integrating over λ, we obtain by using Fubini's theorem that

$$\begin{aligned}
1 &\geq \int_{-\infty}^{\infty} E\frac{y}{\sqrt{2\pi}}\exp\left(\lambda A - \frac{\lambda^2}{2}B^2\right)\exp\left(-\frac{\lambda^2y^2}{2}\right)d\lambda \\
&= E\left[\frac{y}{\sqrt{B^2+y^2}}\exp\left\{\frac{A^2}{2(B^2+y^2)}\right\}\right. \\
&\quad \left.\times \int_{-\infty}^{\infty}\frac{\sqrt{B^2+y^2}}{\sqrt{2\pi}}\exp\left\{-\frac{B^2+y^2}{2}\left(\lambda^2 - 2\frac{A}{B^2+y^2}\lambda + \frac{A^2}{(B^2+y^2)^2}\right)\right\}d\lambda\right] \\
&= E\left[\frac{y}{\sqrt{B^2+y^2}}\exp\left\{\frac{A^2}{2(B^2+y^2)}\right\}\right],
\end{aligned}$$

proving (12.3). By the Cauchy-Schwarz inequality and (12.3),

$$\begin{aligned}
&E\exp\left\{\frac{A^2}{4(B^2+y^2)}\right\} \\
&\leq \left\{\left(E\frac{y\exp\left\{A^2/\left(2(B^2+y^2)\right)\right\}}{\sqrt{B^2+y^2}}\right)\left(E\sqrt{\frac{B^2+y^2}{y^2}}\right)\right\}^{1/2} \\
&\leq \left(E\sqrt{\frac{B^2}{y^2}+1}\right)^{1/2} \leq \left(E\left(\frac{B}{y}+1\right)\right)^{1/2} \qquad \text{for } y = EB.
\end{aligned}$$

To prove (12.4) and (12.5), we assume without loss of generality that $EB < \infty$. Using the inequality $ab \le \frac{a^2+b^2}{2}$ with $a = \sqrt{2c}|A|/\sqrt{B^2+(EB)^2}$ and $b = x/\sqrt{2c}$, we obtain $x|A|/\sqrt{B^2+(EB)^2} \le \frac{cA^2}{B^2+(EB)^2} + \frac{x^2}{4c}$, which in the case $c = 1/4$ yields

$$E\exp\left\{\frac{x|A|}{\sqrt{B^2+(EB)^2}}\right\} \le E\exp\left\{\frac{cA^2}{B^2+(EB)^2} + \frac{x^2}{4c}\right\} \le \sqrt{2}\exp(x^2),$$

proving (12.4). Moreover, by Markov's inequality, $P(|A|/\sqrt{B^2+(EB)^2} \ge x) \le \sqrt{2}\exp(-x^2/4)$ for all $x > 0$. Combining this with the formula $EU^p = \int_0^\infty px^{p-1}P(U > x)dx$ for any nonnegative random variable U, we obtain

$$E\left(|A|\Big/\sqrt{B^2+(EB)^2}\right)^p \le \sqrt{2}\int_0^\infty px^{p-1}\exp(-x^2/4)dx = 2^{p-1/2}p\Gamma(p/2). \quad \square$$

Another application of the basic inequality (12.3) is the following.

Corollary 12.5. *Under the same assumption as in Theorem 12.4, for all $x \ge \sqrt{2}$, $y > 0$ and $p > 0$,*

$$P\left(|A|\Big/\sqrt{(B^2+y)\left(1+\frac{1}{2}\log\left(\frac{B^2}{y}+1\right)\right)} \ge x\right) \le \exp\left(-\frac{x^2}{2}\right), \tag{12.6}$$

$$E\left(|A|\Big/\sqrt{(B^2+y)\left(1+\frac{1}{2}\log\left(\frac{B^2}{y}+1\right)\right)}\right)^p \le 2^{p/2} + 2^{(p-2)/2}p\Gamma\left(\frac{p}{2}\right). \tag{12.7}$$

Proof. Note that for $x \ge \sqrt{2}$ and $y > 0$,

$$\begin{aligned}
&P\left\{\frac{A^2}{2(B^2+y)} \ge \frac{x^2}{2}\left(1+\frac{1}{2}\log\left(\frac{B^2}{y}+1\right)\right)\right\} \\
&\le P\left\{\frac{A^2}{2(B^2+y)} \ge \frac{x^2}{2}+\frac{1}{2}\log\left(\frac{B^2}{y}+1\right)\right\} \\
&\le \exp\left(-\frac{x^2}{2}\right)E\frac{\sqrt{y}\exp\{A^2/(2(B^2+y))\}}{\sqrt{B^2+y}} \le \exp\left(-\frac{x^2}{2}\right),
\end{aligned}$$

in which the last inequality follows from (12.3). The proof of (12.7) makes use of (12.6) and is similar to that of (12.5). $\square$

12.2.2 The Mixing Density Functions in Sect. 11.2

The function g defined in (11.10) *and the class of functions L satisfying* (11.3)–(11.5) *will be used throughout this section*, and therefore reference to where they are introduced will be omitted. Special cases and concrete examples of the moment inequalities obtained by the method of mixtures using the mixing density (11.7) will be given in Sect. 12.3. By making use of Lemmas 11.2 and 11.3, we first derive the following inequality due to de la Peña et al. (2000).

Theorem 12.6. *Let $A, B > 0$ be two random variables satisfying* (10.6) *for all $\lambda > 0$. Let $0 < \delta < 1$ and let $h : [0,\infty) \to [0,\infty)$ be nondecreasing such that $\limsup_{y\to\infty} yh(y)/g^{\delta}(y) < \infty$. Then*

$$Eh\left(\frac{A^+}{B}\right) \le 4\sup_{y\ge 1}\frac{y\left(L(y)\vee g^{1-\delta}(y)\right)h(y)}{g(y)} + Eh\left(1\vee\sqrt{\frac{2}{1-\delta}\log^+ L\left(B\vee\frac{1}{B}\right)}\right), \tag{12.8}$$

$$Eh\left(\frac{A^+/B}{\sqrt{1\vee\log^+ L\left(B\vee\frac{1}{B}\right)}}\right) \le h\left(\sqrt{\frac{2}{1-\delta}}\right) + \sup_{y\ge 1}\frac{y\left(L(y)\vee g^{1-\delta}(y)\right)h(y)}{g(y)}. \tag{12.9}$$

Proof. Let $Q_\delta = \{g^{1-\delta}(A/B) \le L(B\vee B^{-1})\}$. By Lemma 11.3,

$$\begin{aligned} Eh\left(\frac{A^+}{B}\right) &\le Eh(1)I\left(\frac{A}{B}\le 1\right) \\ &\quad + Eh\left(\sqrt{\frac{1}{1-\delta}\log^+ L\left(B\vee\frac{1}{B}\right)}\right)I\left(\frac{A}{B}>1, Q_\delta\right) \\ &\quad + E\frac{g\left(\frac{A}{B}\right)}{4\frac{A}{B}\left(L\left(\frac{A}{B}\right)\vee L\left(B\vee\frac{1}{B}\right)\right)} \\ &\quad \times \frac{4\left(\frac{A}{B}\right)\left(L\left(\frac{A}{B}\right)\vee g^{1-\delta}\left(\frac{A}{B}\right)\right)h\left(\frac{A}{B}\right)}{g\left(\frac{A}{B}\right)}I\left(\frac{A}{B}>1, Q_\delta^c\right) \\ &\equiv I + II + III. \end{aligned}$$

Dropping the event Q_δ from II yields

$$I + II \le Eh\left(1\vee\sqrt{\frac{2}{1-\delta}\log^+ L\left(B\vee\frac{1}{B}\right)}\right).$$

By Lemma 11.2,

$$III \le \sup_{y\ge 1} \frac{4y\left(L(y)\vee g^{1-\delta}(y)\right)h(y)}{g(y)},$$

proving (12.8). To prove (12.9) we use an analogous three-term decomposition

$$\begin{aligned}
&Eh\left(\frac{A^+/B}{\sqrt{1\vee\log^+ L\left(B\vee\frac{1}{B}\right)}}\right)\\
&\le Eh\left(\frac{1}{\sqrt{1\vee\log^+ L\left(B\vee\frac{1}{B}\right)}}\right) I\left(\frac{A^+}{B}\le 1\right)\\
&\quad + Eh\left(\frac{\sqrt{\frac{2}{1-\delta}\log^+ L\left(B\vee\frac{1}{B}\right)}}{\sqrt{1\vee\log^+ L\left(B\vee\frac{1}{B}\right)}}\right) I\left(\frac{A^+}{B}>1, Q_\delta\right)\\
&\quad + E\left\{\frac{g(A/B)}{4\frac{A}{B}\left(L\left(\frac{A}{B}\right)\vee L\left(B\vee\frac{1}{B}\right)\right)}\right.\\
&\qquad \left.\times\frac{4\frac{A}{B}\left(L\left(\frac{A}{B}\right)\vee g^{1-\delta}\left(\frac{A}{B}\right)\right)h\left(\frac{A}{B}\right)}{g\left(\frac{A}{B}\right)} I\left(\frac{A}{B}\ge 1, Q_\delta^c\right)\right\}\\
&\le h\left(\sqrt{\frac{2}{1-\delta}}\right) + 4\sup_{y\ge 1}\frac{y\left(L(y)\vee g^{1-\delta}(y)\right)}{g(y)}h(y).
\end{aligned}$$

□

We next consider the canonical assumption (10.5) instead of (10.6) and obtain inequalities similar to (12.8) and (12.9) but with A^+/B replaced by $\sup_{t\le\tau} A_t/\sqrt{B_t^2+1}$, where τ is a nonnegative random variable. We shall change g to g_r defined in Lemma 11.3, choosing $r<1$ so that we can use the following result of Shao (1998).

Lemma 12.7. *Let T_k, $k\ge 1$, be a nonnegative supermartingale. Then, for all $0<r<1$,*

$$E\left(\sup_{k\ge 1} T_k\right)^r \le \frac{(ET_1)^r}{1-r}.$$

Theorem 12.8. *Let $T=\{0,1,2,\dots\}$ or $T=[0,\infty)$. Suppose that $\{(A_t,B_t),\ t\in T\}$ satisfies the canonical assumption (10.5) for all $\lambda>0$, and that B_t is positive and nondecreasing in $t>0$, and $A_0=B_0=0$. In the case $T=[0,\infty)$, A_t and B_t are also assumed right-continuous. Let $0<\delta$, $r<1$ and $h:[0,\infty)\to[0,\infty)$ be nondecreasing such that $\limsup_{y\to\infty}\left(yh(y)/g_r^\delta(y)\right)<\infty$, where $g_r(y)=(\exp(ry^2/2))I(y\ge 1)$. Then for any random time τ,*

$$
\begin{aligned}
Eh\left(\sup_{t\le\tau}\frac{A_t}{\sqrt{B_t^2+1}}\right) &\le \sup_{y\ge 1}\frac{4y(L(y)\vee g_r^{1-\delta}(y))h(y)}{g_r(y)}\\
&\quad + Eh\left(1\vee\left(\sqrt{\frac{2}{r(1-\delta)}\log^+ L\left(\sqrt{B_\tau^2+1}\right)}\right)\right),
\end{aligned}\tag{12.10}
$$

$$
\begin{aligned}
&Eh\left(\sup_{t\le\tau}\frac{A_t}{\sqrt{B_t^2+1}\sqrt{1\vee\log^+ L\left(\sqrt{B_t^2+1}\right)}}\right)\\
&\qquad\le \sup_{y\ge 1}\frac{4y(L(y)\vee g_r^{1-\delta}(y))h(y)}{g_r(y)} + h\left(\sqrt{\frac{2}{r(1-\delta)}}\right).
\end{aligned}\tag{12.11}
$$

Proof. By considering $\tau\wedge t$ and applying the monotone convergence theorem in letting $t\to\infty$, we can assume without loss of generality that τ is bounded. There exists a sequence of random times $t_n\le\tau$ such that

$$
\lim_{n\to\infty}\frac{A_{t_n}^+}{\sqrt{B_{t_n}^2+1}} = \sup_{0\le t\le\tau}\frac{A_t^+}{\sqrt{B_t^2+1}} = \sup_{0\le t\le\tau}\frac{A_t}{\sqrt{B_t^2+1}},
$$

recalling that $A_0=0$. Since $0<r<1$,

$$
\begin{aligned}
E\left(\exp\left\{\lambda A_{t_n}-\frac{\lambda^2}{2}(B_{t_n}^2+1)\right\}\right)^r &\le E\left(\exp\left\{\lambda A_{t_n}-\frac{\lambda^2}{2}B_{t_n}^2\right\}\right)^r\\
&\le E\left(\sup_{0\le s\le\tau}\exp\left\{\lambda A_s-\frac{\lambda^2}{2}B_s^2\right\}\right)^r\\
&= E\left(\sup_{s\ge 0}\exp\left\{\lambda A_{\tau\wedge s}-\frac{\lambda^2}{2}B_{\tau\wedge s}^2\right\}\right)^r\\
&\le \frac{1}{1-r}\left(E\exp\left\{\lambda A_0-\frac{\lambda^2}{2}B_0^2\right\}\right)^r = \frac{1}{1-r},
\end{aligned}
$$

where the last inequality follows from Lemma 12.7 and the last equality follows from $A_0=B_0=0$. For notational simplicity, let $A=A_{t_n}$ and $B=B_{t_n}$. Multiplying by the mixing density $f(\lambda)$ defined in (11.7) and integrating, $\frac{1}{1-r}$ is bounded from below by

$$
\begin{aligned}
&\int_0^\infty E\exp\left\{r\left(\lambda A-\frac{\lambda^2}{2}\{B^2+1\}\right)\right\}\frac{1}{\lambda L(\lambda\vee\lambda^{-1})}d\lambda\\
&= E\int_0^\infty\frac{\exp\left\{r\left(\frac{Ay}{\sqrt{B^2+1}}-\frac{y^2}{2}\right)\right\}}{yL\left(\frac{y}{\sqrt{B^2+1}}\vee\frac{\sqrt{B^2+1}}{y}\right)}dy \qquad \left(\text{by Fubini, letting } y=\lambda\sqrt{B^2+1}\right)
\end{aligned}
$$

$$= E\left\{\exp\left(\frac{rA^2}{2(B^2+1)}\right)\int_{-\frac{A}{\sqrt{B^2+1}}}^{\infty}\frac{\exp\left(-rx^2/2\right)}{\left(x+\frac{A}{\sqrt{B^2+1}}\right)L\left(\frac{x+\frac{A}{\sqrt{B^2+1}}}{\sqrt{B^2+1}}\vee\frac{\sqrt{B^2+1}}{x+\frac{A}{\sqrt{B^2+1}}}\right)}dx\right\}$$

$$\geq E\left\{\frac{I\left(\frac{A}{\sqrt{B^2+1}}\geq 1\right)\exp\left(\frac{rA^2}{2(B^2+1)}\right)\int_{-1}^{0}\exp\left(-rx^2/2\right)dx}{3\frac{A}{\sqrt{B^2+1}}\left(L\left(\frac{A}{B^2+1}\right)\vee L\left(\sqrt{B^2+1}\vee\frac{1}{\sqrt{B^2+1}}\right)\right)}\right\} \quad \text{by Lemma 11.1}$$

$$\geq E\frac{g_r\left(\frac{A}{\sqrt{B^2+1}}\right)}{4\frac{A}{\sqrt{B^2+1}}\left(L\left(\frac{A}{\sqrt{B^2+1}}\right)\vee L(\sqrt{B^2+1})\right)},$$

where the second equality is obtained via the change of variables $x = y - \frac{A}{\sqrt{B^2+1}}$ and the last one uses the fact that $\sqrt{B^2+1} \geq 1/\sqrt{B^2+1}$. Replacing $g(x)$ by $g_r(x)$ and using the same argument as that in the proof of Theorem 12.6,

$$\begin{aligned}
Eh\left(\sup_{0\leq t\leq\tau}\frac{A_t^+}{\sqrt{B_t^2+1}}\right) &= \lim_{n\to\infty} Eh\left(\frac{A_{t_n}^+}{\sqrt{B_{t_n}^2+1}}\right)\\
&\leq \sup_{y\geq 1}\frac{4y\left(L(y)\vee g_r^{1-\delta}(y)\right)}{(1-r)g_r(y)}h(y)\\
&\quad + \lim_{n\to\infty} Eh\left(1\vee\sqrt{\frac{2}{r(1-\delta)}\log^+ L(\sqrt{B_{t_n}^2+1})}\right)\\
&\leq \sup_{y\geq 1}\frac{4y\left(L(y)\vee g_r^{1-\delta}(y)\right)}{(1-r)g_r(y)}h(y)\\
&\quad + Eh\left(1\vee\sqrt{\frac{2}{r(1-\delta)}\log^+ L(\sqrt{B_\tau^2+1})}\right),
\end{aligned}$$

where the last inequality follows because $t_n \leq \tau$ and B_t is increasing in t, giving (12.10). The proof of (12.11) is similar and follows by replacing $\sqrt{B_t^2+1}$ with $(\sqrt{B_s^2+1})\sqrt{1\vee\log^+ L(\sqrt{B_s^2+1})}$. □

In Theorems 12.6 and 12.8, the assumed growth rate of h is related to g (or g_r) and not to L. The next two theorems, due to de la Peña et al. (2004), relate the growth rate to both g and L and give analogs of (12.8) and (12.10) in this case.

Theorem 12.9. *Let h be a nondecreasing function on $[0,\infty)$ such that for some $x_0 \geq 1$ and $c > 0$,*

$$0 < h(x) \leq cg(x)/L(x) \qquad \textit{for all } x \geq x_0. \tag{12.12}$$

Let q be a strictly increasing, continuous function on $[0,\infty)$ such that for some $\bar{c} \geq c$,

$$L(x) \leq q(x) \leq \frac{\bar{c}g(x)}{h(x)} \qquad \text{for all } x \geq x_0. \tag{12.13}$$

Let $B > 0$ and A be random variables satisfying (10.6) for all $\lambda > 0$. Then

$$Eh(A^+/B) \leq 4\bar{c} + h(x_0) + Eh\left(q^{-1}\left(L(B \vee B^{-1})\right)\right). \tag{12.14}$$

Consequently, $Eh(A^+/B) < \infty$ if $Eh(q^{-1}(L(B \vee B^{-1}))) < \infty$.

Proof. By Lemma 11.2,

$$E\frac{g(A^+/B)}{L(A/B) \vee L(B \vee 1/B)} \leq 4.$$

Let $Q = \{L(B \vee B^{-1}) \leq q(A/B)\}$. Then $Eh(A^+/B)$ is majorized by

$$\begin{aligned}
&h(x_0) + E\left\{\frac{h(A^+/B)I(Q)I(A/b \geq x_0)}{g(A/B)/\left(L(A/B) \vee L(B \vee 1/B)\right)}\right.\\
&\quad \left. \times \left(\frac{g(A/B)}{L(A/B) \vee L(B \vee 1/B)}\right)\right\} + Eh\left\{\left(\frac{A^+}{B}\right) I(Q^c) I\left(\frac{A}{B} \geq x_0\right)\right\}\\
&\leq h(x_0) + \sup_{y \geq x_0} \frac{h(y)\left(L(y) \vee q(y)\right)}{g(y)}\\
&\quad \times E\left(\frac{g(A/B)}{L(A/B) \vee L(B \vee 1/B)}\right) + Eh\left(q^{-1}\left(L\left(B \vee \frac{1}{B}\right)\right)\right)\\
&\leq h(x_0) + 4\sup_{y \geq x_0} \frac{h(y)q(y)}{g(y)} + Eh\left(q^{-1}\left(L\left(B \vee \frac{1}{B}\right)\right)\right).
\end{aligned}$$

□

We next consider the case where the canonical assumption (10.6) holds only for $0 < \lambda < \lambda_0$. Some new ideas are needed because A^+/B may fall outside this range. In this connection we also generalize the canonical assumption (10.6) by replacing the quadratic function $\lambda^2 B^2/2$ and the upper bound 1 by a convex function $\Phi(\lambda B)$ and a finite positive constant c. Unlike Theorem 12.9 that involves a single function q to give the upper bound (12.14), a new idea to handle the restricted range $0 < \lambda < \lambda_0$ is to use a family of functions q_b.

Theorem 12.10. *Suppose that $\Phi(\cdot)$ is a continuous function with $\Phi'(x)$ strictly increasing, continuous and positive for $x > 0$, with $\lim_{x\to\infty} \Phi(x) = \infty$ and $\sup_{x>0} \Phi''(x) < \infty$. Let $B > 0$ and A be random variables such that there exists $c > 0$ for which*

$$E\exp\{\lambda A - \Phi(\lambda B)\} \leq c \qquad \text{for all } 0 < \lambda < \lambda_0. \tag{12.15}$$

For $w > \Phi'(1)$, define y_w by the equation $\Phi'(y_w) = w$, and let

$$g_\Phi(w) = y_w^{-1} \exp\{wy_w - \Phi(y_w)\}. \tag{12.16}$$

Let $\eta > \tilde{\eta} > 0$. Let $h : [0,\infty) \to (0,\infty)$ be a nondecreasing function. For $b \geq \eta$, let q_b be a strictly increasing, continuous function on $(0,\infty)$ such that for some $\tilde{c} > 0$ and $w_0 > \Phi'(2)$,

$$q_b(w) \leq \tilde{c}\{g_\Phi(w)I(y_w \leq \lambda_0 b) + e^{\lambda_0 \tilde{\eta} w} I(y_w > \lambda_0 b)\}/h(w) \tag{12.17}$$

for all $w \geq w_0$. Then there exists a constant C depending only on $\lambda_0, \eta, \tilde{\eta}, c, \tilde{c}$ and Φ such that

$$Eh\left(A^+/(B \vee \eta)\right) \leq C + h(w_0) + Eh\left(q_{B\vee\eta}^{-1}\left(L(B \vee \eta)\right)\right). \tag{12.18}$$

Proof. The proof uses two variants of Lemma 11.2. We split $A/B \geq w_0$ into two cases: $y_{A/B} > \lambda_0 B$ and $y_{A/B} \leq \lambda_0 B$. Since $\Phi(x)$ is increasing in $x > 0$, (12.15) holds with B replaced by $B \vee \eta$ and, therefore, we shall assume without loss of generality that $B \geq \eta$. Integrating (12.15) with respect to the probability density function (11.7) yields

$$c \geq E \int_0^{\lambda_0} \frac{\exp\{\lambda A - \Phi(\lambda B)\}}{\lambda L(\lambda \vee \lambda^{-1})} d\lambda = E \int_0^{\lambda_0 B} \frac{\exp\{xA/B - \Phi(x)\}}{xL(x/B \vee B/x)} dx \tag{12.19}$$

The first variant of Lemma 11.2, given in (12.20), provides an exponential bound for A/B when $\lambda_0 B < y_{A/B}$. Observe that using the definition of y_w, we have that $x\frac{A}{B} - \Phi(x)$ increases in x for $x \leq y_{A/B}$, and decreases in x for $x \geq y_{A/B}$. Take any $0 < \tilde{\eta} < \eta$, and let $\lambda_1 = \lambda_0 \vee \lambda_0^{-1} \vee \tilde{\eta}$. Since $B \geq \eta > \tilde{\eta}$, it follows from (12.19) and (11.3) that

$$\begin{aligned}
c \geq & E \int_{\lambda_0 \eta}^{\lambda_0 \tilde{\eta}} \frac{\exp\{xA/B - \Phi(x)\}}{xL(x/B \vee B/x)} dx I\left(\frac{A}{B} \geq w_0\right) I(y_{A/B} > \lambda_0 B) \\
\geq & E \int_{\lambda_0 \eta}^{\lambda_0 \tilde{\eta}} \frac{\exp\{\lambda_0 \tilde{\eta} A/B - \Phi(\lambda_0 \eta)\}}{L(\lambda_0 \vee B/(\lambda_0 \tilde{\eta}))} \frac{dx}{x} I\left(\frac{A}{B} \geq w_0\right) I(y_{A/B} > \lambda_0 B) \\
\geq & \frac{e^{-\Phi(\lambda_0 \tilde{\eta})}}{3\lambda_1/\tilde{\eta}} \log\left(\frac{\eta}{\tilde{\eta}}\right) E\left\{\frac{e^{\lambda_0 \tilde{\eta} A/B}}{L(B)} I\left(\frac{A}{B} \geq w_0\right) I(y_{A/B} > \lambda_0 B)\right\}.
\end{aligned} \tag{12.20}$$

The second variant of Lemma 11.2, given in (12.22), bounds A/B when $\lambda_0 B \geq y_{A/B}$. Since $w_0 > \Phi'(2), y_{w_0} > 2$. Define

$$a_* = \sup\{a \leq 1 : a^2 \Phi''(x) \leq 1 \text{ for all } x > y_{w_0} - a\}. \tag{12.21}$$

Note that $a_* > 0$ and $y_{w_0} - a_* > 1$. Since $\Phi'(y_w) - w = 0$, a two-term Taylor expansion for $w \geq w_0$ and $x \in (y_w - a_*, y_w)$ yields

$$\begin{aligned} wx - \Phi(x) &= wy_w - \Phi(y_w) - \frac{(x-y_w)^2}{2}\Phi''(\xi^*) \\ &\geq wy_w - \Phi(y_w) - \frac{(x-y_w)^2}{2a_*^2}, \end{aligned}$$

in which ξ^* lies between x and y_w. The last inequality follows from (12.17) and (12.21), noting that $\xi^* > x > y_w - a_* \geq y_{w_0} - a_*$. It then follows from (12.19) that

$$\begin{aligned} c &\geq E\Bigg[I\left(y_{A/B} \leq \lambda_0 B, \frac{A}{B} \geq w_0 \right) \\ &\quad \times \int_{y_{A/B}-a_*}^{y_{A/B}} \frac{\exp\{(A/B)y_{A/B} - \Phi(y_{A/B}) - (x - y_{A/B})^2/(2a_*^2)\}}{xL(x/B \vee B/x)} dx \Bigg] \\ &\geq E\Bigg[I\left(y_{A/B} \leq \lambda_0 B, \frac{A}{B} \geq w_0 \right) \\ &\quad \times \frac{\exp\{(A/B)y_{A/B} - \Phi(y_{A/B})\}}{y_{A/B}\{L(\lambda_0 \vee B)\}} \int_{y_{A/B}-a_*}^{y_{A/B}} \exp\left\{ -\frac{(x - y_{A/B})^2}{2a_*^2} \right\} dx \Bigg], \end{aligned}$$

using $x > y_{A/B} - a_* \geq y_{w_0} - a_* > 1$ so that $\frac{B}{x} < B$. From Lemma 11.1 and the fact that $B \geq \eta$, we have $L(\lambda_0 \vee B) \leq 3(1 \vee \frac{\lambda_0}{\eta})L(B)$. Hence,

$$\begin{aligned} c &\geq E\Bigg[I\left(y_{A/B} \leq \lambda_0 B, \frac{A}{B} \geq w_0 \right) \\ &\quad \times \frac{\exp\{(A/B)y_{A/B} - \Phi(y_{A/B})\}}{3y_{A/B}(1 \vee (\lambda_0/\eta))L(B)} a_* \int_0^1 \exp\left(-\frac{z^2}{2} \right) dz \Bigg] \\ &\geq \frac{a_*}{4(1 \vee (\lambda_0/\eta))} E \frac{g_\Phi(y_{A/B}) I(y_{A/B} \leq \lambda_0 B, A/B \geq w_0)}{L(B)}. \end{aligned} \tag{12.22}$$

Let $Q = \{L(B) \leq q_B(A/B)\}$. Then rewriting (12.17) as an upper bound for h and using the definition of Q, we can majorize $Eh(A^+/B)$ by

$$\begin{aligned} &h(w_0) + \tilde{c}E\Bigg[I(Q) \Bigg\{ \frac{g_\Phi(A/B)}{L(B)} I\left(\frac{A}{B} \geq w_0, y_{A/B} \leq \lambda_0 B \right) \\ &\quad + \frac{e^{\lambda_0 \tilde{\eta} A/B}}{L(B)} I\left(\frac{A}{B} \geq w_0, y_{A/B} > \lambda_0 B \right) \Bigg\} \Bigg] \\ &\quad + Eh\left(\frac{A}{B} \right) I\left(Q^c \cap \left\{ \frac{A}{B} \geq w_0 \right\} \right) \leq h(w_0) + C + Eh\left(q_B^{-1}(L(B)) \right), \end{aligned}$$

in which the inequality follows from (12.17), (12.20) and (12.22). □

While Theorem 12.10 provides an analog of (12.8) when the canonical assumption (10.6) holds only for the restricted range $0 < \lambda < \lambda_0$ and generalizes (10.6) to (12.5) in this connection, de la Peña et al. (2004) have also provided an analog of (12.11) when the canonical assumption (10.5) holds only for $0 < \lambda < \lambda_0$. They use ideas similar to those in the proof of Theorem 12.10 and have accordingly generalized (10.5) to (10.23) in the following.

Theorem 12.11. *Let $T = \{0,1,2,\dots\}$ or $T = [0,\infty)$, $1 < r \le 2$, and $\Phi_r(x) = x^r/r$ for $x > 0$. Let A_t, B_t be stochastic processes (on the same probability space) such that B_t is positive and nondecreasing in $t > 0$, with $A_0 = B_0 = 0$, and*

$$\{\exp(\lambda A_t - \Phi_r(\lambda B_t)),\ t \in T\} \text{ is a supermartingale for } 0 < \lambda < \lambda_0. \tag{12.23}$$

In the case $T = [0,\infty)$, assume furthermore that A_t and B_t are right-continuous. Let $\eta > 0$, $\lambda_0\eta > \varepsilon > 0$ and $h : [0,\infty) \to (0,\infty)$ be a nondecreasing function such that $h(x) \le e^{\varepsilon x}$ for all large x. Then there exists a constant C depending only on $\lambda_0, \eta, r, \varepsilon, h$ and L such that

$$Eh\left(\sup_{t\ge 0}\left\{A_t(B_t \vee \eta)^{-1}\left[1 \vee \log^+ L(B_t \vee \eta)\right]^{-(r-1)/r}\right\}\right) \le C. \tag{12.24}$$

12.3 Applications and Examples

12.3.1 Proof of Lemma 8.11

Let $A = \sum_{i=1}^n \xi_i$, $B^2 = \sum_{i=1}^n \xi_i^2 + \sum_{i=1}^n E(\xi_i^2|\mathscr{F}_{i-1})$. By Theorem 9.21, (A,B) satisfies the canonical assumption (10.6) for all $\lambda \in \mathbb{R}$. Therefore we can apply (12.4) to conclude that

$$E\exp\left\{\theta|A|\Big/\sqrt{B^2+(EB)^2}\right\} \le \sqrt{2}\exp(\theta^2) \qquad \text{for all } \theta > 0.$$

Noting that $EB^2 \ge (EB)^2$ and setting $\theta = x/2$, application of Markov's inequality then yields

$$\begin{aligned}
&P\left\{\frac{|\sum_{i=1}^n \xi_i|}{\left[\sum_{i=1}^n\left(\xi_i^2 + E(\xi_i^2|\mathscr{F}_{i-1}) + 2E\xi_i^2\right)\right]^{1/2}} \ge x\right\} \\
&\le P\left\{|A|\Big/\sqrt{B^2+(EB)^2} \ge x\right\} \le e^{-\theta x}E\exp\left\{\theta|A|\Big/\sqrt{B^2+(EB)^2}\right\} \\
&\le \sqrt{2}\exp(-\theta x + \theta^2) = \sqrt{2}e^{-x^2/4}.
\end{aligned}$$

12.3.2 Generalizations of Theorems 12.1, 12.2 and 12.3

Suppose $B > 0$ and A satisfy (10.6) for all $\lambda \in \mathbb{R}$. Let $p > 0$. Then by Theorem 12.6 with $h(x) = |x|^p$, there exist constants C_p, $C_{1,p}$ and $C_{2,p}$, depending only on p, such that

$$E\left|\frac{A^+}{B}\right|^p \leq C_{1,p} + C_{2,p} E\left(\log^+ L\left(B \vee \frac{1}{B}\right)\right)^{p/2}, \tag{12.25}$$

which is a consequence of (12.8), and

$$E\left(\frac{|A|}{B\sqrt{1 \vee \log^+ L(B \vee B^{-1})}}\right)^p \leq C_p, \tag{12.26}$$

which is a consequence of (12.9). Note that (12.25) addresses the case $q = p$ in Theorem 12.1. Moreover, the upper bound (12.2) for a continuous local martingale M_t is the special case, corresponding to $p = 1$, of Theorem 12.8 with $h(x) = |x|^p$, $A_t = M_t$ (and also $A_t = -M_t$), $B_t^2 = \langle M \rangle_t$ and $L(x) = 2(\log x e^e)(\log\log x e^e)^2 I(x \geq 1)$, for which (12.10) reduces to

$$E\left(\sup_{0 \leq t \leq \tau}\left(\frac{|M_t|}{\sqrt{1 + \langle M \rangle_t}}\right)^p\right) \leq C_{1,p} + C_{2,p} E\left(\log\left(1 + \log(1 + \tau)\right)\right)^{p/2}. \tag{12.27}$$

Example 12.12. Let $\{Y_i\}$ be a sequence of i.i.d. random variables with $P(Y_i = 1) = P(Y_i = -1) = \frac{1}{2}$ and $T = \{\inf n \geq e^e : \sum_{j=1}^n Y_j \geq \sqrt{2n \log\log n}\}$ with $T = \infty$ if no such n exists. By a result of Erdös (1942), $P(T < \infty) = 1$. Let $X_{n,j} = Y_j I(T \geq j)$ for $1 \leq j \leq n$ with $X_{n,j} = 0$ when $j > n$. Then

$$\frac{X_{n,1} + \cdots + X_{n,n}}{\sqrt{X_{n,1}^2 + \cdots + X_{n,n}^2}} = \frac{\sum_{i=1}^{T \wedge n} Y_i}{\sqrt{T \wedge n}} \to \sqrt{2 \log\log T}. \tag{12.28}$$

This shows that (12.25) is sharp.

Whereas the preceding applications have followed Theorems 12.1 and 12.2 and focused on the L_p-norms of self-normalized processes, we can choose h to be an exponential function in the results of Sect. 12.2.2, as illustrated in the following.

Example 12.13. Supposed $B > 0$ and (A, B) satisfies the canonical assumption (10.6) for all $\lambda > 0$. Let $0 < \theta < 1$ and $h(x) = \exp(\theta x^2/2)$ for $x \geq 0$. With this choice of h and with L defined by (11.6), it follows from Theorem 12.9 that

$$E \exp\left[\frac{\theta}{2}\left(\frac{A^+}{B}\right)^2\right] < \infty$$

$$\text{if } E\left\{(\log \tilde{B})(\log\log \tilde{B})^{3/2}(\log\log\log \tilde{B})^{1+\delta}\right\}^{\theta/(1-\theta)} < \infty \tag{12.29}$$

for some $\delta > 0$, where $\tilde{B} = B \vee B^{-1} \vee e^3$; see Problem 12.2.

We next consider the canonical assumption (10.5) for all $\lambda > 0$ and extend Theorem 12.3 due to Kikuchi (1991) by applying Theorem 12.8.

Example 12.14. Consider the case of continuous local martingales A_t. We can apply Theorem 12.8 with $B_t = \sqrt{\langle A \rangle_t}$, in view of Lemma 10.2. Putting $h(x) = \exp(\alpha x^2)$, with $0 < \alpha < \frac{1}{2}$, in (12.11) yields an absolute constant $C(\alpha)$ such that

$$E\left[\sup_{t\geq 0} \exp\left(\frac{\alpha A_t^2}{(\langle A \rangle_t + 1)\log\log(\langle A \rangle_t + e^2)}\right)\right] \leq C(\alpha)$$

which can be regarded as an extension to $p = 0$ of Theorem 12.3.

12.3.3 Moment Inequalities Under Canonical Assumption for a Restricted Range

Section 10.2 has described a number of models that satisfy (10.5) or (10.6) only for the restricted range $0 < \lambda < \lambda_0$. The following example applies Theorem 12.10 to handle this case.

Example 12.15. Suppose $B > 0$ and (A, B) satisfies (10.6) only for $0 \leq \lambda \leq \lambda_0$. Thus, (12.15) holds with $\Phi(x) = x^2/2$ and g_Φ reduce to the function g defined by (11.10) in this case, noting that $y_w = w$. Let $p > 0$ and $h(x) = x^p$ for $x \geq 0$. For $b \geq \eta > \tilde{\eta} > 0$, let q_b be a strictly increasing function on $(0, \infty)$ such that for all large b,

$$\begin{aligned} q_b(w) &= e^{w^2/2}/w^{p+1} && \text{if } w \leq \lambda_0(\tilde{\eta} b)^{1/2}, \\ &\leq e^{w^2/2}/w^{p+1} && \text{if } \lambda_0(\tilde{\eta} b)^{1/2} < w \leq \lambda_0, \\ &\leq e^{\lambda_0 \tilde{\eta} w}/w^p && \text{if } w > \lambda_0 b. \end{aligned} \tag{12.30}$$

The inequality (12.13) holds with $\tilde{c} = 1$. Using (11.6) as the choice of L, it follows from (12.30) that $q_b^{-1}(L(b)) \sim (2\log\log b)^{1/2}$ as $b \to \infty$. Therefore, by Theorem 12.10,

$$E\left(A^+/(B \wedge \eta)\right)^p < \infty \text{ if } E\left\{\log\left(|\log(B \wedge \eta)| \wedge e\right)\right\}^{p/2} < \infty. \tag{12.31}$$

Similarly, letting $h(x) = e^{\xi x}$ with $0 < \xi < \lambda_0 \tilde{\eta}$, it follows from Theorem 12.10 that

$$\begin{aligned} &E\exp\left(\xi A^+/(B \vee \eta)\right) < \infty \\ &\text{if } E\exp\left\{\xi\left[2(\log\log\tilde{B})(\log\log\log\tilde{B})^{1+\delta}\right]^{1/2}\right\} \end{aligned} \tag{12.32}$$

for some $\delta > 0$, where $\tilde{B} = B \vee e^3$.

12.4 Supplementary Results and Problems

1. Prove Lemma 12.7.
2. Explain how (12.29) can be derived from Theorem 12.9.
3. One choice of q_b that satisfies (12.30) for sufficiently large b is to let $q_b(w) = w^{-p}\exp\left(f^2(w)\right)$ for $\lambda_0(\tilde{\eta}b)^{1/2} < w \le \lambda_0 b$, where f is linear on $[\lambda_0(\tilde{\eta}b)^{1/2}, \lambda_0 b]$ and is uniquely determined by requiring q_b to be continuous. Show that in this case $f^2(w) \le w^2/2 - \log w$ for $\lambda_0(\tilde{\eta}b)^{1/2} \le w \le \lambda_0 b$ if b is sufficiently large.

Chapter 13
Laws of the Iterated Logarithm for Self-Normalized Processes

In this chapter we first give Stout's (1973) generalization of Kolmogorov's law of the iterated logarithm (LIL) for sums of independent zero-mean random variables with finite variances (see (2.2)) to martingales that are self-normalized by the conditional variances. We then consider self-normalization by a function of the sum of squared martingale differences as in de la Peña et al. (2004). This self-normalization yields a universal upper LIL that is applicable to all adapted sequences. In the case of martingales satisfying certain boundedness assumptions, a compact LIL is then derived.

13.1 Stout's LIL for Self-Normalized Martingales

Stout's LIL for martingales involves normalization by a function of the conditional variance $s_n^2 = \sum_{i=1}^n E(d_i^2|\mathscr{F}_{i-1})$. In the case where the d_i's are independent zero-mean random variables, the conditional variance is equal to the variance and hence Stout's result, which is stated in the following theorem, is a generalization of Kolmogorov's LIL. Let $m_n = \sqrt{2\log\log(e^2 \vee s_n^2)}$.

Theorem 13.1. *Let $M_n = \sum_{i=1}^n d_i$ be a martingale with respect to a filtration $\{\mathscr{F}_n\}$. There exists a function $\varepsilon(\cdot)$ such that $\varepsilon(x)$ decreases to 0 as $x \downarrow 0$ and for some constants $0 < K \leq \frac{1}{2}$ and $\varepsilon(K) < 1$,*

$$\limsup \frac{M_n}{s_n m_n} \leq 1 + \varepsilon(K) \quad a.s.$$

whenever:

(1) $s_n^2 < \infty$ a.s. for each $n \geq 1$ and $s_n^2 \to \infty$ a.s.
(2) $d_i \leq K_i s_i / m_i$ and $\limsup K_i < K$ a.s., where K_i is $\mathscr{F}_{i-1}$-measurable.

Before proceeding with the proof, we present a simple corollary.

V.H. de la Peña et al., *Self-Normalized Processes: Limit Theory and Statistical Applications*, Probability and its Applications,

© Springer-Verlag Berlin Heidelberg 2009

Corollary 13.2. *Let K_i be $\mathscr{F}_{i-1}$–measurable for all $i \geq 1$. Assume that $K_i \to 0$ a.s. If $s_n^2 < \infty$, $s_n^2 \to \infty$ a.s. and*

$$Y_i \leq \frac{K_i s_i}{m_i} \quad a.s. \qquad \text{for all } i \geq 1, \tag{13.1}$$

then

$$\limsup \frac{M_n}{s_n m_n} \leq 1 \ a.s. \tag{13.2}$$

Stout (1970) has shown that equality in fact holds in (13.2) if Y_i in (13.1) is replaced by its absolute value.

Lemma 13.3. *Let $f_K(x) = (1+x)^2[1 - K(1+x)/2] - 1$ for $0 < x \leq 1$, $0 < K \leq \frac{1}{2}$. Then $f_K(x)$ is an increasing function satisfying $f_K(0) < 0$ and $f_K(1) > 0$ for each $0 < K \leq \frac{1}{2}$. Let $\varepsilon(K)$ be the zero of $f_K(\cdot)$ for each $0 < K \leq \frac{1}{2}$. Then $1 > \varepsilon(K)$ for each $0 < K \leq \frac{1}{2}$ and $\varepsilon(K)$ decreases to 0 as $K \to 0$.*

Proof (of Theorem 13.1). We use a truncation argument so that Theorem 9.18 can be applied. Let $d_i' = d_i I(K_i \leq K)$ for $i \geq 1$ and $M_n' = \sum_{i=1}^n d_i'$ for $n \geq 1$. Since $\limsup_i K_i < K$, it suffices to show that

$$\limsup \frac{M_n'}{s_n m_n} \leq 1 + \varepsilon(K) \ a.s. \tag{13.3}$$

To prove (13.3), we introduce a sequence of stopping times that allow us to replace the random quantities s_i, m_i by constants. Once this is done we are able to use the supermartingale in Theorem 9.18.

Take any $p > 1$. For $k \geq 1$, let τ_k be the smallest n for which $s_{n+1}^2 \geq p^{2k}$. Note that s_{n+1}^2 is $\mathscr{F}_n$-measurable and hence τ_k is a stopping time. Let $M_n^{(k)} = M'_{\tau_k \wedge n}$. Then $\{M_n^{(k)}, \mathscr{F}_n, n \geq 1\}$ is a supermartingale. Note that

$$\frac{s_{\tau_{k-1}+1} + m_{\tau_{k-1}+1}}{2p^{2k} \log\log(e^2 \vee p^{2k})} \geq \frac{p^{-2} \log\log(e^2 \vee p^{2(k-1)})}{\log\log(e^2 \vee p^{2k})} \approx p^{-2}.$$

Let $\delta' > 0$ and pick $\delta > 0$ in such a way that $(1+\delta)p^{-1} > 1 + \delta'$. Then $P(M_n' > (1+\delta) s_n m_n \ i.o.)$ is bounded above by

$$\begin{aligned}
&P\left(\sup_{\tau_k \geq n \geq 0} M_n' > (1+\delta) s_{\tau_{k-1}+1} m_{\tau_{k-1}+1} \ i.o. \right) \\
&= P\left(\sup_{n \geq 0} M_n^{(k)} > (1+\delta) s_{\tau_{k-1}+1} m_{\tau_{k-1}+1} \ i.o. \right) \\
&\leq P\left(\sup_{n \geq 0} M_n^{(k)} > (1+\delta') \left[2p^{2k} \log\log(e^2 \vee p^{2k}) \right]^{1/2} \ i.o. \right),
\end{aligned}$$

in which "i.o." means "for infinitely many k's." Therefore, by the Borell–Cantelli lemma, it suffices to show that for all $\delta' > \varepsilon(K)$,

$$\sum_{k=1}^{\infty} P\left(\sup_{n\geq 0} M_n^{(k)} > (1+\delta')\left[2p^{2k}\log\log(e^2 \wedge p^{2k})\right]^{1/2}\right) < \infty. \tag{13.4}$$

For k large enough, using the definition of τ_k and the fact that s_n/m_n is non-decreasing and the bound on the d_i's, we obtain

$$M_{n+1}^{(k)} - M_n^{(k)} \leq \frac{Ks_{\tau_k}}{m_{\tau_k}} \leq \frac{Kp^k}{[2\log\log(e^2 \wedge p^{2k})]^{1/2}} \quad a.s.$$

Let $(s_n^{(k)})^2 = \sum_{i=1}^n E[(M_i^{(k)} - M_{i-1}^{(k)})^2 | \mathscr{F}_{i-1}]$ for all $n \geq 1$. We apply Theorem 9.18 to the martingale $\{M_n^{(k)}, \mathscr{F}_n, n \geq 1\}$ for k large enough, with

$$\begin{aligned} c &= Kp^k/[2\log\log(e^2 \wedge p^{2k})]^{1/2}, \\ \lambda &= (1+\delta')[2\log\log(e^2 \wedge p^{2k})]^{1/2}/p^k. \end{aligned} \tag{13.5}$$

Set

$$T_n = \exp\left(\lambda M_n^{(k)}\right)\exp\left[-\frac{\lambda^2}{2}\left(1+\frac{\lambda c}{2}\right)\left(s_n^{(k)}\right)^2\right]$$

for $n \geq 1$ and $T_0 = 0$. If we choose $\delta' \leq 1$, then $\lambda c = (1+\delta')K \leq 1$ by (13.5) and the assumption $K \leq \frac{1}{2}$. Finally, note that $\sup_{n\geq 1}(s_n^{(k)})^2 \leq s_{\tau_k}^2$ *a.s.*

$$\begin{aligned} &P\left(\sup_{n\geq 0} M_n^{(k)} > (1+\delta')\left[2p^{2k}\log\log(e^2 \wedge p^{2k})\right]^{1/2}\right) \\ &\quad = P\left(\sup_{n\geq 0} M_n^{(k)} > \lambda p^{2k}\right) \\ &\quad = P\left(\sup_{n\geq 0} \exp(\lambda M_n^{(k)}) > \exp(\lambda^2 p^{2k})\right) \\ &\quad \leq P\left(\sup_n T_n > \exp\left[\lambda^2 p^{2k} - \lambda^2\left(1+\frac{\lambda c}{2}\right)s_{\tau_k}^2\right]\right) \\ &\quad \leq P\left(\sup_n T_n > \exp\left[\lambda^2 p^{2k} - \lambda^2\left(1+\frac{\lambda c}{2}\right)p^{2k}\right]\right) \\ &\quad \leq \exp\left(-\lambda^2 p^{2k} + \lambda^2\left(1+\frac{\lambda c}{2}\right)p^{2k}\right), \end{aligned}$$

by Theorem 9.18. Putting in the values of λ and c given in (13.5), the above upper bound reduces to

$$\exp\left\{-(1+\delta')^2\left[1-\frac{1}{2}K(1+\delta')\right]\log\log(e^2 \vee p^{2k})\right\}.$$

Note that $(1+\delta')^2[1-K(1+\delta')/2] - 1 > 0$ for all $1 \geq \delta' > \varepsilon(K)$. Pick such δ'. Then there exists $\beta > 1$ such that

$$P\left(\sup_{n\geq 0} M_n^{(k)} > (1+\delta')\left[2p^{2k}\log\log(e^2\vee p^{2k})\right]^{1/2}\right) \leq \exp\left[-\beta\log\log(e^2\wedge p^{2k})\right]$$
$$= (2k\log p)^{-\beta}$$

for k large enough. Therefore $\sum_{k=1}^{\infty}(2k\log p)^{-\beta} < \infty$, completing the proof. □

13.2 A Universal Upper LIL

In this section we describe a universal upper LIL (Theorem 13.5) developed by de la Peña et al. (2004). When a partial sum of random variables $X_1, X_2, \ldots$ is centered and normalized by a sequence of constants, only under rather special conditions does the usual LIL hold even if the variables are i.i.d. In contrast, Theorem 13.5 shows that there is a universal upper bound of LIL type for the almost sure rate at which such sums can grow after centering by a sum of conditional expectations of suitably truncated variables and normalizing by the square root of the sum of squares of the X_j's. Specifically, let $S_n = X_1 + X_2 + \cdots + X_n$ and $V_n^2 = X_1^2 + X_2^2 + \cdots + X_n^2$, where $\{X_i\}$ is adapted to the filtration $\{\mathscr{F}_i\}$. In Theorem 13.5 we prove that given any $\lambda > 0$, there exist positive constants a_λ and b_λ such that $\lim_{\lambda\to\infty} b_\lambda = \sqrt{2}$ and

$$\limsup \frac{\{S_n - \sum_{i=1}^n \mu_i(-\lambda v_n, a_\lambda v_n)\}}{V_n(\log\log V_n)^{1/2}} \leq b_\lambda \ \ a.s. \tag{13.6}$$

on $\{\lim V_n = \infty\}$, where $v_n = V_n(\log\log V_n)^{1/2}$ and $\mu_i(c,d) = E\{X_i I(c \leq X_i < d) \mid \mathscr{F}_{i-1}\}$ for $c < d$. Note that (13.6) is "universal" in the sense that it is applicable to *any* adapted sequence $\{X_i\}$. In particular, suppose $\{S_n, \mathscr{F}_n, n \geq 1\}$ is a supermartingale such that $X_n \geq -m_n$ *a.s.* for some $\mathscr{F}_{n-1}-$ measurable random variable m_n satisfying $P\{0 \leq m_n \leq \lambda v_n$ for all large $n\} = 1$. Then (13.6) yields

$$\limsup \frac{S_n}{V_n(\log\log V_n)^{1/2}} \leq b_\lambda \ \ a.s. \qquad \text{on } \{\lim V_n = \infty\}. \tag{13.7}$$

We derive in Sect. 13.3 the lower half counterpart of (13.7) for the case where $\{S_n, \mathscr{F}_n, n \geq 1\}$ is a martingale such that $|X_n| \leq m_n$ *a.s.* for some $\mathscr{F}_{n-1}$-measurable m_n with $v_n \to \infty$ and $m_n/v_n \to 0$ *a.s.* Combining this with (13.7) (with $\lim_{\lambda\to 0} b_\lambda = \sqrt{2}$) then yields

$$\limsup \frac{S_n}{V_n(\log\log V_n)^{1/2}} = \sqrt{2} \ \ a.s. \tag{13.8}$$

To prove (13.6) for any adapted sequence $\{X_i\}$, one basic technique pertains to upper-bounding the probability of an event of the form $E_k = \{t_k \leq \tau_k < t_{k+1}\}$ in which t_j and τ_j are stopping times defined in (13.10) below. Sandwiching τ_k between t_k and t_{k+1} enables us to replace both the random exceedance and truncation levels in (13.10) by constants. Then the event E_k can be re-expressed in terms of two simultaneous inequalities, one involving centered sums and the other involving a sum of squares. Using these inequalities, we come up with a supermartingale that is then used to bound $P(E_k)$. The supermartingale is already mentioned without proof in Lemma 10.5. Apart from finite mean constraints, Lemma 10.4 gives the basic idea underlying the construction of this supermartingale. Lemma 10.4 corresponds to the case $r = 2$ in the following.

Lemma 13.4. *let* $0 < \gamma < 1 < r \leq 2$. *Define*

$$\begin{aligned} c_r &= \inf\{c > 0 : \exp(x - cx^r) \leq 1 + x \textit{ for all } x \geq 0\}, \\ c_r^{(\gamma)} &= \inf\{c > 0 : \exp(x - c|x|^r) \leq 1 + x \textit{ for all } -\gamma \leq x \leq 0\}, \\ c_{\gamma,r} &= \max\{c_r, c_r^{(\gamma)}\}. \end{aligned}$$

(a) For all $x \geq -\gamma$, $\exp\{x - c_{\gamma,r}|x|^r\} \leq 1 + x$. *Moreover,* $c_r \leq (r-1)^{(r-1)}(2-r)^{(2-r)}/r$ *and*

$$c_r^{(\gamma)} = \frac{-(\gamma + \log(1-\gamma))}{\gamma^r} = \sum_{j=2}^{\infty} \frac{\gamma^{j-r}}{j}.$$

(b) Let $\{d_n\}$ *be a sequence of random variables adapted to the filtration* $\{\mathscr{F}_n\}$ *such that* $E(d_n \mid \mathscr{F}_{n-1}) \leq 0$ *and* $d_n \geq -M$ *a.s. for all n and some nonrandom positive constant M. Let* $A_n = \sum_{i=1}^n d_i$, $B_n^r = rc_{\gamma,r}\sum_{i=1}^n |d_i|^r$, $A_0 = B_0 = 0$. *Then* $\{\exp(\lambda A_n - (\lambda B_n)^r/r), \mathscr{F}_n, n \geq 0\}$ *is a supermartingale for every* $0 \leq \lambda \leq \gamma M^{-1}$.

Proof. The first assertion of (a) follows from the definition of $c_{\gamma,r}$. For $c > 0$, define $g_c(x) = \log(1+x) - x + c|x|^r$ for $x > -1$. Then $g_c'(x) = |x|^{r-1}\{|x|^{2-r}(1-|x|)^{-1} - cr\}$ for $-1 < x < 0$. Since $|x|^{2-r}/(1-|x|)$ is decreasing in $-1 < x < 0$, g_c' has at most one 0 belonging to $(-1,0)$. Let $c^* = -\{\gamma + \log(1-\gamma)\}/\gamma^r$. Then $g_{c^*}(-\gamma) = 0 = g_{c^*}(0)$. It then follows that $g_{c^*}(x) > 0$ for all $-\gamma < x < 0$ and, therefore, $c^* \geq c_r^{(\gamma)}$. If $c^* > c_r^{(\gamma)}$, then $g_{c_r^{(\gamma)}}(-\gamma) < g_{c^*}(-\gamma) = 0$, contradicting the definition of $c_r^{(\gamma)}$. Hence, $c_r^{(\gamma)} = c^*$. Take any $c \geq (r-1)^{(r-1)}(2-r)^{(2-r)}/r$. Then for all $x > 0$,

$$\begin{aligned} g_c'(x) &= \frac{1}{1+x} - 1 + crx^{r-1} \geq \frac{x}{1+x}\left\{-1 + cr\inf_{y>0}\left(y^{r-2} + y^{r-1}\right)\right\} \\ &= \frac{x}{1+x}\left\{-1 + \frac{cr}{(r-1)^{(r-1)}(2-r)^{(2-r)}}\right\} \geq 0. \end{aligned}$$

Since $g_c(0) = 0$, it then follows that $g_c(x) \geq 0$ for all $x \geq 0$. Hence, $c_r \leq (r-1)^{(r-1)}(2-r)^{(2-r)}/r$.

To prove (b), note that since $\lambda d_n \geq -\lambda M \geq -\lambda$ *a.s.* for $0 \leq \lambda \leq \gamma M^{-1}$, (a) yields

$$E\left[\exp\{\lambda d_n - c_{\gamma,r}|\lambda d_n|^r \mid \mathscr{F}_{n-1}\}\right] \leq E[1 + \lambda d_n \mid \mathscr{F}_{n-1}] \leq 1 \ \ a.s.$$

$\square$

Lemma 10.5 is a refinement of Lemma 10.4, in which C_γ corresponds to $c_{\gamma,2}$ in Lemma 13.4(b), by removing the assumptions on the integrability and lower bound on Y_i. Noting that $\exp\{y - y^2/\lambda_i\} \le 1$ if $y \ge \lambda_i$ or if $y < -\gamma_i$, and letting $X_i = Y_i I(-\gamma_n \le Y_n < \lambda_i)$ so that $\mu_i = E(X_i|\mathscr{F}_{i-1})$, we have

$$\begin{aligned} E\left\{\exp(Y_i - \mu_i - \lambda_i^{-1}Y_i^2)|\mathscr{F}_{i-1}\right\} &\le E\left\{\exp(X_i - \mu_i - \lambda_i^{-1}X_i^2)|\mathscr{F}_{i-1}\right\} \\ &\le E\left\{\exp((1+X_i)e^{-\mu_i})|\mathscr{F}_{i-1}\right\} = (1+\mu_i)e^{-\mu_i}, \end{aligned}$$

proving Lemma 10.5 since $(1+\mu_i)e^{-\mu_i} \le 1$. One reason why de la Peña et al. (2004) consider more general $1 < r \le 2$ instead of only $r = 2$ is related to the more general form (12.23) of the canonical assumption, in which $\Phi_r(x) = x^r/r$.

The centering constants in (13.6) involve sums of expectations conditioned on the past which are computed as functions of the endpoints of the interval on which the associated random variable is truncated. The actual endpoints used, however, are neither knowable nor determined until the future. Thus the sequence of centered sums that result is not a martingale. Nevertheless, by using certain stopping times, the random truncation levels can be replaced by non-random ones, thereby yielding a supermartingale structure for which Lemma 10.5 applies, enabling us to establish the following result.

Theorem 13.5. *Let X_n be measurable with respect to $\mathscr{F}_n$, an increasing sequence of σ-fields. Let $\lambda > 0$ and $h(\lambda)$ be the positive solution of*

$$h - \log(1+h) = \lambda^2. \tag{13.9}$$

Let $b_\lambda = h(\lambda)/\lambda$, $\gamma = h(\lambda)/\{1+h(\lambda)\}$ and $a_\lambda = \lambda/(\gamma C_\gamma)$, where C_γ is defined by Lemma 10.4. Then (13.6) *holds on* $\{\lim_{n\to\infty} V_n = \infty\}$ *and* $\lim_{\lambda\to 0} b_\lambda = \sqrt{2}$.

Proof. Recall that $V_n^2 = X_1^2 + \cdots + X_n^2$, $v_n = V_n(\log\log V_n)^{-1/2}$. Let $e_k = \exp(k/\log k)$. Define

$$\begin{aligned} t_j &= \inf\{n : V_n \ge e_j\}, \qquad\qquad (13.10)\\ \tau_j &= \inf\left\{n \ge t_j : S_n - \sum_{i=1}^n \mu_i(-\lambda v_n, a_\lambda v_n) \ge (1+3\varepsilon)b_\lambda V_n(\log\log V_n)^{1/2}\right\}, \end{aligned}$$

letting $\inf\emptyset = \infty$. To prove (13.6), it suffices to show that for all sufficiently small $\varepsilon > 0$,

$$\lim_{K\to\infty} \sum_{k=K}^{\infty} P\{\tau_k < t_{k+1}\} = 0.$$

Note that $\tau_k \ge t_k$ and that t_k may equal t_{k+1}, in which case $\{\tau_k < t_{k+1}\}$ becomes the empty set. Moreover, on $\{\lim_{n\to\infty} V_n = \infty\}$, $t_j < \infty$ for every j and $\lim_{j\to\infty} t_j = \infty$. Since $y(\log\log y)^{-1/2}$ is increasing in $y \ge e_3$, we have the following inequalities on $\{t_k \le \tau_k < t_{k+1}\}$ with $k \ge 3$:

$$e_k \leq \left(\sum_{i=1}^{\tau_k} X_i^2 \right)^{1/2} < e_{k+1}, \tag{13.11}$$

$$d_k := e_k (\log\log e_k)^{-1/2} \leq v_{t_k} \leq v_{\tau_k} < d_{k+1}, \tag{13.12}$$

$$\mu_i(-\lambda v_{\tau_k}, \alpha_\lambda v_{\tau_k}) \geq \mu_i(-\lambda d_{k+1}, \alpha_\lambda d_k) \qquad \text{for } 1 \leq i \leq \tau_k. \tag{13.13}$$

Let $\mu_{i,k} = \mu_i(-\lambda d_{k+1}, \alpha_\lambda d_k)$. We shall replace X_i (for $1 \leq i \leq \tau_k$) by $Y_{i,k} := (\lambda d_{k+1})^{-1} \gamma X_i$ and $\mu_{i,k}$ by $\tilde{\mu}_{i,k} := (\lambda d_{k+1})^{-1} \gamma \mu_i(-\lambda d_{k+1}, \alpha_\lambda d_k)$. Since $\lambda^{-1}\gamma\alpha_\lambda = C_\gamma^{-1}$,

$$\tilde{\mu}_{i,k} = E\left\{ Y_{i,k} 1(-\gamma \leq Y_{i,k} < C_\gamma^{-1} d_k / d_{k+1}) | \mathscr{F}_{i-1} \right\}. \tag{13.14}$$

Since $e_k/d_k = (\log\log e_k)^{1/2}$ and $d_k/d_{k+1} \to 1$ as $k \to \infty$, it follows from (13.11)–(13.13) that for all sufficiently large k, the event $\{\tau_k < t_{k+1}\}$ is a subset of

$$\begin{aligned}
&\left\{ \sum_{i=1}^{\tau_k} (\lambda d_{k+1})^{-1} (X_i - \mu_{i,k}) \geq (1+2\varepsilon) \lambda^{-1} b_\lambda \log\log e_k,\ \tau_k < \infty \right\} \\
&\subset \left\{ \sum_{i=1}^{\tau_k} \left[(\lambda d_{k+1})^{-1} \gamma (X_i - \mu_{i,k}) - C_\gamma (d_{k+1}/d_k)(\lambda d_{k+1})^{-2} \gamma^2 X_i^2 \right] \right. \\
&\qquad \left. \geq (1+2\varepsilon)\gamma\lambda^{-1} b_\lambda \log\log e_k - C_\gamma (d_{k+1}/d_k)(\gamma/\lambda)^2 \log\log e_{k+1},\ \tau_k < \infty \right\} \\
&\subset \left\{ \sup_{n \geq 1} \exp\left[\sum_{i=1}^{n} (Y_{i,k} - \tilde{\mu}_{i,k} - C_\gamma d_k^{-1} d_{k+1} Y_{i,k}^2) \right] \right. \\
&\qquad \left. \geq \exp\left[(1+\varepsilon)(\gamma\lambda^{-1} b_\lambda - C_\gamma \gamma^2 \lambda^{-2})(\log k) \right] \right\}.
\end{aligned}$$

In view of (13.14), we can apply Lemma 10.5 to conclude that the last event above involves the supremum of a non-negative supermartingale with mean ≤ 1. Therefore, application of the supermartingale inequality to this event yields

$$P\{\tau_k < t_{k+1}\} \leq \exp\left\{ -(1+\varepsilon)(\gamma\lambda^{-1} b_\lambda - C_\gamma \gamma^2 \lambda^{-2})(\log k) \right\},$$

which implies (13.6) since

$$\gamma\lambda^{-1} b_\lambda - \lambda^{-2}\gamma^2 C_\gamma = \lambda^{-2} \{\gamma h(\lambda) + \gamma + \log(1-\gamma)\} = 1. \tag{13.15}$$

The first equality in (13.15) follows from $\gamma^2 C_\gamma = -\{\gamma + \log(1-\gamma)\}$ and $b_\lambda = h(\lambda)/\lambda$, and the second equality from $\gamma = h(\lambda)/(1+h(\lambda))$ and (13.9). Moreover, (13.9) implies that $h^2(\lambda) \sim 2\lambda^2$ and, therefore, $b_\lambda \to \sqrt{2}$ as $\lambda \to 0$. □

Remark 13.6. The choice of γ in Theorem 13.5 actually comes from minimizing $\gamma\lambda^{-1} b_\lambda - \lambda^{-2}\gamma^2 C_\gamma$ over $0 < \gamma < 1$, whereas b_λ is employed to make this minimizing value equal to 1, leading to (13.9) that defines $h(\lambda)$.

Another reason why de la Peña et al. (2004) consider more general $1 < r \le 2$ in Lemma 13.4 is to use it to extend Theorem 13.5 to the following result, in which we self-normalize the suitably centered S_n by the more general $(\sum_{i=1}^n |X_i|^r)^{1/2}$.

Theorem 13.7. *Let X_n be measurable with respect to the filtration $\{\mathscr{F}_n\}$. For $1 < r \le 2$, let $V_{n,r} = (\sum_{i=1}^n |X_i|^r)^{1/r}$, $v_{n,r} = V_{n,r}\{\log\log(V_{n,r} \vee e^2)\}^{-1/r}$. Then for any $0 < \gamma < 1$, there exists a positive constant $b_{\gamma,r}$ such that*

$$\limsup_{n\to\infty} \frac{\left\{S_n - \sum_{i=1}^n \mu_i(-\gamma v_{n,r}, c_{\gamma,r}^{-1/(r-1)} v_{n,r})\right\}}{\{V_{n,r}(\log\log V_{n,r})^{(r-1)/r}\}} \le b_{\gamma,r} \quad a.s.$$

on $\{\lim_{n\to\infty} V_{n,r} = \infty\}$, where $c_{\gamma,r}$ is given in Lemma 13.4.

13.3 Compact LIL for Self-Normalized Martingales

Although Theorem 13.5 gives an upper LIL for any adapted sequence $\{X_i\}$, the upper bound in (13.6) may not be attained. A simple example is given in Problem 13.3. In this section we consider the case of martingales $\{S_n, \mathscr{F}_n, n \ge 1\}$ self-normalized by V_n and prove the lower half counterpart of (13.7) when the increments of S_n do not grow too fast, thereby establishing (13.8). This is the content of the following.

Theorem 13.8. *Let $\{X_n\}$ be a martingale difference sequence with respect to the filtration $\{\mathscr{F}_n\}$ such that $|X_n| \le m_n$ a.s. for some $\mathscr{F}_{n-1}$-measurable random variable m_n, with $V_n \to \infty$ and $m_n/\{V_n(\log\log V_n)^{-1/2}\} \to 0$ a.s. Then (13.8) holds.*

Proof. Take $0 < b < \beta < \tilde{\beta} < \sqrt{2}$. Since $1 - \Phi(x) = \exp\{-(\frac{1}{2} + o(1))x^2\}$ as $x \to \infty$, we can choose λ sufficiently large such that

$$\left\{1 - \Phi(\beta\sqrt{\lambda})\right\}^{1/\lambda} \ge \exp(-\tilde{\beta}^2/2), \tag{13.16}$$

where Φ is the standard normal distribution function. Take $a > 1$ and define for $j \ge 2$ and $k = 0, 1, \ldots, [\lambda^{-1}\log j]$,

$$\begin{aligned} a_{j,k} &= a^j + k(a^{j+1} - a^j)/[\lambda^{-1}\log j], \\ t_j(k) &= \inf\{n : V_n^2 \ge a_{j,k}\}. \end{aligned}$$

Let $t_j = \inf\{n : V_n^2 \ge a^j\}$, so $t_j(0) = t_j$, $t_j(\lambda^{-1}\log j) = t_{j+1}$. Since $X_n^2 = o(V_n^2(\log\log V_n)^{-1})$ a.s. and $a_{j,k} \le V_{t_j(k)}^2 < a_{j,k} + X_{t_j(k)}^2$,

$$V_{t_j(k)}^2 = a_{j,k}\left\{1 + o\left((\log j)^{-1}\right)\right\} \quad a.s. \tag{13.17}$$

It will be shown that

$$\frac{\sum_{t_j(k)<n\le t_j(k+1)} X_n^2}{\sum_{t_j(k)<n\le t_j(k+1)} E(X_n^2|\mathscr{F}_{n-1})} \to 1 \tag{13.18}$$

in probability under $P(\cdot|\mathscr{F}_{t_j(k)})$ as $j\to\infty$, uniformly in $0\le k<[\lambda^{-1}\log j]$.

Let $S_{m,n}=\sum_{m<i\le n} X_i$, $V_{m,n}^2=\sum_{m<i\le n} X_i^2$. In view of (13.17),

$$V_{t_j(k),t_j(k+1)}^2 \sim a^j(a-1)/[\lambda^{-1}\log j], \tag{13.19}$$

$$V_{t_j,t_{j+1}}^2 \sim a^j(a-1) \;\; a.s. \tag{13.20}$$

Since X_n^2 is bounded by the $\mathscr{F}_{n-1}$-measurable random variable m_n^2, which is $o(V_n^2(\log\log V_n)^{-1})$ *a.s.*, the conditional Lindeberg condition holds and, in view of (13.18) and (13.19), the martingale central limit theorem (see Sect. 15.3.1) can be applied to yield

$$P\left\{S_{t_j(k),t_j(k+1)} \ge \beta\sqrt{\lambda}V_{t_j(k),t_j(k+1)} \,|\, \mathscr{F}_{t_j(k)}\right\} \to 1-\Phi(\beta\sqrt{\lambda}) \;\; a.s. \tag{13.21}$$

as $j\to\infty$, uniformly in $0\le k<[\lambda^{-1}\log j]$. Since

$$S_{t_j,t_{j+1}} = \sum_{0\le k<[\lambda^{-1}\log j]} S_{t_j(k),t_j(k+1)}$$

and

$$V_{t_j,t_{j+1}}(\log j)^{1/2} = \left(\sqrt{\lambda}+o(1)\right) \sum_{0\le k<[\lambda^{-1}\log j]} V_{t_j(k),t_j(k+1)} \;\; a.s.$$

by (13.19), it follows from (13.21) that as $j\to\infty$,

$$\begin{aligned}
&P\left\{S_{t_j,t_{j+1}} \ge bV_{t_j,t_{j+1}}(\log j)^{1/2} \,|\, \mathscr{F}_{t_j(k)}\right\}\\
&\quad\ge P\left\{S_{t_j(k),t_j(k+1)} \ge \beta\sqrt{\lambda}V_{t_j(k),t_j(k+1)} \text{ for all } 0\le k<[\lambda^{-1}\log j] \,|\, \mathscr{F}_{t_j(k)}\right\}\\
&\quad= \left(1-\Phi(\beta\sqrt{\lambda})+o(1)\right)^{[\lambda^{-1}\log j]}\\
&\quad\ge \exp\left\{-\left(\tilde{\beta}^2/2+o(1)\right)\log j\right\} \;\; a.s.,
\end{aligned}$$

in view of (13.16). Since $\tilde{\beta}^2/2<1$, the conditional Borel–Cantelli lemma (Lemma 9.1(a)) then yields

$$\limsup_{j\to\infty} \frac{S_{t_j,t_{j+1}}}{V_{t_j,t_{j+1}}(\log j)^{1/2}} \ge b \;\; a.s. \tag{13.22}$$

Recalling that $V_n \to \infty$ and $m_n = o(V_n(\log\log V_n)^{-1/2})$ *a.s.*, we obtain from (13.7) that

$$\limsup_{n\to\infty} \frac{S_n}{V_n(\log\log V_n)^{1/2}} \le \sqrt{2} \ a.s. \tag{13.23}$$

and the same conclusion still holds with S_n replaced by $-S_n$ (which is a martingale). Combining this with (13.19) and (13.22) yields

$$\limsup_{j\to\infty} \frac{S_{t_{j+1}}}{V_{t_{j+1}}(\log\log V_{t_{j+1}})^{1/2}} \ge ba^{-1/2}(a-1)^{1/2} - \sqrt{2}a^{-1/2} \ a.s. \tag{13.24}$$

Since a can be chosen arbitrarily large and b arbitrarily close to $\sqrt{2}$ in (13.24),

$$\limsup_{j\to\infty} \frac{S_{t_{j+1}}}{V_{t_{j+1}}(\log\log V_{t_{j+1}})^{1/2}} \ge \sqrt{2} \ a.s.$$

Combining this with the upper half result (13.23) yields (13.8).

It remains to prove (13.18). Let $\alpha_j = a^j(a-1)/[\lambda^{-1}\log j]$. In view of (13.19), we need to show that given any $0<\rho<\frac{1}{2}$ and $\delta>0$,

$$\limsup_{n\to\infty} \left[P\left\{ \sum_{t_j(k)<n\le t_j(k+1)} E(X_n^2|\mathscr{F}_{n-1}) \ge (1+\rho)\alpha_j | \mathscr{F}_{t_j(k)} \right\} + P\left\{ \sum_{t_j(k)<n\le t_j(k+1)} E(X_n^2|\mathscr{F}_{n-1}) \ge (1-\rho)\alpha_j | \mathscr{F}_{t_j(k)} \right\} \right] \le \delta \ a.s. \tag{13.25}$$

Choose $\varepsilon>0$ such that $2\{\max[(I+\rho)e^{-\rho},(1-\rho)e^{\rho}]\}^{1/\varepsilon} < \delta$. Let $\widetilde{X}_n = X_n I(m_n^2 \le \varepsilon\alpha_j)$ and note that since m_n is $\mathscr{F}_{n-1}$-measurable and $X_n^2 \le m_n^2$,

$$0 \le E(X_n^2|\mathscr{F}_{n-1}) - E(\tilde{X}_n^2|\mathscr{F}_{n-1}) \le m_n^2 I(m_n^2 > \varepsilon\alpha_j).$$

Moreover, $P\{m_n^2 \le \varepsilon\alpha_j \text{ for all } t_j(k) < n \le t_j(k+1) \,|\, \mathscr{F}_{t_j(k)}\} \to 1$ *a.s.* Hence, it suffices to consider $E(\tilde{X}_n^2|\mathscr{F}_{n-1})$ instead of $E(X_n^2|\mathscr{F}_{n-1})$ in (13.25). Since $\tilde{X}_n^2 \le \varepsilon\alpha_j$, we can apply Lemma 9.1(b) to conclude that

$$P\left\{ \sum_{t_j(k)<n\le t_j(k+1)} E(X_n^2|\mathscr{F}_{n-1}) \ge (1+\rho)\alpha_j | \mathscr{F}_{t_j(k)} \right\} + P\left\{ \sum_{t_j(k)<n\le t_j(k+1)} E(X_n^2|\mathscr{F}_{n-1}) \ge (1-\rho)\alpha_j | \mathscr{F}_{t_j(k)} \right\}$$
$$\le (1+\rho)e^{-\rho/\varepsilon} + (1-\rho)e^{\rho/\varepsilon} + o(1) < \delta,$$

completing the proof. □

Replacing X_n by $-X_n$ in Theorem 13.8 yields

$$\liminf_{n\to\infty} S_n/\left\{V_n(\log\log V_n)^{1/2}\right\} = -\sqrt{2} \quad a.s.$$

Theorem 13.8 can therefore be strengthened into the following compact LIL by a standard argument; see Proposition 2.1 of Griffin and Kuelbs (1989).

Corollary 13.9. *With the same notation and assumptions as in Theorem 13.8, the cluster set of the sequence* $\{S_n/[V_n(\log\log(V_n\wedge e^2))^{1/2}]\}$ *is the interval* $[-\sqrt{2},\sqrt{2}]$.

Note that Theorem 13.8 and Corollary 13.9 pertain to martingale difference sequences X_n. This means that given an integrable sequence $\{X_n\}$, one should first consider centering X_n at its conditional expectation given $\mathscr{F}_{n-1}$ before applying the theorems to $\tilde{X}_n = X_n - E(X_n|\mathscr{F}_{n-1})$ and $V_n = \left(\sum_{i=1}^n \tilde{X}_i^2\right)^{1/2}$. Although Theorem 13.8 requires $\tilde{X}_n$ to be bounded by $\mathscr{F}_{n-1}$-measurable $m_n = o(V_n(\log\log V_n)^{-1/2})$, we can often dispense with such boundedness assumptions; see Problem 13.2. In the more general context of Theorem 13.5, the X_n may not even be integrable, so Theorem 13.5 centers the X_n at certain truncated conditional expectations. Using $(\sum_{i=1}^n X_i^2)^{1/2}$ for the norming factor, however, may be too large since it involves uncentered X_i's. To alleviate this problem, we can first center X_n at its conditional median before applying Theorem 13.5 to $\tilde{X}_n = X_n - \mathrm{med}(X_n|\mathscr{F}_{n-1})$, as illustrated in the following example.

Example 13.10. Let $0<\alpha<1$, $d_1\ge 0$, $d_2\ge 0$ with $d_1+d_2>0$. Let $Y, Y_1, Y_2, \ldots$ be i.i.d random variables such that

$$\begin{aligned} P\{Y\ge y\} &= (d_1+o(1))y^{-\alpha}, \\ P\{Y\le -y\} &= (d_2+o(1))y^{-\alpha}, \qquad \text{as } y\to\infty. \end{aligned} \tag{13.26}$$

Let $\hat{S}_n = \sum_{i=1}^n Y_i$, $\hat{V}_n^2 = \sum_{i=1}^n Y_i^2$, $\hat{v}_n = \hat{V}_n(\log\log\hat{V}_n)^{-1/2}$. Then by Theorem 6.14,

$$\limsup_{n\to\infty} \frac{\hat{S}_n}{\hat{V}_n(\log\log n)^{1/2}} = \{\beta(\alpha,d_1,d_2)\}^{-1/2} \quad a.s. \tag{13.27}$$

Moreover, $E\{YI(-\lambda y\le Y<a_\lambda y)\} = (d_1a_\lambda - d_2\lambda + o(1))\alpha y^{1-\alpha}/(1-\alpha)$ as $y\to\infty$ and

$$\frac{n\hat{v}_n^{1-\alpha}}{\hat{V}_n(\log\log\hat{V}_n)^{1/2}} = \frac{n}{\hat{V}_n^{\alpha}(\log\log\hat{V}_n)^{(2-\alpha)/2}} = O(1) \quad a.s. \tag{13.28}$$

since $\log\log\hat{V}_n \sim \log\log n$ and

$$\liminf_{n\to\infty} \frac{\sum_{i=1}^n Y_i^2}{n^{1/\tilde{\alpha}}(\log\log n)^{-(1-\tilde{\alpha})/\tilde{\alpha}}} > 0 \quad a.s.$$

with $\tilde{\alpha} = \alpha/2$ by the so-called delicate LIL (see Breiman, 1968).

Now let $X_n = n^r + Y_n$ with $r > 1/\alpha$ and let $S_n = \sum_{i=1}^n X_i$, $V_n^2 = \sum_{i=1}^n X_i^2$. Since $Y_n = o(n^s)$ $a.s.$ for any $s>1/\alpha$, it follows that $S_n \sim V_n \sim n^{r+1}/(r+1)$ and

$\mu_i(-\lambda v_n, a_\lambda v_n) = i^r + o(n^{(r+1)(1-\alpha)}) = i^r + o(n^r)$ *a.s.*, recalling that $r\alpha > 1$. Therefore, although (13.6) still holds in this case, it is too crude as the nonrandom location shift n^r is the dominant term in X_n causing V_n to swamp the centered S_n. Centering the X_n first at its median will remove this problem. Specifically, if we apply (13.6) to $\tilde{X}_n = X_n - \text{med}(X_n)$ and $\tilde{V}_n^2 = \sum_{i=1}^n \tilde{X}_i^2$, then $\tilde{X}_n = Y_n - \text{med}(Y)$ and (13.27) still holds with $\hat{S}_n$ replaced by $\tilde{S}_n$.

13.4 Supplementary Results and Problems

1. Show that (13.6) implies (13.7) when $S_n = \sum_{i=1}^n X_i$ is a supermartingale such that $X_n \geq -m_n$ *a.s.*, and m_n is $\mathscr{F}_{n-1}$-measurable and satisfies

$$P\{0 \leq m_n \leq \lambda v_n \text{ for all large } n\} = 1. \tag{13.29}$$

2. The following example shows that we cannot dispense with the boundedness assumption $|X_n| \leq m_n$ with $\mathscr{F}_{n-1}$-measurable $m_n = o(v_n)$ for (13.8) to hold for martingales. Let $X_1 = X_2 = 0, X_3, X_4, \ldots$ be independent random variables such that

$$\begin{aligned} P\{X_n = -n^{-1/2}\} &= 1/2 - n^{-1/2}(\log n)^{1/2} - n^{-1}(\log n)^{-2}, \\ P\{X_n = -m_n\} &= n^{-1}(\log n)^{-2}, \\ P\{X_n = n^{-1/2}\} &= 1/2 + n^{-1/2}(\log n)^{1/2}, \end{aligned}$$

for $n \geq 3$, where $m_n \sim 2(\log n)^{5/2}$ is chosen so that $EX_n = 0$. Show that $P\{X_n = -m_n \ i.o.\} = 0$ and that with probability 1, $V_n^2 = \log n + O(1)$.

Since $\tilde{X}_i := X_i I(|X_i| \leq 1) - EX_i I(|X_i| \leq 1)$ are independent bounded random variables with zero means and $\text{Var}(\tilde{X}_i) \sim i^{-1}$, Kolmogorov's LIL yields

$$\limsup_{n\to\infty} \frac{\sum_{i=1}^n \tilde{X}_i}{\{2(\log n)(\log\log\log n)\}^{1/2}} = 1 \ \ a.s. \tag{13.30}$$

Show that $\sum_{i=1}^n EX_i I(|X_i| \leq 1) \sim \frac{4}{3}(\log n)^{3/2}$ and therefore

$$\frac{\sum_{i=1}^n X_i}{V_n(\log\log V_n)^{1/2}} \sim \frac{4(\log n)^{3/2}}{3\{(\log n)(\log\log\log n)\}^{1/2}} \to \infty \ \ a.s. \tag{13.31}$$

Since $m_n(\log\log V_n)^{1/2}/V_n \to \infty$, this shows that without the boundedness condition $X_n \geq -\lambda V_n(\log\log V_n)^{-1/2}$, the upper LIL need not hold for martingales self-normalized by V_n. It also shows the importance of the centering in Theorem 13.5 because subtracting $EX_i 1(|X_i| \leq 1)$ from X_i gives the LIL in view of (13.30).

3. Let $X_1, X_2, \ldots$ be independent random variables with $P(X_i = i!) = \frac{1}{2} = P(X_i = -i!)$. Let $S_n = \sum_{i=1}^n X_i$ and $V_n^2 = \sum_{i=1}^n X_i^2$, $v_n = V_n(\log\log V_n)^{-1/2}$, and define $\mu_i(c,d)$ as in (13.6). Prove

$$\left\{S_n - \sum_{i=1}^n \mu_i(-\lambda v_n, \alpha_\lambda v_n)\right\}/V_n = O(1) \quad a.s.,$$

which shows that although Theorem 13.5 gives an upper LIL for any adapted sequence $\{X_n\}$, the upper bound in (13.6) may not be attained.
4. The following example illustrates the difference between Stout's LIL (Theorem 13.1) and Theorem 13.8. Let X_n be the same as in Problem 2. Note that X_n satisfies the boundedness condition of Theorem 13.1. Show that $\mathrm{Var}(X_i) \sim 4(\log i)^3/i$ and that $s_n^2 = \sum_{i=1}^n E(X_i^2|\mathscr{F}_{i-1}) \sim (\log n)^4$, and therefore

$$\frac{\sum_{i=1}^n X_i}{s_n(\log\log s_n)^{1/2}} \sim \frac{4(\log n)^{3/2}}{3(\log n)^2(\log\log\log n)^{1/2}} \to 0 \quad a.s. \qquad (13.32)$$

which is consistent with Theorem 13.1. Contrasting (13.32) with (13.31) shows the difference between self-normalizing by s_n and V_n for martingales.

Chapter 14
Multivariate Self-Normalized Processes with Matrix Normalization

The general framework of self-normalization in Chap. 10 and the method of mixtures in Chap. 11 has been extended by de la Peña et al. (2008) to the multivariate setting in which A_t is a vector and B_t is a positive definite matrix. Section 14.1 describes the basic concept of matrix square roots and the literature on its application to self-normalization. Section 14.2 extends the moment and exponential inequalities in Chap. 13 to multivariate self-normalized processes. Section 14.3 describes extensions of the boundary crossing probabilities in Sect. 11.3 and the law of the iterated logarithm in Sect. 13.3 to multivariate self-normalized processes with matrix normalization.

14.1 Multivariate Extension of Canonical Assumptions

14.1.1 Matrix Sequence Roots for Self-Normalization

Let C be a symmetric $m \times m$ matrix. Then all the eigenvalues $\lambda_1, \ldots, \lambda_m$ are real. Assume that C is nonnegative definite (i.e., $x'Vx \geq 0$ for all $x \in \mathbb{R}^m$). Then the λ_i are non-negative and so are $\sqrt{\lambda_i}$. Let e_i be an eigenvector associated with λ_i, normalized so that $e_i'e_i = 1$. The eigenvectors corresponding to distinct eigenvalues are orthogonal, and in the case where the eigenvalue λ has multiplicity p, its associated linear space of eigenvectors has dimension p and is orthogonal to the eigenvectors associated with the other eigenvalues. Let Q be the $m \times m$ orthogonal matrix (i.e., $Q^{-1} = Q'$) whose column vectors are $e_1, \ldots, e_m$. Since e_i is the eigenvector associated with λ_i, $Ce_i = \lambda_i e_i$ for $1 \leq i \leq m$ and therefore $CQ = Q\,\mathrm{diag}(\lambda_1, \ldots, \lambda_m)$, yielding the *singular value decomposition*

$$C = Q\,\mathrm{diag}(\lambda_1, \ldots, \lambda_m)Q'. \tag{14.1}$$

© Springer-Verlag Berlin Heidelberg 2009

Since $Q'Q = I$, we can define $C^{1/2}$ (so that $C^{1/2}C^{1/2} = C$) by

$$C^{1/2} = Q\operatorname{diag}(\sqrt{\lambda_1}, \ldots, \sqrt{\lambda_m})Q', \tag{14.2}$$

which is often called the *symmetric square root* of C. When C is positive definite, the eigenvalues are positive and C^{-1} and $C^{-1/2}$ can be evaluated by

$$C^{-1} = Q\operatorname{diag}(\lambda_1^{-1}, \ldots, \lambda_m^{-1})Q',\ C^{-1/2} = Q\operatorname{diag}(1/\sqrt{\lambda_1}, \ldots, 1/\sqrt{\lambda_m})Q'. \tag{14.3}$$

The *Cholesky decomposition* of C is of the form $C = PP'$, where P is an $m \times m$ lower-triangular matrix (and therefore P' is upper-triangular); see Problem 14.1. The *left Cholesky square root* of C is P, and P' is the right Cholesky square root. For the problem of self-normalizing an m-dimensional statistic (e.g., sample mean vector) so that the self-normalized vector converges weakly to a spherically symmetric distribution (such as $N(0,I)$), Vu et al. (1996) have shown that the symmetric and the left Cholesky square roots of the sample estimate of the asymptotic covariance matrix can be used for self-normalization. They point out that the Cholesky square root "is favoured in the older statistical literature because of its computational convenience, an important consideration before the advent of computers." This advantage, however, disappears with the availability of software packages. The `R` function `svd` returns the singular value decomposition $M = U\Lambda V'$ of a general $m \times p$ matrix M with real entries, where U and V are $m \times (m \wedge p)$ and $p \times (m \wedge p)$ matrices with orthonormal columns and Λ is a diagonal matrix. The `R` function `chol` returns the Cholesky decomposition $C = PP'$ of a nonnegative definite symmetric matrix.

In this and subsequent chapters we use the symmetric square root of a positive definite matrix C for self-normalization. As will be shown in Sect. 14.2, the singular value decomposition of C that is used in (14.2) to define $C^{1/2}$ also provides linear transformations that orthogonalize the variables in the integrals associated with the method of mixtures for pseudo-maximization.

14.1.2 Canonical Assumptions for Matrix-Normalized Processes

We first extend the canonical assumption (10.6) to the setting of a random vector A and the canonical assumption on a random vector A and a symmetric, positive definite random matrix C:

$$E\exp\{\theta' A - \theta' C\theta/2\} \le 1 \qquad \text{for all } \theta \in \mathbb{R}^d. \tag{14.4}$$

We then relax (14.4) to the form

$$E\exp\{\theta' A - \Phi(C^{1/2}\theta)\} \le \gamma \qquad \text{if } \|\theta\| < \varepsilon, \tag{14.5}$$

for some $\gamma > 0$ and $\varepsilon > 0$, where $\Phi : \mathbb{R}^d \to [0,\infty)$ is isotropic, strictly convex in $\|\theta\|$ such that $\Phi(0) = 0$, $\lim_{\|\theta\|\to\infty} \Phi(\theta) = \infty$ and $\Phi(\theta)$ has bounded second derivatives for large $\|\theta\|$. An important special case is $\Phi_q(\theta) = \|\theta\|^q/q$ with $1 < q \le 2$. These exponential and L_p-bounds are then strengthened into corresponding maximal inequalities for self-normalized processes under the canonical assumption

$$\left\{\exp(\theta' A_t - \Phi_q(C_t^{1/2}\theta)),\ t \in T\right\} \quad \text{is a supermartingale for } \|\theta\| < \varepsilon, \tag{14.6}$$

where T is either $\{0,1,2,\dots\}$ or $[0,\infty)$.

The following lemmas, which give important special cases of these canonical assumptions, are extensions of corresponding results in Chaps. 9, 10 and 13.

Lemma 14.1. *Let M_t be a continuous martingale taking values in $\mathbb{R}^d$, with $M_0 = 0$. Then $\exp\{\theta' M_t - \theta'\langle M\rangle_t \theta/2\}$ is a supermartingale with mean ≤ 1, for all $\theta \in \mathbb{R}^d$.*

Lemma 14.2. *Let $\{M_t : t \ge 0\}$ be a locally square-integrable martingale taking values in $\mathbb{R}^d$, with $M_0 = 0$. Then*

$$\exp\left\{\theta' M_t - \frac{1}{2}\theta'\langle M\rangle_t^c \theta - \sum_{s \le t}[(\theta'\Delta M_s)^+]^2 - [\sum_{s \le t}((\theta'\Delta M_s)^-)^2]_t^{(p)}\right\}$$

is a supermartingale with mean ≤ 1, for all $\theta \in \mathbb{R}^d$, where the superscript (p) denotes the dual predictable projection process.

Lemma 14.3. *Let $\{d_n\}$ be a sequence of random vectors adapted to a filtration $\{\mathscr{F}_n\}$ such that d_i is conditionally symmetric (i.e., $\mathscr{L}(\theta' d_n|\mathscr{F}_{i-1}) = \mathscr{L}(-\theta' d_n|\mathscr{F}_{n-1})$. Then $\exp\{\theta'\sum_{i=1}^n d_i - \theta'\sum_{i=1}^n d_i d_i'\theta/2\}$, $n \ge 1$, is a supermartingale with mean ≤ 1, for all $\theta \in \mathbb{R}^d$.*

Lemma 14.4. *Let $\{d_n\}$ be a sequence of random vectors adapted to a filtration $\{\mathscr{F}_n\}$ such that $E(d_n|\mathscr{F}_{n-1}) = 0$ and $\|d_n\| \le M$ a.s. for all n and some non-random positive constant M. Let $0 < \varepsilon \le M^{-1}$, $A_n = \sum_{i=1}^n d_i$, $C_n = (1 + \frac{1}{2}\varepsilon M)\sum_{i=1}^n E(d_i d_i'|\mathscr{F}_{i-1})$. Then (14.6) holds.*

Lemma 14.5. *Let $\{d_n\}$ be a sequence of random vectors adapted to a filtration $\{\mathscr{F}_n\}$ such that $E(d_n|\mathscr{F}_{n-1}) = 0$ and $\sigma_n^2 = E(\|d_n\|^2|\mathscr{F}_{n-1}) < \infty$. Assume that there exists a positive constant M such that $E(\|d_n\|^k|\mathscr{F}_{n-1}) \le (k!/2)\sigma_n^2 M^{k-2}$ a.s. or $P(\|d_n\| \le M|\mathscr{F}_{n-1}) = 1$ a.s. for all $n \ge 1$, $k > 2$. Let $A_n = \sum_{i=1}^n d_i$, $V_n = \sum_{i=1}^n E(d_i d_i'|\mathscr{F}_{i-1})$. Then for $\|\theta\| \le 1/M$, $\{\exp(\theta' A_n - \frac{1}{2}\theta' V_n \theta/(1 - M\|\theta\|)$, $\mathscr{F}_n$, $n \ge 0\}$ is a supermartingale with mean ≤ 1.*

Lemma 14.6. *Let $\{d_n\}$ be a sequence of random vectors adapted to a filtration $\{\mathscr{F}_n\}$ such that $E(d_n|\mathscr{F}_{n-1}) = 0$ and $\|d_n\| \le M$ a. s. for all n and some nonrandom positive constant M. Let $0 < \gamma < 1, a_\gamma = -\{\gamma + \log(1-\gamma)\}/\gamma^2, A_n = \sum_{i=1}^n d_i$, $C_n = 2a_\gamma\sum_{i=1}^n d_i d_i'$. Then (14.6) holds with $\varepsilon = \gamma M^{-1}$.*

14.2 Moment and Exponential Inequalities via Pseudo-Maximization

Consider the canonical assumption (14.4). If the random function $\exp\{\theta' A - \theta' C\theta/2\}$ could be maximized over θ inside the expectation, taking the maximizing value $\theta = C^{-1}A$ in (14.4) would yield $E\exp\{A'C^{-1}A/2\} \le 1$. This in turn would give the exponential bound $P(\|C^{-1/2}A\| > x) \le \exp(-x^2/2)$. Although we cannot interchange the order of $\max_\lambda$ and E that is needed in the above argument, we can integrate both sides of (14.4) with respect to a probability measure F on θ and use Fubini's theorem to interchange the order of integration with respect to P and F. To achieve an effect similar to maximizing the random function $\exp\{\theta' A - \theta' C\theta/2\}$, F would need to assign positive mass to and near $\theta = C^{-1}A$ that maximizes $\exp\{\theta' A - \theta' C\theta/2\}$, for all possible realizations of (A, C). This leads us to choose probability measures of the form $dF(\theta) = f(\theta)d\theta$, with f positive and continuous. Note that

$$\int_{\mathbb{R}^d} e^{\theta' A - \theta' C\theta/2} f(\theta)d\theta = e^{A'C^{-1}A/2} \int_{\mathbb{R}^d} e^{-(\theta - C^{-1}A)'C(\theta - C^{-1}A)/2} f(\theta)d\theta. \tag{14.7}$$

Let $\lambda_{\max}(C)$ and $\lambda_{\min}(C)$ denote the maximum and minimum eigenvalues of C, respectively. Since $(\theta - C^{-1}A)'C(\theta - C^{-1}A) \ge \lambda_{\min}(C)\|\theta - C^{-1}A\|^2$, it follows that as $\lambda_{\min}(C) \to \infty$,

$$\int_{\mathbb{R}^d} e^{-(\theta - C^{-1}A)'C(\theta - C^{-1}A)/2} f(\theta)d\theta \sim \frac{(2\pi)^{m/2}}{\sqrt{\det C}} f(C^{-1}A). \tag{14.8}$$

Combining (14.7) with (14.8) yields Laplace's asymptotic formula that relates the integral on the left-hand side of (14.7) to the maximum value $\exp(A'C^{-1}A/2)$ of $\exp\{\theta' A - \theta' C\lambda/2\}$. Thus integration of $\exp(\theta' A - \theta' C\theta/2)$ with respect to the measure F provides "pseudo-maximization" of the integrand over θ when $\lambda_{\min}(C) \to \infty$. By choosing f appropriately to reflect the growth rate of $C^{-1/2}A$, we can extend the moment and exponential inequalities in Sect. 12.2 to the multivariate case. In particular, we shall prove the following two theorems and a related lemma.

Theorem 14.7. *Let A and C satisfy the canonical assumption* (14.4). *Let V be a positive definite nonrandom matrix. Then*

$$E\left[\sqrt{\frac{\det(V)}{\det(C+V)}} \exp\left\{\frac{1}{2}A'(C+V)^{-1}A\right\}\right] \le 1, \tag{14.9}$$

$$E\exp\{A'(C+V)^{-1}A/4\} \le \left\{E\sqrt{\det(I+V^{-1}C)}\right\}^{\frac{1}{2}}. \tag{14.10}$$

Proof. Put $f(\theta) = (2\pi)^{-d/2}\sqrt{\det V}\exp(-\theta' V\theta/2)$, $\theta \in \mathbb{R}^d$, in (14.7) after multiplying both sides of (14.4) by $f(\theta)$ and integrating over θ. By Fubini's theorem,

$$1 \geq E\left[\frac{\sqrt{\det(V)}}{(2\pi)^{d/2}} e^{A'(C+V)^{-1}A/2} \int_{\mathbb{R}^d} e^{-\{\theta-(C+V)^{-1}A\}'(C+V)\{\theta-(C+V)^{-1}A\}} d\theta\right]$$
$$= E\sqrt{\frac{\det(V)}{\det(C+V)}} e^{A'(C+V)^{-1}A/2},$$

proving (14.9). To prove (14.10), apply (14.9) to the upper bound in the Cauchy–Schwarz inequality

$$E\exp\{A'(C+V)^{-1}A/4\}$$
$$\leq \left\{\left(E\sqrt{\frac{\det(V)}{\det(C+V)}}\exp\left(\frac{1}{2}A'(C+V)^{-1}A\right)\right)\left(E\sqrt{\frac{\det(C+V)}{\det(V)}}\right)\right\}^{1/2}. \qquad \square$$

Note that (14.10) is of the form $Eh(A'(C+V)^{-1}A) \leq EH(V^{-1}C)$, where H is a function that depends on h and V is a positive definite matrix used to shift C away from 0 (the matrix with zero entries). For $d = 1$, de la Peña et al. (2000) and de la Peña et al. (2004) also consider the case without shifts, for which they obtain inequalities of the form $Eh(A/B) \leq EH(B \vee B^{-1})$, where $B = C^{1/2}$. The pseudo-maximization technique can be used to generalize these inequalities to the multivariate case, for which we replace $B \vee B^{-1}$ in the case $d = 1$ by $\lambda_{\max}(B) \vee \lambda_{\min}^{-1}(B)$ for $d \times d$ positive definite matrices B.

As in Sects. 11.2 and 12.2, a key idea in this generalization is to choose the density function f in (14.7) to be

$$f(\theta) = \tilde{f}(\|\theta\|)/\|\theta\|^{d-1} \quad \text{for } \theta \in \mathbb{R}^d, \qquad \text{with } \tilde{f}(r) = \frac{1}{rL(r \vee r^{-1})} \quad \text{for } r > 0, \tag{14.11}$$

where $L : (0,\infty) \to [0,\infty)$ is a nondecreasing function satisfying (11.3)–(11.5), in which (11.5) is now modified to $\int_1^\infty \tilde{f}(r)dr = 1/[2\text{vol}(\mathbb{S}^d)]$, where $\text{vol}(\mathbb{S}^d)$ denotes the volume of the unit sphere $\mathbb{S}^d = \{\theta \in \mathbb{R}^d : \|\theta\| = 1\}$. Since f is isotropic,

$$\int_{\mathbb{R}^d} f(\lambda)d\lambda = \text{vol}(\mathbb{S}^d)\int_0^\infty \tilde{f}(r)dr = \text{vol}(\mathbb{S}^d)\left\{\int_0^1 \tilde{f}(r)dr + \int_1^\infty \tilde{f}(r)dr\right\} = 1.$$

The following properties of L play an important role in applying the pseudo-maximization technique to derive inequalities for self-normalized vectors from the canonical assumption (14.4).

Lemma 14.8.
(a) For $x \neq 0$ and positive definite matrix B,

$$L(\|B^{-1}x\| \vee \|B^{-1}x\|^{-1}) \leq 3\{L(\|x\| \vee 1) \vee L(\lambda_{\max}(B) \vee \lambda_{\max}(B^{-1}))\}.$$

(b) Under (14.4) *for A and C, let $B = C^{1/2}$ and define $g : (0,\infty) \to [0,\infty)$ by*

$$g(r) = \frac{e^{r^2/2}}{r^d} I(r \geq 1). \tag{14.12}$$

Then

$$E \frac{g(\|B^{-1}A\|)}{L(\|B^{-1}A\|) \vee L(\lambda_{\max}(B) \vee \lambda_{\min}^{-1}(B))} \left\{ \frac{\lambda_{\min}(B)}{\lambda_{\max}(B)} \right\}^{d-1}$$
$$\leq 18 \left(\int_{-1/2}^{1/2} e^{-z^2/2} dz/2 \right)^{-d}.$$

Proof. The proof of (a) is a straightforward modification of that of Lemma 11.1, noting that $\lambda_{\min}(B^{-1})\|y\| \leq \|B^{-1}y\| \leq \lambda_{\max}(B^{-1})\|y\|$, and that $\lambda_{\min}(B^{-1}) = 1/\lambda_{\max}(B)$. To prove (b), application of (14.4) and (14.7) to the density function (14.11) yields

$$E\left\{ e^{-A'C^{-1}A/2} I\left(\|B^{-1}A\| \geq 1\right) \int_{\mathbb{R}^d} \frac{\exp\{-(\theta - C^{-1}A)'C(\theta - C^{-1}A)/2\}}{\|\theta\|^d L(\|\theta\| \vee \|\theta\|^{-1})} d\theta \right\} \leq 1. \tag{14.13}$$

To evaluate the integral in (14.13), we use the singular value decomposition that gives $C = Q' \operatorname{diag}(\lambda_1, \ldots, \lambda_d)Q$, where the λ_i are eigenvalues of the positive definite matrix C and Q is an orthogonal matrix. Noting that the Euclidean norm $\|\cdot\|$ is invariant under orthogonal transformations, we use the change of variables $x = Q\theta$ to rewrite the integral as

$$\int_{\mathbb{R}^d} \frac{\exp\{-\sum_{i=1}^d \lambda_i (x_i - \tilde{a}_i)^2/2\}}{\|x\|^d L(\|x\| \vee \|x\|^{-1})} dx \geq \int_I \frac{\exp\{-\sum_{i=1}^d \lambda_i (x_i - \tilde{a}_i)^2/2\}}{\|x\|^d L(\|x\| \vee \|x\|^{-1})} dx, \tag{14.14}$$

where I is the rectangle $\prod_{i=1}^d [\tilde{a}_i - (2\sqrt{\lambda_i})^{-1}, \tilde{a}_i + (2\sqrt{\lambda_i})^{-1}]$ and $\tilde{a} = (\tilde{a}_1, \ldots, \tilde{a}_d)' = QC^{-1}A$. Note that $B = C^{1/2} = Q' \operatorname{diag}(\sqrt{\lambda_1}, \ldots, \sqrt{\lambda_d})Q$. Next use the change of variables $y_i = \sqrt{\lambda_i} x_i$ $(i = 1, \ldots, d)$ for the integral in the right-hand side of (14.14) and apply part (1) of the lemma, so that (14.14) is bounded below by

$$\int_{I^*} \frac{(\lambda_1 \ldots \lambda_d)^{-1/2} \exp(-\sum_{i=1}^d (y_i - \sqrt{\lambda_i}\tilde{a}_i)^2/2)}{3(\sum_{i=1}^d y_i^2/\lambda_i)^{d/2} \{L(\|y\| \vee 1) \vee L(\max_{1 \leq i \leq d} \sqrt{1/\lambda_i} \vee \max_{1 \leq i \leq d} \sqrt{\lambda_i})\}} dy, \tag{14.15}$$

where $I^* = \prod_{i=1}^d [\sqrt{\lambda_i}\tilde{a}_i - 1/2, \sqrt{\lambda_i}\tilde{a}_i + 1/2]$. Note that

$$\begin{aligned} \tilde{a} = QB^{-1}B^{-1}A &= \operatorname{diag}(1/\sqrt{\lambda_1}, \ldots, 1/\sqrt{\lambda_d})QB^{-1}A \\ &= (a_1^*/\sqrt{\lambda_1}, \ldots, a_d^*/\sqrt{\lambda_d})', \end{aligned} \tag{14.16}$$

where $a^* = (a_1^*, \dots, a_d^*)' = QB^{-1}A$. Therefore $I^* = \prod_{i=1}^d [a_i^* - \frac{1}{2}, a_i^* + \frac{1}{2}]$, and $\|y\| < 2\|a^*\|$ for $y \in I^*$ and $\|a^*\| \geq 1$. Hence (14.15) can be bounded below by

$$\frac{(\lambda_1, \dots, \lambda_d)^{-1/2} \left(\int_{-1/2}^{1/2} e^{-z^2/2} dz \right)^d}{3 \left(\min_{1 \leq i \leq d} \lambda_i\right)^{-d/2} (2\|B^{-1}A\|^d) \{6L(\|B^{-1}A\|) \vee L(\lambda_{\max}(B^{-1}) \vee \lambda_{\max}(B))\}} \tag{14.17}$$

in view of (11.3) and $\|a^*\| = \|B^{-1}A\|$. Combining (14.13) with (14.14)–(14.17) gives the desired conclusion, noting that the eigenvalues of B are $\sqrt{\lambda_1}, \dots, \sqrt{\lambda_d}$. □

Replacing $L(B \vee B^{-1})$ in the one-dimensional case by

$$\ell(B) = L(\lambda_{\max}(B) \vee \lambda_{\min}^{-1}(B)) \{\lambda_{\max}(B)/\lambda_{\min}(B)\}^{d-1} \tag{14.18}$$

for $d \times d$ matrices B, we can use the same argument as that of Theorem 12.9 to derive the following result from Lemma 14.8(b).

Theorem 14.9. *Let h be a nondecreasing function on $[0, \infty)$ such that for some $x_0 \geq 1$ and $\alpha > 0$,*

$$0 < h(x) \leq \alpha g(x)/L(x) \quad \text{for all } x \geq x_0, \tag{14.19}$$

where g is defined by (14.12) and $L : (0, \infty) \to (0, \infty)$ is a nondecreasing function satisfying (11.3)–(11.5). Let q be a strictly increasing, continuous function on $[0, \infty)$ such that for some $\tilde{\alpha} \geq \alpha$,

$$L(x) \leq q(x) \leq \tilde{\alpha} g(x)/h(x) \quad \text{for all } x \geq x_0. \tag{14.20}$$

Let A and C satisfy the canonical assumption (14.4) and let $B = C^{1/2}$. Then there exists a positive constant ζ_d (depending only on d) such that

$$Eh(\|B^{-1}A\|) \leq \zeta_d \tilde{\alpha} + h(x_0) + Eh(q^{-1}(\ell(B))), \tag{14.21}$$

where ℓ is defined in (14.18).

As a corollary of Theorem 14.9, we obtain that under the canonical assumption (14.4), there exist universal constants $\zeta_{d,p}$ and $\tilde{\zeta}_{d,p}$ for any $p > 0$ such that

$$\begin{aligned} E\|B^{-1}A\|^p \leq & \zeta_{d,p} + \tilde{\zeta}_{d,p} E\{\log^+ \log(\lambda_{\max}(B) \vee \lambda_{\min}^{-1}(B)) \\ & + [\log \lambda_{\max}(B) - \log \lambda_{\min}(B)]\}^{p/2}. \end{aligned} \tag{14.22}$$

In the univariate case $d = 1$, the term $\log \lambda_{\max}(B) - \log \lambda_{\min}(B)$ disappears and (14.22) reduces to

$$E|A/B|^p \leq \zeta_p + \tilde{\zeta}_p E\{\log^+ \log(B \vee B^{-1})\}^{p/2}. \tag{14.23}$$

The following example shows that for $d > 1$, the term $\log \lambda_{\max}(B) - \log \lambda_{\min}(B)$ in (14.22) cannot be removed.

Example 14.10. Lai and Robbins (1981, p. 339) consider the simple linear regression model $y_i = \alpha + \beta u_i + \varepsilon_i$, in which ε_i are i.i.d. random variables with $E\varepsilon_i = 0$ and $E\varepsilon_i^2 = 1$ and the u_i are sequentially determined regressors defined by

$$u_1 = 0, \quad u_{n+1} = \bar{u}_n + c\bar{\varepsilon}_n \tag{14.24}$$

so that u_{n+1} is $\mathscr{F}_n$-measurable, where $\mathscr{F}_n$ is the σ-field generated by $\{\varepsilon_1, \ldots, \varepsilon_n\}$ and $c \neq 0$ is nonrandom. They have shown that

$$\sum_{i=1}^n (u_i - \bar{u}_n)^2 = c^2 \sum_{i=2}^n (i-1)\bar{\varepsilon}_{i-1}^2/i \sim c^2 \log n \ \ a.s., \tag{14.25}$$

$$\sum_{i=1}^n (u_i - \bar{u}_n)\varepsilon_i \Big/ \sum_{i=1}^n (u_i - \bar{u}_n)^2 \to -c^{-1} \ \ a.s. \tag{14.26}$$

Example 1 of Lai and Wei (1982) uses (14.25) to prove that

$$\begin{aligned} \lambda_{\max}\left(\sum_{i=1}^n x_i x_i'\right) &\sim n\left\{1 + c^2\left(\sum_{i=1}^\infty i^{-1}\varepsilon_i\right)^2\right\} \ \ a.s.,\\ \lambda_{\min}\left(\sum_{i=1}^n x_i x_i'\right) &\sim c^2(\log n)\Big/\left\{1 + c^2\left(\sum_{i=1}^\infty i^{-1}\varepsilon_i\right)^2\right\} \ \ a.s., \end{aligned} \tag{14.27}$$

where $x_i = (1, u_i)'$. Standard projection calculations associated with the simple linear regression model can be used to show that

$$\begin{aligned} W_n &:= \left(\sum_{i=1}^n x_i\varepsilon_i\right)'\left(\sum_{i=1}^n x_i x_i'\right)^{-1}\left(\sum_{i=1}^n x_i\varepsilon_i\right)\\ &= \frac{(\sum_1^n \varepsilon_i)^2}{n} + \frac{[\sum_1^n (u_i - \bar{u}_n)\varepsilon_i]^2}{\sum_1^n (u_i - \bar{u}_n)^2}. \end{aligned} \tag{14.28}$$

Whereas the LIL yields $(\sum_1^n \varepsilon_i)^2/n = O(\log\log n)$ *a.s.*, the last term in (14.28) is of order $\log n$ (rather than $\log\log n$) since

$$\frac{[\sum_1^n (u_i - \bar{u}_n)\varepsilon_i]^2}{\sum_1^n (u_i - \bar{u}_n)^2} = \left\{\frac{\sum_1^n (u_i - \bar{u}_n)^2\varepsilon_i}{\sum_1^n (u_i - \bar{u}_n)^2}\right\}^2 \sum_{i=1}^n (u_i - \bar{u}_n)^2 \sim \log n \ \ a.s., \tag{14.29}$$

by (14.25) and (14.26). By Fatou's lemma, $\liminf_{n\to\infty} E(W_n/\log n) \geq 1$, showing that the term $\log\lambda_{\max}(\sum_{i=1}^n x_i x_i')$ cannot be dropped from (14.22).

For other extensions of the results in Sects. 12.2 and 12.3, see de la Peña et al. (2008).

14.3 LIL and Boundary Crossing Probabilities for Multivariate Self-Normalized Processes

Theorem 13.8 has the following multivariate extension, details of which can be found in de la Peña et al. (2008).

Theorem 14.11. *Let $\{M_n, \mathscr{F}_n, n \geq 0\}$ be a martingale taking values in $\mathbb{R}^d$, with $M_0 = 0$. Let $d_i = M_i - M_{i-1}$ and define V_n either by $V_n = \sum_{i=1}^n E(d_i d_i' | \mathscr{F}_{i-1})$ or by $V_n = \sum_{i=1}^n d_i d_i'$ for all n. Assume that*

$$\|d_n\| \leq m_n \ a.s. \quad \text{for some } \mathscr{F}_{n-1}\text{-measurable } m_n, \tag{14.30}$$

$$\operatorname{tr}(V_n) \to \infty \quad \text{and} \quad m_n(\log\log m_n)^{1/2}/\operatorname{tr}(V_n) \to 0 \ a.s., \tag{14.31}$$

$$\lim_{n\to\infty} V_n/\operatorname{tr}(V_n) = \Gamma \ a.s. \tag{14.32}$$

for some positive definite nonrandom matrix Γ. Define $W_n(t) = V_n^{-1/2} M_i / \{2\log\log \operatorname{tr}(V_n)\}^{1/2}$ for $t = \operatorname{tr}(V_i)/\operatorname{tr}(V_n)$, $W_n(0) = 0$, and extend by linear interpolation to $W_n : [0,1] \to \mathbb{R}^d$. Then with probability 1, $\{W_n, n \geq 1\}$ is relatively compact in $C^d[0,1]$ and its set of limit points in $C^d[0,1]$ is

$$\Big\{ f = (f_1, \dots, f_d) : f_i(0) = 0, f_i \text{ is absolutely continuous and } \sum_{i=1}^d \int_0^1 \left(\frac{d}{dt} f_i(t)\right)^2 dt \leq 1 \Big\}. \tag{14.33}$$

Consequently, $\limsup_{n\to\infty} (M_n' V_n^{-1} M_n) / \{2\log\log \operatorname{tr}(V_n)\} = 1$ *a.s.*

In the case $d = 1$, (14.32) clearly holds with $\Gamma = 1$, and (14.30) and (14.31) are Stout's assumptions for the martingale LIL in Theorem 13.1. Theorem 14.11 in this case can be regarded as the Strassen-type version (see Problem 2.8) of Theorem 13.1.

Let $f : \mathbb{R}^d \to [0,\infty)$ be an isotropic function such that $\int_{\|\theta\|<\varepsilon} f(\theta) d\theta < \infty$. Under (14.6), $\{\psi(A_t, C_t^{1/2}), t \in T\}$ is a nonnegative supermartingale, where

$$\psi(A,B) = \int_{\|\theta\|<\varepsilon} f(\theta) \exp(\theta' A - \Phi_q(B\theta)) d\theta. \tag{14.34}$$

Let $B_t = C_t^{1/2}$, $A_0 = B_0 = 0$. Therefore by the supermartingale inequality, for any $a > 0$,

$$P\{\psi(A_t, B_t) \geq a \text{ for some } t \geq 0\} \leq \int_{\|\theta\|<\varepsilon} f(\theta) d\theta / a. \tag{14.35}$$

Let $\lambda(B)$ be the $d \times 1$ vector of ordered eigenvalues (not necessarily distinct) of a positive definite matrix B. It will be shown that

$$\psi(A,B) < a \Longleftrightarrow B^{-1}A \in \Gamma_{a,\lambda(B)}, \tag{14.36}$$

where $\Gamma_{a,\lambda}$ is a convex subset of $\mathbb{R}^d$ depending only on $a > 0$ and a parameter $\lambda \in \mathbb{R}^d$. Therefore (14.35) can be re-expressed as

$$P\{B_t^{-1}A_t \notin \Gamma_{a,\lambda(B_t)} \text{ for some } t \geq 0\} \leq \int_{\|\theta\|<\varepsilon} f(\theta)d\theta/a. \tag{14.37}$$

In the case $d = 1$, $\lambda(B) = B \in (0,\infty)$ and the convex set is an interval $(-\infty, \gamma_a(B))$, so the probability in (14.36) is the boundary crossing probability $P\{A_t/B_t \geq \gamma_a(B_t)$ for some $t \geq 0\}$.

To prove (14.36), use the transformation $x = B\theta$ to rewrite the integral in (14.34) as

$$\frac{1}{\det(B)} \int_{\|B^{-1}x\|<\varepsilon} f(B^{-1}x)\exp\{x'B^{-1}A - \Phi_q(x)\}dx. \tag{14.38}$$

Let $\lambda(B) = (\sqrt{\lambda_1}, \ldots, \sqrt{\lambda_d})$ and use the singular value decomposition $B^2 = Q'$ diag $(\lambda_1, \ldots, \lambda_d)Q$, where Q is an orthogonal matrix, to express $\|B^{-1}x\|$ in terms of $\lambda(B)$:

$$\|B^{-1}x\|^2 = x'B^{-2}x = \sum_{i=1}^{d} (Qx)_i^2/\lambda_i.$$

Moreover, $\det(B) = \prod_{i=1}^d \sqrt{\lambda_i}$. Since f and Φ_q are isotropic, applying a further change of variables $z = Qx$ to the integral in (14.38) can be used to express (14.34) as a function $\tilde{\psi}(B^{-1}A, \lambda(B))$ of $B^{-1}A$ and the eigenvalues of B. For fixed λ, the function $\tilde{\psi}(w,\lambda)$ is a convex function of $w \in \mathbb{R}^d$, and therefore $\Gamma_{a,\lambda} := \{w : \tilde{\psi}(w,\lambda) < a\}$ is convex. Since $\psi(A,B) = \tilde{\psi}(B^{-1}A, \lambda(B))$, (14.36) follows.

14.4 Supplementary Results and Problems

1. The decomposition $V = LDL'$, with diagonal matrix $D = \text{diag}(d_1, \ldots, d_p)$ and lower-triangular matrix L whose diagonal elements are 1, is called the *modified Cholesky decomposition* of V. Show that for a positive definite matrix V, the elements of L and D in its modified Cholesky decomposition can be computed inductively, one row at a time, beginning with $d_1 = V_{11}$:

$$L_{ij} = \left(V_{ij} - \sum_{k=1}^{j-1} L_{ik}d_kL_{jk}\right) \Big/ d_j, \qquad j = 1, \ldots, i-1;$$

$$d_i = V_{ii} - \sum_{k=1}^{i-1} d_kL_{ik}^2, \qquad i = 2, \ldots, p.$$

Moreover, show that the d_i are positive and that $P := LD^{1/2}$ is a lower-triangular matrix, thereby yielding the Cholesky decomposition $V = PP'$.

2. Prove Lemmas 14.1–14.6 by noting that for fixed $\theta \in \mathbb{R}^d$, $\theta' A_t$ and $\theta' C_t \theta$ are scalars to which the corresponding results in Chaps. 9, 10 and 13 are applicable.
3. The following example illustrates the subtleties of matrix normalization in self-normalized LIL. Suppose that in Example 14.10 the ε_i are symmetric Bernoulli random variables. Let $x_i = (1, u_i)'$ and

$$A_n = \sum_{i=1}^{n} x_i \varepsilon_i, \qquad C_n = \sum_{i=1}^{n} x_i x_i' = \sum_{i=1}^{n} \mathrm{Cov}(x_i \varepsilon_i | \mathscr{F}_{i-1}),$$

noting that $\varepsilon_i^2 = 1$. By Lemma 14.3, the canonical assumption (14.6) holds with $q = 2$ and $\varepsilon = \infty$. In view of (14.27), $\log \lambda_{\max}(C_n) \sim \log n$ and $\log \lambda_{\min}(C_n) \sim \log \log n$ *a.s.*

(a) Show that $\|C_n^{-1/2} A_n\|^2 \sim 2 \log \log n + \log n$ *a.s.* and that

$$\|C_n^{-1/2} A_n\|^2 / \{\log \log \lambda_{\max}(C_n)\}^{1/2} \to \infty \quad a.s. \tag{14.39}$$

(b) The components of A_n are $\sum_{i=1}^{n} \varepsilon_i$ and $\sum_{i=1}^{n} u_i \varepsilon_i$, which are martingales with bounded increments and satisfy the univariate LIL. Therefore it may be somewhat surprising that the LIL fails to hold for the self-normalized $C_n^{-1/2} A_n$ as (14.39) shows. However, the components of $C_n^{-1/2} A_n$ are $n^{-1/2} \sum_1^n \varepsilon_i$ and $\{\sum_1^n (u_i - \bar{u}_n) \varepsilon_i\} / \{\sum_1^n (u_i - \bar{u}_n)^2\}^{\frac{1}{2}}$. Explain why $(u_i - \bar{u}_n)\varepsilon_i$ is not even $\mathscr{F}_i$-measurable for $i \leq n-1$.

Part III
Statistical Applications

Chapter 15
The t-Statistic and Studentized Statistics

This chapter first describes in Sect. 15.1 the t-distribution introduced by Gosset (1908) and its multivariate extensions, in the form of the multivariate t-distribution, Hotelling's T^2-statistic and the F-distribution, all derived from sampling theory of a normal (or multivariate normal) distribution with unknown variance (or covariance matrix). It then develops the asymptotic distributions of these self-normalized sample means even when the population has infinite second moment. Related results such as the law of the iterated logarithm (LIL) for these self-normalized statistics are also described.

Self-normalized statistics, with matrix normalization as in the T^2-statistic, are ubiquitous in statistical applications. Section 15.2 describes these general *Studentized statistics*; the term "Studentized" refers to Gosset's (Student's) basic approach that divides $\hat{\theta}_n - \theta$ by the estimated standard error $\widehat{\text{se}}_n$ of the sample estimate $\hat{\theta}_n$ (which is the sample mean in Gosset's case) of a population parameter θ. In the multivariate case, $1/\widehat{\text{se}}_n$ is replaced by $C_n^{-1/2}$, where C_n is typically a consistent estimator of the covariance matrix of $\hat{\theta}_n$ when the latter exists or of the covariance matrix in the asymptotic normal distribution of $\hat{\theta}_n$. This principle extends far beyond the setting of i.i.d. observations, and Sect. 15.3 shows that the asymptotic theory of these extensions to time series and control systems is typically related to self-normalized martingales.

15.1 Distribution Theory of Student's t-Statistics

Let $X_1, X_2, \ldots, X_n$ be i.i.d. normal random variables with mean μ and variance $\sigma^2 > 0$. The sample mean $\bar{X}_n = n^{-1}\sum_{i=1}^n X_i$ is normal with mean μ and variance σ^2/n and therefore $\sqrt{n}(\bar{X}_n - \mu)/\sigma$ is standard normal. In practice, σ is unknown and $\sqrt{n}(\bar{X}_n - \mu)/\sigma$ cannot be used as a *pivot* (i.e., a quantity whose distribution does not depend on unknown parameters) for inference on μ. The sample variance $s_n^2 = (n-1)^{-1}\sum_{i=1}^n (X_i - \bar{X}_n)^2$ is an unbiased estimate of σ^2, and replacing σ in $\sqrt{n}(\bar{X}_n - \mu)/\sigma$ by s_n gives a pivot since the distribution of

V.H. de la Peña et al., *Self-Normalized Processes: Limit Theory and Statistical Applications*, 207
Probability and its Applications,
© Springer-Verlag Berlin Heidelberg 2009

$$T_n := \frac{\sqrt{n}(\bar{X}_n - \mu)}{s_n} = \sqrt{\frac{n-1}{n}} \frac{\sum_{i=1}^n (X_i - \mu)/\sigma}{\left\{\sum_{i=1}^n \left[(X_i - \mu) - (\bar{X}_n - \mu)\right]^2 / \sigma^2\right\}^{1/2}} \tag{15.1}$$

does not depend on (μ, σ), noting that $(X_i - \mu)/\sigma$ is $N(0,1)$. The distribution of the pivot (15.1) was first derived by W. S. Gosset in 1908 under the name Student as he was working at a brewery at that time. It has since been called Student's t-distribution with $\nu = n-1$ degrees of freedom, and has density function

$$f(t) = \frac{\Gamma((\nu+1)/2)}{\sqrt{\pi}\Gamma(\nu/2)} \left(1 + \frac{t^2}{\nu}\right)^{-(\nu+1)/2}, \qquad -\infty < t < \infty. \tag{15.2}$$

The t-distribution with ν degrees of freedom converges to the standard normal distribution as $\nu \to \infty$. Without assuming normality, T_n still converges in distribution to a standard normal random variable even though its exact distribution is not Student t, provided that σ^2 is finite and positive so that the central limit theorem can be applied to $\sqrt{n}(\bar{X}_n - \mu)$ and the law of large numbers to s_n^2. Remarkably, T_n still has a limiting distribution when $EX_i^2 = \infty$ and the distribution of X_i belongs to the domain of attraction of a normal or stable law.

15.1.1 Case of Infinite Second Moment

As pointed out in (1.1), (15.1) in the case $\mu = 0$ can be expressed as $T_n = U_n\{(n-1)/(n-U_n^2)\}^{1/2}$, where $U_n = (\sum_{i=1}^n X_i)/(\sum_{i=1}^n X_i^2)^{1/2}$ is the self-normalized sum. Let $S_n = \sum_{i=1}^n X_i$, $V_n^2 = \sum_{i=1}^n X_i^2$. When X_i is symmetric, Efron (1969) has shown that U_n has a limiting distribution. Logan et al. (1973) have derived the limiting distribution of U_n when X_i belongs to the domain of attraction of a stable law with index $0 < \alpha < 2$. They first note that for appropriately chosen constants a_n, $a_n S_n$ and $a_n^2 V_n^2$ have the same joint limiting distribution as that in the case when X_i is stable with index α. They then restrict to exactly stable X_i having density function g that satisfies

$$x^{\alpha+1} g(x) \to r, \quad x^{\alpha+1} g(-x) \to \ell \qquad \text{as } x \to \infty, \tag{15.3}$$

with $r + \ell > 0$. The characteristic function of $(S_n/n^{1/\alpha}, V_n^2/n^{2/\alpha})$ can be written as

$$E\exp\left(\frac{it_1 S_n}{n^{1/\alpha}} + \frac{it_2 V_n^2}{n^{2/\alpha}}\right) = \left\{1 + \int_{-\infty}^{\infty} \left[\exp\left(\frac{ixt_1}{n^{1/\alpha}} + \frac{ix^2 t_2}{n^{2/\alpha}}\right) - 1\right] g(x)dx\right\}^n. \tag{15.4}$$

First consider the case $0 < \alpha < 1$. Let

$$K(y) = \begin{cases} r & \text{if } y > 0, \\ \ell & \text{if } y < 0. \end{cases} \tag{15.5}$$

Using the change of variables $x = n^{1/\alpha}y$, we can rewrite the integral in (15.4) as

$$\begin{aligned}
&\int_{-\infty}^{\infty} \{\exp(iyt_1 + iy^2t_2) - 1\} \frac{(n^{1/\alpha}|y|)^{1+\alpha}}{n|y|^{1+\alpha}} g(n^{1/\alpha}y)dy \\
&\sim \frac{1}{n}\int_{-\infty}^{\infty} \{\exp(iyt_1 + iy^2t_2) - 1\} \frac{K(y)}{|y|^{1+\alpha}}dy
\end{aligned} \tag{15.6}$$

as $n \to \infty$, noting that the second integral in (15.6) converges and applying (15.3) and the dominated convergence theorem. From (15.4) and (15.5) it follows that the characteristic function of $(S_n/n^{1/\alpha}, V_n^2/n^{2/\alpha})$ converges to

$$\Psi(t_1,t_2) := \exp\left\{\int_{-\infty}^{\infty} \left[\exp(iyt_1 + iy^2t_2) - 1\right] [K(y)/|y|^{1+\alpha}]dy\right\}. \tag{15.7}$$

Since $V_n^2/n^{2/\alpha}$ has a limiting distribution concentrated on $(0,\infty)$, it then follows that S_n/V_n has a limiting distribution which is the distribution of S/V, where (S,V^2) has characteristic function (15.7).

Let $c(t) = Ee^{itS/V}$ be the characteristic function of S/V. To derive a formula for c from the characteristic function (15.7) of (S,V^2), let $a > 0$ and set $t_1 = u$ and $t_2 = iau^2$ in (15.7), in which the integral can be extended to all real t_1 and $\mathrm{Im}(t_2) > 0$. Using (15.5) and the change of variables $x = yu$ yields $\Psi(u, iau^2) = \exp(u^\alpha \psi(a))$, where

$$\psi(a) = \int_{-\infty}^{\infty} \left[\exp(ix - ax^2) - 1\right] [K(x)/|x|^{1+\alpha}]dx, \qquad a > 0. \tag{15.8}$$

A key idea is to relate c to ψ via

$$\begin{aligned}
&\int_0^{\infty} u^{-1} \{\exp(u^\alpha \psi(a)) - \exp(u^\alpha \psi(b))\} du \\
&= \int_0^{\infty} u^{-1} \{\Psi(u, iau^2) - \Psi(u, ibu^2)\} du \\
&= E\int_0^{\infty} u^{-1} e^{iuS} (e^{-au^2V^2} - e^{-bu^2V^2}) du \\
&= E\int_0^{\infty} t^{-1} e^{itS/V} (e^{-at^2} - e^{-bt^2}) dt \qquad \text{(with } t = Vu\text{)} \\
&= \int_0^{\infty} t^{-1} c(t) (e^{-at^2} - e^{-bt^2}) dt.
\end{aligned}$$

Differentiating the equation relating the first and last terms above with respect to a then yields

$$\int_0^{\infty} tc(t)e^{-at^2} dt = -\psi'(a)\int_0^{\infty} u^{\alpha-1} \exp(u^\alpha \psi(a)) du = \alpha^{-1}\psi'(\alpha)/\psi(\alpha); \tag{15.9}$$

see Problem 15.1. By multiplying (15.9) by $(\pi a)^{-1/2}\exp(-s^2/4a)$, integrating over a from 0 to ∞, and using the identity $t\int_0^\infty (\pi a)^{-1/2}\exp[-s^2/4a - at^2]da = e^{-st}$ for positive s and t, Logan et al. (1973, pp. 795–797) derive from (15.9) a formula for $\int_0^\infty c(t)e^{-st}dt$, which is the Laplace transform of the Fourier transform of the distribution function F of S/V. They then invert this transform to compute the density function of F numerically.

For the case $1 < \alpha < 2$, instead of using (15.4), note that $\int_{-\infty}^{\infty} xg(x)dx = EX_1 = 0$ and write $E\exp(it_1 n^{-1/\alpha}S_n + it_2 n^{-2/\alpha}V_n^2)$ as

$$\left\{1 + \int_{-\infty}^{\infty}\left[\exp\left(\frac{ixt_1}{n^{1/\alpha}} + \frac{ix^2t_2}{n^{2/\alpha}}\right) - 1 - \frac{ixt_1}{n^{1/\alpha}}\right] g(x)dx\right\}^{1/n}.$$

We can then proceed as before after modifying (15.8) as

$$\psi(a) = \int_{-\infty}^{\infty}\left[\exp(ix - ax^2) - 1 - ix\right]\left[K(x)/|x|^{1+\alpha}\right]dx.$$

For the case $\alpha = 1$, Logan et al. (1973, pp. 798–799) note that U_n has a proper limiting distribution only if $r = \ell$ (i.e., only if X_i has a symmetric Cauchy distribution), and indicate how the preceding arguments can be modified for the symmetric Cauchy distribution.

As pointed out in Chap. 4, there is a stochastic representation of the distribution of S/V in terms of i.i.d. symmetric Bernoulli and exponent random variables; see Theorem 4.5. Giné et al. (1997) have proved that the t-statistic has a limiting normal distribution if and only if X_1 is in the domain of attraction of a normal law. This is the content of Theorem 4.1 which settles one of the conjectures of Logan et al. (1973, pp. 789). Theorem 4.5, which is due to Chistyakov and Götze (2004a), settles the other conjecture of Logan et al. (1973) that the "only possible nontrivial limiting distributions" are those when X_1 follows a stable law. Mason and Zinn (2005) have also given an elementary proof of this conjecture when X_1 is symmetric.

15.1.2 Saddlepoint Approximations

As shown in the preceding section, the limiting distribution of the t-statistic is normal when X_i belongs to the domain of attraction of a normal law and is a complicated distribution, which depends on α and is specified by the Laplace transform of its characteristic function, when X_i belongs to the domain of attraction of a stable law with index α. Jing et al. (2004) have developed simpler *saddlepoint approximations* for the density function and the tail probability of the self-normalized mean $\sqrt{n}\bar{X}_n/V_n$. These saddlepoint approximations do not require X_i to belong to the domain of attraction of a normal or stable law and are in fact applicable to *all* distributions satisfying

$$\int_{-\infty}^{\infty}\int_{-\infty}^{\infty}|E\exp(itX_1 + isX_1^2)|^\rho\,dt\,ds < \infty \qquad \text{for some } \rho \geq 1. \tag{15.10}$$

This corresponds to the case of an integrable characteristic function in Sect. 2.3.2; because the characteristic function of a sum of n i.i.d. random variables is the nth power of the individual characteristic function, we only need the ρth absolute power of the characteristic function to be integrable. For $0 < b < 1$, the numerical results in Sect. 5 of their paper compare the probability $P(\sqrt{n}\bar{X}_n/V_n \geq b)$, estimated by 1 million Monte Carlo simulations, with the saddlepoint approximation, the normal approximation that is valid only when X_i belongs to the domain of attraction of a normal law, and the Edgeworth expansion (see Chap. 16) that requires finiteness of $E|X_1|^r$ for $r \geq 3$. They study four different underlying distributions, ranging from the normal to the Cauchy, for the case $n = 5$ and for $b = 0.05, 0.10, \ldots, 0.90, 0.95$, and have found that the saddlepoint approximations are remarkably accurate. In contrast, the normal and Edgeworth approximations perform much worse.

A key ingredient of the saddlepoint approximation to the density function f_n of $\sqrt{n}\bar{X}_n/V_n$ is the cumulant generating function $K(s,t) = \log E\exp(sX + tX^2)$, which is finite if $t < 0$. Note that in the notation of Sect. 10.1.1, $K(\theta, -\rho\theta) = \psi(\theta, \rho)$. To approximate $f_n(b)$ for $|b| < 1$, note that $x/\sqrt{y} = b \Leftrightarrow y = x^2/b^2$ and define

$$\Lambda(a,b) = \sup_{t<0, s\in\mathbb{R}} \left\{sa + ta^2/b^2 - K(s,t)\right\} = \hat{s}a + \hat{t}a^2/b^2 - K(\hat{s},\hat{t}), \tag{15.11}$$

where $\hat{s}$ and $\hat{t} < 0$ are the solutions of

$$(\partial K/\partial s)(\hat{s},\hat{t}) = a, \qquad (\partial K/\partial t)(\hat{s},\hat{t}) = a^2/b^2. \tag{15.12}$$

Let a_b be the minimizer of $\Lambda(a,b)$ given by the solution of the equation $\Lambda_a(a,b) = 0$, where we use Λ_a to denote $\partial\Lambda/\partial a$ and Λ_{aa} to denote $\partial^2\Lambda/\partial a^2$. Then under (15.10), Jing et al. (2004) have shown that the saddlepoint approximation to f_n is

$$\hat{f}_n(b) \approx \sqrt{\frac{n}{2\pi}} \frac{2a_b^2/|b|^3}{\{\det(\nabla^2\Lambda(a_b,b))\Lambda_{aa}(a_b,b)\}^{1/2}} e^{-n\Lambda(a_b,b)}, \tag{15.13}$$

where $\nabla^2\Lambda$ denotes the Hessian matrix of second partial derivatives of Λ. Moreover, they also showed that

$$P\left\{\frac{\sqrt{n}\bar{X}_n}{V_n} \geq b\right\} = 1 - \Phi\left(\sqrt{n}w\right) - \frac{\phi\left(\sqrt{n}w\right)}{\sqrt{n}} \left\{\frac{1}{w} - \frac{1}{v} + O(n^{-1})\right\}, \tag{15.14}$$

where Φ and ϕ are the standard normal distribution and density functions and

$$w = \{2\Lambda(a_b,b)\}^{1/2}, \qquad v = -\left\{\det\left(\nabla^2\Lambda(a_b,b)\right)\Lambda_{aa}(a_b,b)\right\}^{1/2} \hat{t}(a_b,b),$$

noting that the $\hat{s}$ and $\hat{t}$ in (15.12) are actually functions of a, b and can be denoted by $\hat{s}(a,b)$ and $\hat{t}(a,b)$.

Daniels and Young (1991) have derived the saddlepoint approximation (15.14) under (15.10) and the assumption $E\exp(sX_1 + tX_1^2) < \infty$ for all (s,t) with $|s| + |t| < \delta$ and some $\delta > 0$. They also consider the equations in (15.12) without requiring $\hat{t} < 0$.

A key observation of Jing et al. (2004) is that the solution $(\hat{s},\hat{t})$ of (15.12) has the property that $\hat{t} < 0$ and therefore we only need the finiteness of $E\exp(sX_1 + tX_1^2)$ for $t < 0$, which always holds.

15.1.3 The t-Test and a Sequential Extension

The saddlepoint approximation (15.14) can be used to calculate the type I error probability of the one-sided t-test that rejects the null hypothesis $H_0 : \mu = 0$ if the t-statistic (15.1), in which $\mu = 0$, exceeds some threshold, for general (not necessarily normal) X_i satisfying (15.10) even when X_i has fat tails that result in infinite $E|X_i|$. When X_i is symmetric, Efron (1969, pp. 1285–1288) notes that the probability in (15.14) is bounded by $P\{N(0,1) \geq b\}$, at least for tail probabilities in the usual hypothesis testing range. Concerning the question as to "why worry about limiting normality if the type I errors tend to be in the conservative direction in any case," he points out that the t-test may have poor power relative to more robust tests such as the sign test and rank tests.

Chan and Lai (2000, pp. 1645, 1647–1649) have developed similar approximations for the type I and type II errors of the t-test and its sequential extension, called the *repeated t-test*, by using a different approach that involves change of measures and geometric integration associated with Laplace's method for tubular neighborhoods of extremal manifolds. The repeated t-test of H_0 stops sampling at stage

$$\tau = \inf\left\{n \geq \delta c : \frac{n}{2}\log\left(1 + \frac{\bar{X}_n^2}{s_n^2}\right) \geq c\right\} \wedge [ac], \tag{15.15}$$

where $0 < \delta < a$ and $c > 0$ are the design parameters of the test. Note that $l_n := (n/2)\log(1 + \bar{X}_n^2/s_n^2) \approx n\bar{X}_n^2/(2s_n^2) \approx T_n^2/2$ in view of (15.1); l_n is the generalized likelihood ratio statistic when the X_i are i.i.d. normal with unknown mean and variance (see Chap. 17). The test rejects H_0 if stopping occurs prior to $n_1 := [ac]$ or if $l_{n_1} \geq c$ when stopping occurs at n_1. Chan and Lai (2000) make use of the finiteness of the moment generating function $e^{\psi(\theta,\rho)} := E\exp(\theta X_1 - \rho\theta^2 X_1^2)$ to embed the distribution of (X_i, X_i^2) in a bivariate exponential family with density functions

$$f_{\theta,\rho}(x,y) = \exp\left\{\theta x - \rho\theta^2 y - \psi(\theta,\rho)\right\}$$

with respect to the probability measure P that corresponds to $\theta = 0$; see Sect. 10.1.1.

Besides the repeated t-test, there are other sequential extensions of the t-test in the literature. These are reviewed in Sects. 18.1, and 18.2 also describes methods for analyzing the error probabilities of these tests.

15.2 Multivariate Extension and Hotelling's T^2-Statistic

15.2.1 Sample Covariance Matrix and Wishart Distribution

Let $Y_1, \ldots, Y_n$ be independent $m \times 1$ random $N(\mu, \Sigma)$ vectors with $n > m$ and positive definite Σ. Define

$$\bar{Y} = \frac{1}{n}\sum_{i=1}^{n} Y_i, \qquad W = \sum_{i=1}^{n}(Y_i - \bar{Y})(Y_i - \bar{Y})'. \tag{15.16}$$

The sample mean vector $\bar{Y}$ and the sample covariance matrix $W/(n-1)$ are independent, generalizing the corresponding result in the case $m = 1$. Suppose $Y_1, \ldots, Y_n$ are independent $N(0, \Sigma)$ random vectors of dimension m. Then the random matrix $\sum_{i=1}^{n} Y_i Y_i'$ is said to have a Wishart distribution, denoted by $W_m(\Sigma, n)$. This definition can be used to derive the density function of W when Σ is positive definite. We begin by considering the case $m = 1$ and noting that χ_n^2, which is the distribution of $Z_1^2 + \cdots + Z_n^2$ with i.i.d. standard normal Z_i, is the same as the gamma$(n/2, 1/2)$ distribution. Therefore $\sigma^2 \chi_n^2$ has the density function $w^{(n-2)/2} e^{-w/(2\sigma^2)} / \left[(2\sigma^2)^{n/2} \Gamma(n/2)\right]$, $w > 0$. The density function of the Wishart distribution $W_m(\Sigma, n)$ generalizes this to

$$f(W) = \frac{\det(W)^{(n-m-1)/2} \exp\left\{-\frac{1}{2}\mathrm{tr}\left(\Sigma^{-1} W\right)\right\}}{[2^m \det(\Sigma)]^{n/2} \Gamma_m(n/2)}, \qquad W > 0, \tag{15.17}$$

in which $W > 0$ denotes that W is positive definite and $\Gamma_m(\cdot)$ denotes the multivariate gamma function

$$\Gamma_m(t) = \pi^{m(m-1)/4} \prod_{i=1}^{m} \Gamma\left(t - \frac{i-1}{2}\right). \tag{15.18}$$

Note that the usual gamma function corresponds to the case $m = 1$. The following properties of the Wishart distribution are generalizations of some well-known properties of the chi-square distribution and the gamma(α, β) distribution:

(a) If $W \sim W_m(\Sigma, n)$, then $E(W) = n\Sigma$.
(b) Let $W_1, \ldots, W_k$ be independently distributed with $W_j \sim W_m(\Sigma, n_j)$, $j = 1, \ldots, k$. Then $\sum_{j=1}^{k} W_j \sim W_m(\Sigma, \sum_{j=1}^{k} n_j)$.
(c) Let $W \sim W_m(\Sigma, n)$ and A be a nonrandom $m \times m$ nonsingular matrix. Then $AWA' \sim W_m(A\Sigma A', n)$. In particular, $a'Wa \sim (a'\Sigma a)\chi_n^2$ for all nonrandom $m \times 1$ vectors $a \neq 0$.

15.2.2 The Multivariate t-Distribution and Hotelling's T^2-Statistic

The multivariate generalization of $\sum_{t=1}^{n}(y_t - \bar{y})^2 \sim \sigma^2 \chi_{n-1}^2$ involves the Wishart distribution and is given by $W := \sum_{i=1}^{n}(Y_i - \bar{Y})(Y_i - \bar{Y})' \sim W_m(\Sigma, n-1)$. As indicated in

Sect. 14.1.1, we can use the singular value decomposition $W = P\text{diag}(\lambda_1, \ldots, \lambda_m)P'$, where P is an orthogonal matrix and $\lambda_1, \ldots, \lambda_m$ are the eigenvalues of W, to define $W^{1/2} = P\text{diag}(\sqrt{\lambda_1}, \ldots, \sqrt{\lambda_m})P'$. Moreover, as noted above, W is independent of $\bar{Y} \sim N(\mu, \Sigma/n)$. Hence the situation is the same as in the univariate ($m = 1$) case. It is straightforward to generalize the t-distribution to the multivariate case as follows. If $Z \sim N(0, \Sigma)$ and $W \sim W_m(\Sigma, k)$ such that the $m \times 1$ vector Z and the $m \times m$ matrix W are independent, then $(W/k)^{-1/2}Z$ is said to have the *m-variate t-distribution* with k degrees of freedom.

By making use of the density function (15.17) of the Wishart distribution, it can be shown that the m-variate t-distribution with k degrees of freedom has the density function

$$f(t) = \frac{\Gamma((k+m)/2)}{(\pi k)^{m/2}\Gamma(k/2)} \left(1 + \frac{t't}{k}\right)^{-(k+m)/2}, \qquad t \in \mathbb{R}^m. \tag{15.19}$$

The square of a t_k random variable has the $F_{1,k}$-distribution. More generally, if T has the m-variate t-distribution with k degrees of freedom such that $k \geq m$, then

$$\frac{k-m+1}{km} T'T \text{ has the } F_{m,k-m+1}\text{-distribution.} \tag{15.20}$$

Applying (15.20) to *Hotelling's T^2-statistic*

$$T^2 = n(\bar{Y} - \mu)' \left(W/(n-1)\right)^{-1} (\bar{Y} - \mu), \tag{15.21}$$

where $\bar{Y}$ and W are defined in (15.16), yields

$$\frac{n-m}{m(n-1)} T^2 \sim F_{m,n-m}, \tag{15.22}$$

noting that

$$\left[W/(n-1)\right]^{-1/2} \left[\sqrt{n}(\bar{Y} - \mu)\right] = \left[W_m(\Sigma, n-1)/(n-1)\right]^{-1/2} N(0, \Sigma)$$

has the m-variate t-distribution.

In the preceding definition of the multivariate t-distribution, it is assumed that Z and W share the same Σ. More generally, we can consider the case where $Z \sim N(0, V)$ instead. By considering $V^{-1/2}Z$ instead of Z, we can assume that $V = I$. Then the density function of $(W/k)^{-1/2}Z$, with independent $Z \sim N(0, I)$ and $W \sim W_m(\Sigma, k)$, has the general form

$$f(t) = \frac{\Gamma((k+m)/2)}{(\pi k)^{m/2}\Gamma(k/2)\sqrt{\det(\Sigma)}} \left(1 + \frac{t'\Sigma^{-1}t}{k}\right)^{-(k+m)/2}, \qquad t \in \mathbb{R}^m. \tag{15.23}$$

15.2.3 Asymptotic Theory in the Case of Non-Normal Y_i

Let $Y_i, \ldots, Y_n$ be i.i.d. m-dimensional random vectors (not necessarily normal) with $EY = 0$. Hahn and Klass (1980) have shown that there exist $m \times m$ nonrandom matrices A_n such that $A_n \sum_{i=1}^n Y_i$ has a limiting $N(0,I)$ distribution if and only if

$$\lim_{y\to\infty} \sup_{\|\theta\|=1} \frac{y^2 P(|\theta' Y| > y)}{E[(\theta' Y)^2 I\{|\theta' Y| \le y\}]} = 0. \tag{15.24}$$

Note that the zero-mean random variable $\theta' Y$ belongs to the domain of attraction of the normal law if and only if

$$y^2 P\left(|\theta' Y| > y\right) / E\left[(\theta' Y)^2 I\{|\theta' Y| \le y\}\right] \to 0 \qquad \text{as } y \to \infty; \tag{15.25}$$

see Sect. 4.2 and Sepanski (1994, 1996). Thus (15.24) requires this convergence to be uniform in θ belonging to the unit sphere. The construction of A_n is quite complicated, even in the two-dimensional case; see Hahn and Klass (1980, p. 269). In contrast, self-normalization simply involves multiplying $\sum_{i=1}^n Y_i$ by V_n^{-1}, where $V_n = \{\sum_{i=1}^n (Y_i - \bar{Y})(Y_i - \bar{Y})'\}^{1/2}$. Making use of this result of Hahn and Klass (1980), Vu et al. (1996) have shown that $V_n^{-1} \sum_{i=1}^n Y_i$ has a limiting $N(0,I)$ distribution if (15.24) holds. They have also shown that the result still holds if the Cholesky square root (see Sect. 14.1.1) is used instead of the symmetric square root V_n. Giné and Götze (2004) have proved the converse that the weak convergence of $V_n^{-1} \sum_{i=1}^n Y_i$ to $N(0,I)$ implies (15.24) when Y_i is symmetric.

Dembo and Shao (2006) have obtained the LIL for the T^2-statistic (15.21) under $EY = 0$, $h(y) := E\|Y\|^2 I\{\|Y\| \le y\}$ is slowly varying, and

$$\liminf_{y\to\infty} \min_{\|\theta\|=1} E\left[(\theta' Y)^2 I\{\|Y\| \le y\}\right] / h(y) > 0. \tag{15.26}$$

They prove this LIL under (15.26) by extending Theorem 6.1, which establishes the self-normalized moderate deviation formula (6.1) in the domain of attraction of a normal distribution, to the multivariate case. This is the content of the following theorem, and they conjecture that the LIL still holds for the T^2-statistic (15.21) under the weaker assumption (15.24).

Theorem 15.1. *Let $Y, Y_1, Y_2, \ldots$ be i.i.d. m-dimensional random vectors such that $EY = 0$, $h(y) := E\|Y\|^2 I(\|Y\| \le y)$ is slowly varying and (15.26) holds. Define $T_n^2 (= T^2)$ by (15.21). Let $\{x_n,\ n \ge 1\}$ be a sequence of positive numbers with $x_n \to \infty$ and $x_n = o(n)$ as $n \to \infty$. Then*

$$\lim_{n\to\infty} x_n^{-1} \log P\left(T_n^2 \ge x_n\right) = -\frac{1}{2}. \tag{15.27}$$

Moreover, the LIL holds for T_n^2:

$$\limsup T_n^2 / \log\log n = \beta \ \ a.s. \tag{15.28}$$

15.3 General Studentized Statistics

The t-statistic $\sqrt{n}(\bar{X}_n - \mu)/s_n$ is a special case of more general *Studentized statistics* $(\widehat{\theta}_n - \theta)/\widehat{\mathrm{se}}_n$ that are of fundamental importance in statistical inference on an unknown parameter θ of an underlying distribution from which the sample observations $X_1, \dots, X_n$ are drawn. In nonparametric inference, θ is a functional $g(F)$ of the underlying distribution function F and $\widehat{\theta}_n$ is usually chosen to be $g(\widehat{F}_n)$, where $\widehat{F}_n$ is the empirical distribution. The standard deviation of $\widehat{\theta}_n$ is often called its *standard error*, which is typically unknown, and $\widehat{\mathrm{se}}_n$ denotes a consistent estimate of the standard error of $\widehat{\theta}_n$. For the t-statistic, μ is the mean of F and $\bar{X}_n$ is the mean of $\widehat{F}_n$. Since $\mathrm{Var}(\bar{X}_n) = \mathrm{Var}(X_1)/n$, we estimate the standard error of $\bar{X}_n$ by $s_n/\sqrt{n}$, where s_n^2 is the sample variance. An important property of a Studentized statistic is that it is an *approximate pivot*, which means that its distribution is approximately the same for all θ. For parametric problems, θ is usually a multidimensional vector and $\widehat{\theta}_n$ is an asymptotically normal estimate (e.g., by maximum likelihood). Moreover, the asymptotic covariance matrix $V_n(\theta)$ of $\widehat{\theta}_n$ depends on the unknown parameter θ, so $V_n^{-1/2}(\widehat{\theta}_n)(\widehat{\theta}_n - \theta)$ is the self-normalized (Studentized) statistic that can be used as an approximate pivot for tests and confidence regions. The theoretical basis for the approximate pivotal property of Studentized statistics lies in the limiting standard normal distribution, or in some other limiting distribution that does not involve θ (or F in the nonparametric case).

The results in Sect. 15.2.3 on the asymptotic normality of the t-statistic and its multivariate extension when the observations are independent (as in Part I) and belong to the domain of attraction of the (multivariate) normal law have been applied to linear regression in errors-in-variables models by Martsynyuk (2007a,b). By Studentizing the generalized least squares estimates appropriately, she obtains statistics that are approximately pivotal in the sense that their asymptotic distributions are independent of unknown parameters in the distributions of the measurement errors, thereby giving weaker conditions on the explanatory variables than previous authors in the construction of asymptotically valid confidence intervals for the regression parameters. We next consider applications of Studentized statistics in the more general settings considered in Part II, where martingale theory plays a basic role.

15.3.1 Martingale Central Limit Theorems and Asymptotic Normality

To derive the asymptotic normality of $\widehat{\theta}_n$, one often uses a martingale M_n associated with the data, and approximates $V_n^{-1/2}(\widehat{\theta}_n)(\widehat{\theta}_n - \theta)$ by $\langle M \rangle_n^{-1/2} M_n$. For example, in the asymptotic theory of the maximum likelihood estimator $\widehat{\theta}_n$, $V_n(\theta)$ is the inverse of the observed Fisher information matrix $I_n(\theta)$, and the asymptotic normality of $\widehat{\theta}_n$ follows by using Taylor's theorem to derive

$$-I_n(\theta)(\widehat{\theta}_n - \theta) \doteq \sum_{i=1}^{n} \nabla \log f_\theta(X_i|X_1,\ldots,X_{i-1}). \tag{15.29}$$

The right-hand side of (15.29) is a martingale whose predictable variation is $-I_n(\theta)$. Therefore the Studentized statistic associated with the maximum likelihood estimator can be approximated by a self-normalized martingale, i.e.,

$$V_n^{-1/2}(\hat{\theta}_n)(\hat{\theta}_n - \theta) \doteq I_n^{1/2}(\theta)(\hat{\theta}_n - \theta) \doteq \langle M \rangle_n^{-1/2} M_n,$$

where $M_n = \sum_{i=1}^n e_i$, with $e_i = \nabla \log f_\theta(X_i|X_1,\ldots,X_{i-1})$. If there exist nonrandom positive definite matrices B_n such that

$$B_n^{-1} \langle M \rangle_n^{1/2} \xrightarrow{P} I \qquad \text{as } n \to \infty \tag{15.30}$$

and if for every $\varepsilon > 0$,

$$\sum_{i=1}^{n} E\left[\|B_n^{-1} e_i\|^2 I\{\|B_n^{-1} e_i\|^2 \geq \varepsilon\} | X_i,\ldots,X_{i-1}\right] \xrightarrow{P} 0, \tag{15.31}$$

then $\langle M \rangle_n^{-1/2} M_n$ converges in distribution to the multivariate standard normal distribution, by applying the following martingale central limit theorem to $x_{ni} = B_n^{-1} e_i$; see Durrett (2005, p. 411). Condition (15.31), or its more general form (15.33) below, is usually referred to as *conditional Lindeberg*.

Theorem 15.2. *Let* $\{x_{n,m}, \mathscr{F}_{n,m}, 1 \leq m \leq n\}$ *be a martingale difference array (i.e.,* $E(x_{n,m}|\mathscr{F}_{n,m-1}) = 0$ *for* $1 \leq m \leq n$*). Let*

$$S_{n,0} = 0, \quad S_{n,k} = \sum_{i=1}^{k} x_{n,m}, \quad V_{n,m} = \sum_{m=1}^{k} E(x_{n,m}^2 | \mathscr{F}_{n,m-1}),$$

and define

$$S_n(t) = \begin{cases} S_{n,m} & \text{if } t = m/n \text{ and } m = 0,1,\ldots,n \\ \text{linear} & \text{for } t \in [(m-1)/n, m/n]. \end{cases} \tag{15.32}$$

Suppose that $V_{n,[nt]} \xrightarrow{P} t$ *for every* $t \in [0,1]$ *and*

$$\sum_{m=1}^{n} E\left[x_{n,m}^2 I\{x_{n,m}^2 \geq \varepsilon\} | \mathscr{F}_{n,m-1}\right] \xrightarrow{P} 0 \qquad \text{for every } \varepsilon > 0. \tag{15.33}$$

Then S_n *converges weakly in* $C[0,1]$ *to Brownian motion as* $n \to \infty$.

15.3.2 Non-Normal Limiting Distributions in Unit-Root Nonstationary Autoregressive Models

The assumption (15.30) is crucial for ensuring the weak convergence of the self-normalized martingale $\langle M \rangle_n^{-1/2} M_n$ to standard normal. Without this condition,

$\langle M \rangle_n^{-1/2} M_n$ may not have a limiting standard normal distribution. A well-known example in the time series literature is related to least squares estimates of β in the autoregressive model $y_t = \beta y_{t-1} + \varepsilon_t$ when $\beta = 1$, the so-called *unit-root nonstationary* model. Here ε_t are i.i.d. unobservable random variables with $E\varepsilon_t = 0$ and $E\varepsilon_t^2 = \sigma^2 > 0$. The least squares estimate

$$\hat{\beta}_n = \frac{\sum_{t=2}^n y_t y_{t-1}}{\sum_{t=2}^n y_{t-1}^2} = \beta + \frac{\sum_{t=2}^n y_{t-1}\varepsilon_t}{\sum_{t=2}^n y_{t-1}^2}$$

is consistent but not asymptotically normal in this case. Note that since $\beta = 1$, $y_t = y_0 + \sum_{i=1}^t \varepsilon_i$ is a zero-mean random walk and that

$$\left(\sum_{t=2}^n \varepsilon_t\right)^2 = \sum_{t=2}^n \varepsilon_t^2 + 2\sum_{t=2}^n \varepsilon_t (y_{t-1} - y_0).$$

Moreover, by Theorem 15.2 and the continuous mapping theorem for weak convergence in $C[0,1]$ (Durrett, 2005, p. 407), it follows that

$$\left(\sum_{t=2}^n y_{t-1}^2\right)^{1/2} (\hat{\beta}_n - \beta) = \frac{1}{2} \frac{\left\{\left(\frac{1}{\sqrt{n}}\sum_{t=2}^n \varepsilon_t\right)^2 - \frac{1}{n}\sum_{t=2}^n \varepsilon_t^2 + \frac{2y_0}{n}\sum_{t=2}^n \varepsilon_t\right\}}{\left\{\frac{1}{n}\sum_{t=2}^n (y_{t-1}/\sqrt{n})^2\right\}^{1/2}}$$

$$\Rightarrow \frac{\sigma}{2} \frac{\left(W^2(1) - 1\right)}{\left\{\int_0^1 W^2(t)dt\right\}^{1/2}}$$

where $\Rightarrow$ denotes weak convergence and $\{W(t), t \geq 0\}$ denotes standard Brownian motion.

In the preceding autoregressive model, the mean level is assumed to be 0. For the more general autoregressive model $y_t = \alpha + \beta y_{t-1} + \varepsilon_t$, there is a similar result for the Studentized statistic $(\hat{\beta} - 1)/\widehat{\text{se}}(\hat{\beta})$, which is called the *Dickey–Fuller test statistic* for unit-root nonstationarity (i.e., $\beta = 1$ and $\alpha = 0$); see Problem 15.3. Extensions to more lagged variables $y_{t-1}, y_{t-2}, \cdots, y_{t-p}$ and to multivariate y_t have also been considered in the literature, leading to the *augmented Dickey–Fuller test* and the *cointegration test*; see Lai and Xing (2008, Sects. 9.4.4 and 9.4.5) for details.

15.3.3 Studentized Statistics in Stochastic Regression Models

The autoregressive model in the previous section is a special case of *stochastic regression* models of the form

$$y_t = \beta' x_t + \varepsilon_t, \tag{15.34}$$

in which $\{\varepsilon_t\}$ is a martingale difference sequence with respect to a filtration $\{\mathscr{F}_t\}$ and x_t is $\mathscr{F}_{t-1}$-measurable. An important class of these models in control systems is the ARMAX model (autoregressive models with moving average errors and exogenous inputs):

$$y_t = a_1 y_{t-1} + \cdots + a_p y_{t-p} + b_1 u_{t-d} + \cdots + b_q u_{t-d-q+1} + \varepsilon_t + c_1 \varepsilon_{t-1} + \cdots + c_r \varepsilon_{t-r}, \tag{15.35}$$

in which y_t represents the output and u_t the input at time t and the ε_t are random disturbances, $\beta = (a_1, \ldots, a_p, b_1, \ldots, b_q, c_1, \ldots, c_r)'$ is a vector of unknown parameters and $d \geq 1$ represents the delay. The regressor

$$x_t = (y_{t-1}, \ldots, y_{t-p}, u_{t-d}, \ldots, u_{t-d-q+1}, \varepsilon_{t-1}, \ldots, \varepsilon_{t-r})' \tag{15.36}$$

is indeed $\mathscr{F}_{t-1}$-measurable but includes the unobservable $\varepsilon_{t-1}, \ldots, \varepsilon_{t-r}$ and therefore ordinary least squares cannot be used to estimate β. In the white-noise case (i.e., $r = 0$), x_t does not contain ε_i and the model is called ARX, for which β can be estimated by least squares.

To begin with, consider the linear regression model $y_t = a + bx_t + \varepsilon_t$. The least squares estimate of b based on $(x_1, y_1), \ldots, (x_n y_n)$ is

$$\hat{b}_n = \frac{\sum_{i=1}^n (x_i - \bar{x}_n) y_i}{\sum_{i=1}^n (x_i - \bar{x}_n)^2} = b + \frac{\sum_{i=1}^n (x_i - \bar{x}_n) \varepsilon_i}{\sum_{i=1}^n (x_i - \bar{x}_n)^2}. \tag{15.37}$$

Even when ε_i are independent with $E\varepsilon_i = 0, E\varepsilon_i^2 < \infty$ and the x_i are nonrandom constants such that $\sum_{i=1}^n (x_i - \bar{x}_n)^2 \to \infty$, strong consistency of $\hat{b}_n$ does not follow directly from the strong law of large numbers since it involves a double array of weights $a_{ni} := x_i - \bar{x}_n$ to form the weighted sum $\sum_{i=1}^n a_{ni} w_i$. By making use of the properties of the double array a_{ni} associated with least squares regression, Lai et al. (1978, 1979) have established the strong consistency of least squares estimates $\hat{\beta}_n := (\sum_{t=1}^n x_t x_t')^{-1} \sum_{t=1}^n x_t y_t$ in the regression model (15.34) under the minimal assumption that $\sum_{i=1}^n (x_i - \bar{x}_n)^2 \to \infty$, when the x_i are nonrandom and $\{\varepsilon_i, \mathscr{F}_i, i \geq 1\}$ is a martingale difference sequence with $\limsup_{n \to \infty} E(\varepsilon_n^2 | \mathscr{F}_{n-1}) < \infty$.

This assumption, however, is not strong enough to ensure strong consistency of $\hat{\beta}_n$ in stochastic regression models (15.34) in which x_t is $\mathscr{F}_{t-1}$-measurable. In this case, letting $\lambda_{\max}(\cdot)$ and $\lambda_{\min}(\cdot)$ denote the maximum and minimum eigenvalue of a symmetric matrix, respectively, Lai and Wei (1982) have shown that

$$\hat{\beta}_n \to \beta \;\; a.s. \text{ on } \left\{ \lambda_{\min}\left(\sum_{t=1}^n x_t x_t' \right) \Big/ \log \lambda_{\max}\left(\sum_{t=1}^n x_t x_t' \right) \to \infty \right\}, \tag{15.38}$$

when $\sup_n E(|\varepsilon_n|^{2+\delta} | \mathscr{F}_{n-1}) < \infty$ for some $\delta > 0$. They also give an example in which $\lambda_{\max}(\sum_{t=1}^n x_t x_t') \sim Un$, $\lambda_{\min}(\sum_{t=1}^n x_t x_t') \sim V \log n$ and $\hat{\beta}_n \to b_{U,V} \neq \beta$ *a.s.*, where $U \neq 0, V \neq 0$ and $b_{U,V}$ are random variables. The proof of (15.38) uses the (squared) Studentized statistic

$$Q_n = (\hat{\beta}_n - \beta)' \left(\sum_{t=1}^n x_t x_t' \right) (\hat{\beta}_n - \beta), \tag{15.39}$$

which is shown to satisfy

$$Q_n = O\left(\log \lambda_{\max}\left(\sum_{t=1}^{n} x_t x_t'\right)\right) \quad a.s. \tag{15.40}$$

Since $Q_n = (\hat{\beta}_n - \beta)'/\sum_{t=1}^{n} x_t x_t')(\hat{\beta}_n - \beta) \geq \lambda_{\min}(\sum_{t=1}^{n} x_t x_t')\|\hat{\beta}_n - \beta\|^2$, (15.38) follows from (15.40).

To prove (15.40), Lai and Wei (1982) make use of the following recursive representations of $\hat{\beta}_n$ and $P_n := (\sum_{t=1}^{n} x_t x_t')^{-1}$:

$$\hat{\beta}_n = \hat{\beta}_{n-1} + P_n x_n (y_n - \hat{\beta}_{n-1}' x_n), \tag{15.41}$$

$$P_n = P_{n-1} - \frac{P_{n-1} x_n x_n' P_{n-1}}{1 + x_n' P_{n-1} x_n}. \tag{15.42}$$

The recursions (15.41) and (15.42) lead to a recursive inequality for Q_n of the following form:

$$Q_n \leq (1 + \alpha_{n-1}) Q_{n-1} + \theta_n - \gamma_n + w_{n-1} \varepsilon_n, \tag{15.43}$$

with $\alpha_n = 0, \theta_n = x_n' P_n x_n \varepsilon_n^2, w_{n-1} = 2\{(\hat{\beta}_{n-1} - \beta)' x_n\}(1 - x_n' P_n x_n)$ and

$$\gamma_n = \left\{(\hat{\beta}_{n-1} - \beta)' x_n\right\}^2 (1 - x_n' P_n x_n) \geq 0. \tag{15.44}$$

Lai (2003, p. 394) calls $Q_n \geq 0$ an *extended stochastic Liapounov function* if it is $\mathscr{F}_n$-measurable and satisfies (15.43), in which $\alpha_n \geq 0, \theta_n \geq 0, \gamma_n \geq 0$ and w_n are $\mathscr{F}_n$-measurable random variables such that $\sum_{n=1}^{\infty} \alpha_n < \infty$ *a.s.* and $\{\varepsilon_n, \mathscr{F}_n, n \geq 1\}$ is a martingale difference sequence such that $\limsup_{n\to\infty} E(\varepsilon_n^2 | \mathscr{F}_{n-1}) < \infty$ *a.s.* He uses (15.43) and strong laws for martingales to show that

$$\max\left(Q_n, \sum_{i=1}^{n} \gamma_i\right) = O\left(\sum_{i=1}^{n} \theta_i + \left(\sum_{i=1}^{n-1} w_i^2\right)^{1/2+\eta}\right) \quad a.s. \tag{15.45}$$

for every $\eta > 0$. Moreover,

$$Q_n \text{ converges and } \sum_{i=1}^{n} E(\gamma_i | \mathscr{F}_{i-1}) < \infty \ \ a.s. \text{ on } \left\{\sum_{i=1}^{\infty} E(\theta_i | \mathscr{F}_{i-1}) < \infty\right\}. \tag{15.46}$$

Applying (15.45) to the α_n, θ_n, w_n and γ_n in (15.44) yields (15.40) since $\sum_{i=2}^{n} w_{i-1}^2 \leq 4\sum_{i=1}^{n} \gamma_i$ and

$$\sum_{i=1}^{n} \theta_i = \sum_{i=1}^{n} x_i' P_i x_i \varepsilon_i^2 = O\left(\sum_{i=1}^{n} x_i' P_i x_i\right) = O\left(\log \lambda_{\max}\left(\sum_{i=1}^{n} x_i x_i'\right)\right) \quad a.s.;$$

see Lemma 2 of Lai and Wei (1982) and recall the assumption $\limsup_{n\to\infty} E(|\varepsilon_n|^{2+\delta} | \mathscr{F}_{n-1}) < \infty$ *a.s.* for the stochastic regression model.

For the ARMAX model (15.35), one can use the extended least squares estimator that replaces the unobserved ε_i in (15.36) by the prediction error $\hat{\varepsilon}_i = y_i - \hat{\beta}'_{i-1}x_i$. Lai and Wei (1986) have modified the preceding argument to prove an analog to (15.38) under certain conditions on $(c_1,\ldots,c_r)$. They have also shown how the inputs u_t can be chosen in the adaptive control problem of keeping the outputs y_t as close as possible to some target values y_t^* when the parameters $a_1,\ldots,a_p,b_1,\ldots,b_q,c_1,\ldots,c_r$ of the ARMAX model are unknown.

15.4 Supplementary Results and Problems

1. Prove the two equalities in (15.9). The argument leading to (15.9) in Sect. 15.1.1 is closely related to the derivation of the following identity in Logan and Shepp (1968, p. 310), according to a conversation with Larry Shepp:
$$\int_0^\infty \frac{f(ax)-f(bx)}{x}dx = f(0)\log\frac{b}{a} \tag{15.47}$$
for $a>0,\ b>0$ and continuously differentiable $f:[0,\infty)\to\mathbb{R}$ such that $\int_1^\infty |f'(x)|\log x < \infty$. Note that both sides of (15.47) are equal when $b=a$. Differentiating the left-hand side of (15.47) with respect to b yields $-\int_0^\infty f'(bx)dx = f(0)/b$, and the derivative of the right-hand side of (15.47) with respect to b is also $f(0)/b$.
2. Derive (15.19) from (15.17), and prove (15.20) and (15.23).
3. Let $\hat{\theta}_n = (\hat{\alpha}_n,\hat{\beta}_n)'$ be the least squares estimate, based on $y_1,\ldots,y_n$, of the parameter vector $\theta=(\alpha,\beta)'$ in the autoregressive model $y_t = \alpha+\beta y_{t-1}+\varepsilon_t$, in which $\{\varepsilon_t, t\ge 1\}$ is a martingale difference sequence with respect to a filtration $\{\mathscr{F}_t\}$ such that y_0 is $\mathscr{F}_0$-measurable and
$$\lim_{t\to\infty} E(\varepsilon_t^2|\mathscr{F}_{t-1}) = \sigma \ a.s., \quad \sup_t E(|\varepsilon_t|^p|\mathscr{F}_{t-1})<\infty \ a.s. \quad \text{for some } p>2. \tag{15.48}$$
Let $x_t = (1,y_{t-1})'$:
   (a) If $|\beta|<1$, show that the Studentized statistic $(\sum_{t=2}^n x_t x_t')^{1/2}(\hat{\theta}_n-\theta)$ has a limiting normal distribution as $n\to\infty$.
   (b) If $\beta=1$ and $\alpha=0$, show that $(\sum_{t=2}^n x_t x_t')^{1/2}(\hat{\theta}_n-\theta)$ still converges weakly as $n\to\infty$, but to a non-normal limiting distribution.

Chapter 16
Self-Normalization for Approximate Pivots in Bootstrapping

An alternative to inference based on the approximate normal distribution theory in Sect. 15.3.1 for Studentized statistics is to make use of bootstrap resampling. Let $X_1,\dots,X_n$ be independent random vectors with common distribution F. The empirical distribution puts probability mass $1/n$ at each X_i, or equivalently has distribution function $\hat{F}(x) = n^{-1}\sum_{i=1}^n I_{\{X_i \le x\}}$. A *bootstrap sample* $(X_1^*,\dots,X_n^*)$ is obtained by sampling with replacement from $\hat{F}$ so that the X_i^* are i.i.d. with common distribution function $\hat{F}$. Let g be a functional of F and $\theta = g(F)$. The "plug-in estimate" of θ is $\hat{\theta} = g(\hat{F})$. For example, the kth sample moment $n^{-1}\sum_{i=1}^n X_i^k = \int x^k d\hat{F}(x)$ is an estimate of the kth moment of F, for which $g(F) = \int x^k dF(x)$. Denoting $g(\hat{F})$ also by $g(X_1,\dots,X_n)$, we use the bootstrap sample to form a *bootstrap replicate* of $\hat{\theta}$ via $\hat{\theta}^* = g(X_1^*,\dots,X_n^*)$. The sampling distribution of $\hat{\theta}$ can be estimated by Monte Carlo simulations involving a large number of bootstrap replicates generated from $\hat{F}$. In particular, the standard error $\mathrm{se}(\hat{\theta})$ of $\hat{\theta}$, which is the standard deviation of the sampling distribution of $\hat{\theta}$, can be estimated by the following steps:

1. Generate B independent bootstrap samples from $\hat{F}$
2. Evaluate $\hat{\theta}_b^*$ for the bth bootstrap sample, $b = 1,\dots,B$
3. Compute the mean $\bar{\theta}^* = B^{-1}\sum_{b=1}^B \hat{\theta}_b^*$ of the bootstrap replicates and estimate the standard error $\mathrm{se}(\hat{\theta})$ by

$$\widehat{\mathrm{se}}(\hat{\theta}) = \left\{ \sum_{b=1}^{B} (\hat{\theta}_b^* - \bar{\theta}^*)^2/(B-1) \right\}^{1/2}$$

16.1 Approximate Pivots and Bootstrap-t Confidence Intervals

The limiting standard normal distribution of the Studentized statistic $(\hat{\theta} - \theta)/\widehat{\mathrm{se}}$ provides an approximate $(1-2\alpha)$-level confidence interval $\hat{\theta} \pm z_{1-\alpha}\widehat{\mathrm{se}}$, where z_p is the pth quantile of the standard normal distribution for which $z_\alpha = -z_{1-\alpha}$. For the

V.H. de la Peña et al., *Self-Normalized Processes: Limit Theory and Statistical Applications*, 223
Probability and its Applications,
© Springer-Verlag Berlin Heidelberg 2009

special case where θ is the mean of a distribution, Gosset derived in 1908 a better approximation that replaces the normal quantile $z_{1-\alpha}$ by the quantile $t_{n-1;1-\alpha}$ of a t-distribution with $n-1$ degrees of freedom; the approximation is exact if the underlying distribution is normal. If the underlying distribution F were known, one could evaluate the αth and $(1-\alpha)$th quantiles of $(\hat{\theta}-\theta)/\widehat{\text{se}}$ by Monte Carlo by sampling from F and thereby obtain the confidence interval $[\hat{\theta}-u_{1-\alpha}\widehat{\text{se}}, \hat{\theta}-u_{\alpha}\widehat{\text{se}}]$, where u_p denotes the pth quantile of $(\hat{\theta}-\theta)/\widehat{\text{se}}$. Since F is actually unknown, the bootstrap approach draws samples from $\hat{F}$ to evaluate its quantiles $\hat{u}_{\alpha}$ and $\hat{u}_{1-\alpha}$, which are then used to form the bootstrap confidence interval $[\hat{\theta}-\hat{u}_{1-\alpha}\widehat{\text{se}}, \hat{\theta}-\hat{u}_{\alpha}\widehat{\text{se}}]$.

The preceding bootstrap confidence interval is often called a *bootstrap-t interval* since it is based on the Studentized statistic (or generalized t-statistic). An important property of Studentized statistics is that they are asymptotic pivots in the sense that their limiting distributions do not depend on F, which suggests that we can approximate the quantiles of the limiting distribution by sampling from $\hat{F}$. An alternative bootstrap interval is the *percentile interval* $[\hat{\theta}-\hat{q}_{1-\alpha}, \hat{\theta}-\hat{q}_{\alpha}]$, which uses the quantiles $\hat{q}_{\alpha}$ and $\hat{q}_{1-\alpha}$ of the bootstrap replicates $\hat{\theta}_1^*-\hat{\theta},\ldots,\hat{\theta}_B^*-\hat{\theta}$ to estimate the quantiles q_{α} and $q_{1-\alpha}$ of the sampling distribution of $\hat{\theta}-\theta$. Hall (1988) uses Edgeworth expansions to compare the accuracy of different types of bootstrap confidence intervals. The next section describes these Edgeworth expansions for the bootstrap-t and percentile intervals.

16.2 Edgeworth Expansions and Second-Order Accuracy

16.2.1 Edgeworth Expansions for Smooth Functions of Sample Means

To derive the Edgeworth expansions rigorously, Hall considers the case of smooth functions of sample means for which Bhattacharya and Ghosh (1978) have developed a complete theory of Edgeworth expansions. Let $h:\mathbb{R}^d \to \mathbb{R}$ be a function that is sufficiently smooth in some neighborhood of the common mean μ of i.i.d. random vectors $X_1,\ldots,X_n$. Taylor expansion of $h(\bar{X})$ around $h(\mu)$ can be combined with the central limit theorem to show that $\sqrt{n}\{h(\bar{X})-h(\mu)\}$ has a limiting normal distribution with mean 0 and variance $\sigma^2=(\nabla h(\mu))'V(\nabla h(\mu))$, where V is the common covariance matrix of the X_i. This is often called the *delta method*.

Theorem 16.1. *Suppose that for some integer $\nu \geq 1$, h has $\nu+2$ continuous derivatives in a neighborhood of μ and that $E(\|X_1\|^{\nu+2})<\infty$. Assume also that the characteristic function of X_1 satisfies Cramér's condition*

$$\limsup_{\|t\|\to\infty} \left|E\exp(it'X_1)\right| < 1. \tag{16.1}$$

Let Φ and ϕ denote the standard normal distribution and density functions. Then

$$P\{\sqrt{n}(h(\bar{X})-h(\mu))/\sigma \leq x\} = \Phi(x) + \sum_{j=1}^{v} n^{-j/2}\pi_j(x)\phi(x) + o(n^{-v/2}) \quad (16.2)$$

uniformly in x, where π_j is a polynomial of degree $3j-1$, odd for even j and even for odd j, with coefficients depending on the moments of X_1 up to order $j+2$ and the derivatives of h at μ.

For the details of the proof, see Bhattacharya and Ghosh (1978). Here we summarize the main ideas. First consider the simplest case $d=1$ and $h(x)=x$. If the characteristic function of X_1 belongs to $L^p(\mathbb{R})$ for some $p \geq 1$, then $\sqrt{n}(\bar{X}-\mu)/\sigma$ has a density function f_n (with respect to Lebesgue measure) for $n \geq p$ and an Edgeworth expansion of f_n can be obtained by Fourier inversion of the characteristic function $Ee^{it\sqrt{n}(\bar{X}-\mu)/\sigma} = (Ee^{i(t/\sqrt{n})(X_1-\mu)/\sigma})^n$, the Fourier transform of f_n. The Edgeworth expansion (16.2) for $h(x)=x$ can then be obtained by integrating the Edgeworth expansion of f_n. For general d and h, we can obtain a linear approximation to $h(\bar{X})-h(\mu)$ and a change of variables involving the Jacobian matrix to derive an Edgeworth expansion of the density function of $\sqrt{n}\{h(\bar{X})-h(\mu)\}/\sigma$. A more delicate Fourier inversion argument can be used when the integrability assumption on the characteristic function of X_1 is replaced by Cramér's condition; see Sect. 2.3.2, Bhattacharya and Ranga Rao (1976, pp. 211–214) and Bhattacharya and Ghosh (1978, p. 445) for details.

16.2.2 Edgeworth and Cornish–Fisher Expansions: Applications to Bootstrap-t and Percentile Intervals

A confidence interval $I = I(X_1,\ldots,X_n)$ for θ, with nominal coverage probability $1-\alpha$, is said to be *second-order accurate* if $P\{\theta \in I\} = 1-\alpha+O(n^{-1})$. It is called *first-order accurate* if $P\{\theta \in I\} = 1-\alpha+O(n^{-1/2})$. We now make use of Edgeworth expansions to show that bootstrap-t intervals for θ are second-order accurate but one-sided percentile intervals are only first-order accurate when θ is a smooth function of the sample mean vector. Closely related to coverage accuracy is the notion of *correctness* of a confidence interval, which refers to how closely the confidence interval matches an exact confidence interval that it tries to mimic. For example, an exact $1-\alpha$ level upper confidence bound for θ is $\hat{\theta} - u_\alpha \widehat{\text{se}}$ and the bootstrap-t upper confidence bound is $\hat{\theta} - \hat{u}_\alpha \widehat{\text{se}}$. An upper confidence bound is said to be *second-order correct* if it differs from the corresponding exact confidence bound by $O_p(n^{-3/2}) = O_p(n^{-1}\widehat{\text{se}})$. If the difference is $O_p(n^{-1})$, the upper confidence bound is called *first-order correct*. Whereas an Edgeworth expansion gives an asymptotic formula for the distribution function of $\sqrt{n}(\hat{\theta}-\theta)$ or $\sqrt{n}(\hat{\theta}-\theta)/\widehat{\text{se}}$, inverting the formula gives a *Cornish–Fisher expansion* that relates the quantile of the sampling distribution to that of the normal distribution. Cornish–Fisher expansions can be used to show that bootstrap-t confidence bounds are second-order accurate but percentile confidence bounds are only first-order accurate. Since $X_1^*, \ldots, X_n^*$ are *i.i.d.* with distribution function $\hat{F}$, we can apply Edgeworth expansions to the bootstrap distribution of $\hat{\theta}^* - \hat{\theta}$ or $(\hat{\theta}^* - \hat{\theta})/\widehat{\text{se}}$ by translating Cramér's condition on F to a similar property for $\hat{F}$; see Problem 16.1 for the precise statement.

To apply Theorem 16.1 to $(\hat{\theta} - \theta)/\text{se}$ or $(\hat{\theta} - \theta)/\widehat{\text{se}}$, where $\hat{\theta} = g(\bar{X})$, we take the function h in the theorem to be of the form $h(x) = (g(x) - g(\mu))/s(\mu)$ or $h(x) = (g(x) - g(\mu))/s(x)$, where $s^2(\mu)/n$ is the asymptotic variance of $\hat{\theta}$. Note that $s(\bar{X})$ is a consistent estimate of $s(\mu)$ and that $\sigma = 1$ for both choices of h. The coefficients of the polynomials π_j, however, differ for the two choices.

Setting (16.2), with $h(x) = (g(x) - g(\mu))/s(x)$ and $\nu = 2$, equal to α and noting that $\Phi(z_\alpha) = \alpha$, we obtain the following Cornish–Fisher expansion for u_α:

$$u_\alpha = z_\alpha + n^{-1/2} p_1(z_\alpha) + n^{-1} p_2(z_\alpha) + O(n^{-3/2}), \tag{16.3}$$

in which the p_j are polynomials of degree at most $j+1$, odd for even j and even for odd j, and depend on moments of X_1 up to order $j+2$. From the corresponding Edgeworth expansion applied to X_i^* instead of X_i, we obtain likewise

$$\hat{u}_\alpha = z_\alpha + n^{-1/2} \hat{p}_1(z_\alpha) + n^{-1} \hat{p}_2(z_\alpha) + O_p(n^{-3/2}). \tag{16.4}$$

Since the sample moments differ from their population counterparts by $O_p(n^{-1/2})$, it follows from (16.3) and (16.4) that $\hat{u}_\alpha - u_\alpha = O_p(n^{-1})$, and therefore the bootstrap-t upper confidence bound $\hat{\theta} - \hat{u}_\alpha \widehat{\text{se}}$ is second-order correct.

If we consider the percentile upper confidence bound instead, then we can apply (16.2) with $\nu = 2$ and $h(x) = (g(x) - g(\mu))/s(\mu)$ instead, yielding similar Cornish–Fisher expansions for the quantile q_α of $\hat{\theta} - \theta$ and that of $\hat{\theta}^* - \hat{\theta}$:

$$q_\alpha = n^{-1/2} s(\mu) \left\{ z_\alpha + n^{-1/2} P_1(z_\alpha) + n^{-1} P_2(z_\alpha) \right\} + o(n^{-1}), \tag{16.5}$$

$$\hat{q}_\alpha = n^{-1/2} s(\bar{X}) \left\{ z_\alpha + n^{-1/2} \hat{P}_1(z_\alpha) + n^{-1} \hat{P}_2(z_\alpha) \right\} + o_p(n^{-1}). \tag{16.6}$$

Since $s(\bar{X}) - s(\mu) = O_p(n^{-1/2})$, it follows that the percentile upper confidence bound $\hat{\theta} - \hat{q}_\alpha$ is only first-order correct.

To study the coverage accuracy of the bootstrap-t confidence bound, we use (16.3) and (16.4) to derive

$$\hat{u}_\alpha = z_\alpha + n^{-1} \hat{d}_1(\bar{Y}) + \sum_{j=1}^{2} n^{-j/2} p_j(z_\alpha) + O_p(n^{-3/2}), \tag{16.7}$$

where $d_1(\bar{Y}) = \sqrt{n}\{\hat{p}_1(z_\alpha) - p_1(z_\alpha)\}$ and $\bar{Y}$ contains $\bar{X}$ as a subvector and other components involved in the sample moments up to order 3. Therefore

$$
\begin{aligned}
&P\left\{g(\mu) \le g(\bar{X}) - n^{-1/2}s(\bar{X})u_\alpha\right\} \\
&= P\left\{\sqrt{n}\,(g(\bar{X}) - g(\mu))/s(\bar{X}) + n^{-1}\hat{d}_1(\bar{Y}) \ge u_\alpha - \sum_{j=1}^{2} n^{-j/2}p_j(z_\alpha)\right\} \\
&\quad + O(n^{-3/2}).
\end{aligned}
\tag{16.8}
$$

An Edgeworth expansion, up to term of the order n^{-1} and with a remainder of the order $O(n^{-3/2})$, still holds for the second probability in (16.8), which then can be shown to be equal to $1-\alpha+O(n^{-1})$.

Since the percentile upper confidence bound $\hat{\theta}-\hat{q}_\alpha$ is only first-order correct, modifying the preceding argument to derive an Edgeworth expansion for the coverage probability of the one-sided percentile interval yields

$$
P\{\sqrt{n}\,(g(\bar{X})-g(\mu))/s(\mu) \ge \sqrt{n}\hat{q}_\alpha/s(\mu)\} = 1-\alpha+O(n^{-1/2}), \tag{16.9}
$$

showing that the one-sided percentile interval is only first-order accurate. This illustrates the importance of using self-normalized statistics for statistical inference.

For two-sided percentile intervals $[\hat{\theta}-\hat{q}_{1-\alpha}, \hat{\theta}-\hat{q}_\alpha]$, the coverage probability is

$$
\begin{aligned}
&P\{\sqrt{n}\hat{q}_\alpha/s(\mu) \le \sqrt{n}\,(g(\bar{X})-g(\mu))/s(\mu) \le \sqrt{n}\hat{q}_{1-\alpha}/s(\mu)\} \\
&\qquad = P\{\sqrt{n}\,(g(\bar{X})-g(\mu))/s(\mu) \le \sqrt{n}\hat{q}_{1-\alpha}/s(\mu)\} \\
&\qquad\quad - P\{\sqrt{n}\,(g(\bar{X})-g(\mu))/s(\mu) \le \sqrt{n}\hat{q}_\alpha/s(\mu)\}.
\end{aligned}
\tag{16.10}
$$

Since $z_\alpha = -z_{1-\alpha}$ and $\pi_1(z_\alpha) = \pi(z_{1-\alpha})$ because π_1 is an even function, it follows from the difference of the Edgeworth expansions for the two probabilities in (16.10) that the coverage probability of $[\hat{\theta}-\hat{q}_{1-\alpha}, \hat{\theta}-\hat{q}_\alpha]$ is $\Phi(z_{1-\alpha})-\Phi(z_\alpha)+O(n^{-1}) = 1-2\alpha+O(n^{-1})$. The two-sided percentile interval is second-order accurate because of such cancellation of the first-order terms. Unlike π_1, π_2 is an odd function for which $\pi_2(z_\alpha) = -\pi_2(z_{1-\alpha})$. Therefore

$$
n^{-1}\{\pi_2(z_{1-\alpha})\phi(z_{1-\alpha}) - \pi_2(z_\alpha)\phi(z_\alpha)\} = 2n^{-1}\pi_2(z_{1-\alpha})\phi(z_{1-\alpha}),
$$

and the two-sided bootstrap-*t* interval remains second-order accurate as there is no cancellation of the second-order terms.

A practical difficulty that sometimes arises in using bootstrap-*t* confidence intervals is that a good estimate $\widehat{\text{se}}$ of the standard error may be difficult to find, especially when n is not large and $g(F)$ is a nonlinear functional of F. A well-known example is the correlation coefficient of a bivariant distribution F; see Efron and Tibshirani (1993, Sect. 12.6) who also introduce a variance-stabilizing transformation to overcome this difficulty. Other methods to improve the bootstrap-*t* method in nonlinear settings include the BC_a interval, the *ABC* approximation and bootstrap iteration; see Efron and Tibshirani (1993, Sects. 22.4–22.8) and Hall (1992, Sects. 3.10, 3.11) for the underlying theory, Edgeworth expansions, and technical details.

16.3 Asymptotic U-Statistics and Their Bootstrap Distributions

[Asymptotic U-Statistics and Their Bootstrap Distributions]

Although smooth functions of sample mean vectors cover a wide range of statistics in applications, many statistics cannot be expressed as $g(\bar{X})$. When the X_i are i.i.d., it is natural to use symmetric statistics; a statistic $S = S(X_1,\dots,X_n)$ is said to be *symmetric* if it is invariant under permutation of the arguments. In particular, $g(\hat{F})$ is a symmetric statistic. Assuming that $ES^2 < \infty$, let $\mu = ES$ and define

$$\begin{aligned}
A(x_i) &= E(S \mid X_i = x_i) - \mu, \\
B(x_i,x_j) &= E(S \mid X_i = x_i, X_j = x_j) - E(S \mid X_i = x_i) - E(S \mid X_j = x_j) + \mu
\end{aligned}$$

for $i \neq j$, etc. Then $B(x,y) = B(y,x)$ and S has the decomposition

$$\begin{aligned}
S - \mu = {} & \sum_{i=1}^{n} A(X_i) + \sum_{1\le i<j\le n} B(X_i,X_j) + \sum_{1\le i<j<k\le n} C(X_i,X_j,X_k) \\
& + \sum_{1\le i<j<k<h\le n} D(X_i,X_j,X_k,X_h) + \cdots + R(X_1,\dots,X_n),
\end{aligned} \tag{16.11}$$

where all $2^n - 1$ random variables on the right-hand side of (16.11) have mean 0 and are mutually uncorrelated with each other. In fact, $E\{B(X_1,X_2) \mid X_1\} = 0$, $E\{C(X_1,X_2,X_3) \mid X_1,X_2\} = 0$, etc. The decomposition (16.11), due to Efron and Stein (1981), is a generalization of Hoeffding's (1948) decomposition for the special case of U-statistics; see Sect. 8.2.1. Using the first three terms of this decomposition, Lai and Wang (1993) call a real-valued statistic $U_n = U_n(X_1,\dots,X_n)$ an *asymptotic U-statistic* if it has the decomposition

$$\begin{aligned}
U_n = {} & \sum_{i=1}^{n} \left\{ \frac{\alpha(X_i)}{\sqrt{n}} + \frac{\alpha'(X_i)}{n^{3/2}} \right\} + \sum_{1\le i<j\le n} \frac{\beta(X_i,X_j)}{n^{3/2}} \\
& + \sum_{1\le i<j<k\le n} \frac{\gamma(X_i,X_j,X_k)}{n^{5/2}} + R_n,
\end{aligned} \tag{16.12}$$

where $\alpha, \alpha', \beta, \gamma$ are non-random Borel functions which are invariant under permutation of the arguments and which satisfy assumptions (A2)–(A4) below, and the R_n are random variables satisfying (A1):

(A1) $P\{|R_n| \ge n^{-1-\varepsilon}\} = o(n^{-1})$ for some $\varepsilon > 0$.
(A2) $E\alpha(X) = E\alpha'(X) = 0$.
(A3) $E\{\beta(X_1,X_2) \mid X_1\} = 0$, $E\{\gamma(X_1,X_2,X_3) \mid X_1,X_2\} = 0$.
(A4) $E\{|\alpha'(X_1)|^3 + |\gamma(X_1,X_2,X_3)|^4\} < \infty$.

To develop an Edgeworth expansion, with an error of order $o(n^{-1})$, for the distribution of an asymptotic U-statistic, Lai and Wang (1993) make the following assumptions on α and β:

(B) $E\alpha^2(X_1) = \sigma^2 > 0, E\alpha^2(X_1) < \infty$ and $\limsup_{|t|\to\infty} |Ee^{it\alpha(X_1)}| < 1$.

(C) $E|\beta(X_1,X_2)|^r < \infty$ for some $r > 2$ and there exist K Borel functions $f_\nu : \mathbb{R}^p \to \mathbb{R}$ such that $Ef_\nu^2(X_1) < \infty (\nu = 1,\dots,K)$ and the covariance matrix of $W_1,\dots,W_K$ is positive definite, where $W_\nu = E\{\beta(X_1,X_2)f_\nu(X_2) \mid X_1\}$, with $K(r-2) > 4r$ if $\gamma(X_1,X_2,X_3) = 0$ *a.s.* and $K(r-2) > 32r - 40$ otherwise.

Condition (B) is natural for Edgeworth expansions of the major term $\sum_{i=1}^n \alpha(X_i)/\sqrt{n}$ in the decomposition (16.12). In the case $\gamma = R_n = 0$, Bickel, Götze and van Zwet (1986) introduced a condition that is equivalent to condition (C) (see their Lemma 4.1). Noting that $\beta(X_1,X_2)$ can be represented in the form $\sum_{i=1}^n c_\nu g_\nu(X_1)g_\nu(X_2)$ when condition (C) fails, Lai and Wang (1993) introduce the following alternative to condition (C):

(D) There exist constants c_ν and Borel functions $g_\nu : \mathbb{R}^p \to \mathbb{R}$ such that $Eg_\nu(X) = 0, E|g_\nu(X)|^r < \infty$ for some $r \geq 5$ and $\beta(X_1,X_2) = \sum_{\nu=1}^K c_\nu g_\nu(X_1)g_\nu(X_2)$ *a.s.*; moreover, for some $0 < \delta < \min\{1, 2(1 - 11r^{-1}/3)\}$,

$$\limsup_{|t|\to\infty} \sup_{|s_1|+\cdots+|s_K| \leq |t|^{-\delta}} \left| E\exp\left(it \left\{ \alpha(X) + \sum_{\nu=1}^K s_\nu g_\nu(X) \right\} \right) \right| < 1.$$

Under these assumptions, they prove the following Edgeworth expansion for asymptotic U-statistics.

Theorem 16.2. *Let U_n be an asymptotic U-statistic defined by* (16.12) *and (A1)–(A4). Suppose α satisfies (B) and either (C) or (D) holds. Let $\sigma = (E\alpha^2(X))^{1/2}$ and define*

$$\begin{aligned}
a_3 &= E\alpha^3(X), \qquad a_4 = E\alpha^4(X), \qquad a' = E\{\alpha(X)\alpha'(X)\}, \\
b &= E\{\alpha(X_1)\alpha(X_2)\beta(X_1,X_2)\}, \\
c &= E\{\alpha(X_1)\alpha(X_2)\alpha(X_3)\gamma(X_1,X_2,X_3)\}, \\
\kappa_3 &= a_3 + 3b, \\
\kappa_4 &= a_4 - 3\sigma^4 + 4c \\
&\quad + 12E\{\alpha^2(X_1)\alpha(X_2)\beta(X_1,X_2) + \alpha(X_1)\alpha(X_2)\beta(X_1,X_3)\beta(X_2,X_3)\}, \\
P_1(z) &= \kappa_3\sigma^{-3}(z^2-1)/6, \\
P_2(z) &= \left\{ a' + \frac{E\beta^2(X_1,X_2)}{4} \right\} \frac{z}{\sigma^2} + \frac{\kappa_4}{24\sigma^4}(z^3 - 3z) + \frac{\kappa_3^2}{72\sigma^6}(z^5 - 10z^3 + 15z).
\end{aligned}$$

Then $P\{U_n/\sigma \leq z\} = \Phi(z) - n^{-1/2}\phi(z)P_1(z) - n^{-1}\phi(z)P_2(z) + o(n^{-1})$, uniformly in $-\infty < z < \infty$.

To prove Theorem 16.2, Lai and Wang (1993) make use of Theorem 2.25 to analyze the distribution function of $\tilde{U}_n := U_n - R_n$ via its characteristic function, noting that in view of (A1) it suffices to consider $\tilde{U}_n$ instead of U_n. Let $f_n(t) = Ee^{it\tilde{U}_n/\sigma}$. Take $2 < s \leq \min(3,r)$. By making use of the Taylor expansion $e^{iu} = 1 + iu - u^2/2 + O(|u|^s)$ as $u \to 0$, Lai and Wang (1993, pp. 531–532) show that $f_n(t) = g_n(t) + o(n^{-(1+2\rho)}|t|)$ uniformly in $|t| \leq u^\rho$, where

$$g_n(t) = \int_{-\infty}^{\infty} e^{itz} d\left\{\Phi(z) - n^{-1/2}\phi(z)P_1(z) - n^{-1}\phi(z)P_2(z)\right\}$$

and $0 < \rho < 1/4$ such that $s/2 - \rho(s-1) > 1 - 2\rho$. Since $\int_{|t|\geq n^\delta} |t|^{-1}|g_n(t)|dt = o(n^{-1})$ for any $\delta > 0$, Lai and Wang (1993, pp. 532–539) complete the proof of Theorem 16.2 by showing that

$$\int_{n^\rho \leq |t| \leq n^{(r-1)/r}(\log n)^{-1}} |t^{-1} f_n(t)|dt = o(n^{-1}),$$
$$\int_{n^{(r-1)/r}(\log n)^{-1} \leq |t| \leq n \log n} |t^{-1} f_n(t)|dt = o(n^{-1}).$$

Lai and Wang (1993, pp. 526–527, 539–541) have also extended the preceding analysis to derive Edgeworth expansions for asymptotic U-statistics, by making use of the following result of Abramovitch and Singh (1985, p. 129) on the empirical characteristic function $\hat{\psi}_n(t) := \int e^{ity} d\hat{F}_n(y)$, where $\hat{F}_n$ is the distribution function of i.i.d. random variables $Y_1, Y_2, \ldots$ that have distribution function F and characteristic function ψ:

$$\sup_{|t|\leq n^a} \left|\hat{\psi}_n\left(t/\sqrt{n}\right) - \psi\left(t/\sqrt{n}\right)\right| \to 0 \ \ a.s. \qquad \text{for any } a > 0. \tag{16.13}$$

Theorem 16.3. *With the same notation and assumptions as in Theorem 16.2, let H denote the distribution of X_1 and $\hat{H}_n(A) = n^{-1}\sum_{i=1}^n I(X_i \in A)$ denote the empirical distribution, and let $X_1^*, \ldots, X_n^*$ be i.i.d. with common distribution $\hat{H}_n$. Suppose that there exist functions $\hat{\alpha}_n, \hat{A}_n, \hat{\beta}_n, \hat{\gamma}_n$, depending on $\hat{H}_n$ and invariant under permutation of the arguments, such that*

$$n^{-1}\sum_{i=1}^n |\hat{A}_n(X_i)|^3 + n^{-3}\sum_{1\leq i<j<k\leq n} |\hat{\gamma}_n(X_i, X_j, X_k)|^4 = O_p(1), \tag{16.14}$$

$$\begin{aligned}\sum_{i=1}^n \hat{\alpha}_n(X_i) = \sum_{i=1}^n \hat{A}_n(X_i) = 0 &= \sum_{i=1}^n \hat{\beta}_n(y_1, X_i)\\ &= \sum_{i=1}^n \hat{\gamma}_n(y_1, y_2, X_i) \qquad \textit{for any } y_1, y_2 \in S(H),\end{aligned} \tag{16.15}$$

$$\sup_{x\in S(H)} \frac{|\hat{\alpha}_n(x) - \alpha(x)|}{1 + |\alpha(x)|} + \sup_{x,y\in S(H)} |\hat{\beta}_n(x,y) - \beta(x,y)| = O_p(n^{-1/2}), \tag{16.16}$$

where $S(H)$ denotes the support of H. Let

$$\begin{aligned}U_n^* = &\sum_{i=1}^n \left\{\frac{\hat{\alpha}_n(X_i^*)}{\sqrt{n}} + \frac{\hat{A}_n(X_i^*)}{n^{3/2}}\right\} + \sum_{1\leq i<j\leq n} \frac{\hat{\beta}_n(X_i^*, X_j^*)}{n^{3/2}}\\ &+ n^{-5/2}\sum_{1\leq i<j<k\leq n} \hat{\gamma}_n(X_i^*, X_j^*, X_k^*) + R_n^*,\end{aligned} \tag{16.17}$$

where $nP\{|\tilde{R}_n^*| \geq n^{-1-\varepsilon} \mid \hat{H}_n\} \xrightarrow{P} 0$ *for some* $\varepsilon > 0$. *Let* $\hat{\sigma}_n^2 = E\{\hat{\alpha}_n^2(X_1^*) \mid \hat{H}_n\}$. *Then* $P\{U_n^* \leq \hat{\sigma}_n z \mid \hat{H}_n\} = \Phi(z) - n^{-1/2}\phi(z)P_1(z) + O_p(n^{-1})$, *uniformly in* $-\infty < z < \infty$. *Consequently,* $\sup_z |P\{U_n/\sigma \leq z\} - P\{U_n^* \leq \hat{\sigma}_n z \mid \hat{H}_n\}| = O_p(n^{-1})$.

Although the conclusion of Theorem 16.3 seems restrictive because it is concerned with the distribution of the standardized statistic U_n/σ, it is straightforward to modify the result so that it is applicable to the Studentized statistic $U_n/\hat{\sigma}_n$. As noted by Gross and Lai (1996), $U_n/\hat{\sigma}_n$ can be expressed as an asymptotic U-statistic with $\sigma = 1$ and therefore Theorem 16.2 is still applicable to $U_n/\hat{\sigma}_n$ (in place of U_n/σ). In this case Theorem 16.3 can be modified to show that

$$\sup_z |P\{U_n/\hat{\sigma}_n \leq z\} - P\{U_n^*/\hat{\sigma}_n^* \leq z \mid \hat{H}_n\}| = O_p(n^{-1}), \tag{16.18}$$

where, similar to U_n^*, $\hat{\sigma}_n^*$ replaces $X_1, \ldots, X_n$ in $\hat{\sigma}_n$ by $X_1^*, \ldots, X_n^*$. Gross and Lai (1996) have applied this result to develop an asymptotic theory of Efron's (1981) "simple" bootstrap method for right censored survival data and have also extended it to right censored and left truncated data described below.

Let $(X_1, T_1, C_1), (X_2, T_2, C_2), \ldots$ be i.i.d. random vectors such that (T_i, C_i) is independent of X_i. The quantities of interest are the X_i which are not completely observable because of the presence of the right censoring variables C_i and left truncation variables T_i. Letting

$$\tilde{X}_i = \min(X_i, C_i), \qquad \delta_i = I(X_i \leq C_i), \tag{16.19}$$

one only observes $(\tilde{X}_i, \delta_i)$ when $\tilde{X}_i \geq T_i$. Thus, the data consist of n observations $(\tilde{X}_{i,o}, \delta_{i,o}, T_{i,o})$ with $\tilde{X}_{i,o} \geq T_{i,o}$, $i = 1, \ldots, n$. Such left truncated and right censored data have wide applications in survival analysis, where X_i represents the failure time of the ith subject in a clinical study. The subject may withdraw from the study, or may be lost to follow-up, or may still survive by the scheduled end of the study. Thus, X_i is right censored. In certain studies of the duration of a disease, patients are followed from an entrance (or left truncation) age to an exit age (due to death or right censoring). When truncation is absent, we can set $T_i = -\infty$. When censoring is absent, multiplying the random variables by -1 converts a left truncated model into a right truncated one, and right truncated data have extensive applications in astronomy and econometrics; see Gross and Lai (1996, p. 509). Let $\hat{\Psi}_n$ denote the empirical distribution that puts probability $1/n$ at each $X_i = (\tilde{X}_{i,o}, \delta_{i,o}, T_{i,o})$, $i = 1, \ldots, n$. The *simple bootstrap* sample consists of i.i.d. random vectors $Z_1^*, \ldots, Z_n^*$ drawn from the distribution $\hat{\Psi}_n$. As noted by Gross and Lai (1996, p. 512), the Z_i are i.i.d. random vectors whose common distribution Ψ is given by

$$\begin{aligned} &P\left\{\delta_{i,o} = \delta, (\tilde{X}_{i,o}, T_{i,o}) \in A\right\} \\ &\qquad = P\{I(X_1 \leq C_1) = \delta, (X_1 \wedge C_1, T_1) \in A\} / P\{X_1 \wedge C_1 \geq T_1\}, \end{aligned}$$

for $\delta = 0$ or 1 and all Borel sets A such if $x \geq t$ if $(x,t) \in A$. Let $S = S(Z_1, \ldots, Z_n)$ be an estimate of the functional $\mu(\Psi)$ and let $\hat{\sigma} = \hat{\sigma}(Z_1, \ldots, Z_n)$ be an estimate of the standard error of S, in which $S/\hat{\sigma}$ can be expressed as an asymptotic

U-statistic. Theorem 16.3 can be used to show that the sampling distribution of $(S - \mu(\Psi))/\hat{\sigma}$ can be approximated by that of $(S^* - \mu(\hat{\Psi}_n))/\hat{\sigma}^*$ with $O_p(n^{-1})$ error, where $S^* = S(Z_1^*, \ldots, Z_n^*)$ and $\hat{\sigma}^* = \hat{\sigma}(Z_1^*, \ldots, Z_n^*)$; see Gross and Lai (1996) for details and examples and Problems 10.3 and 10.4 for the background.

16.4 Application of Cramér-Type Moderate Deviations

Results of the type in Theorem 16.3 are concerned with absolute errors of the bootstrap approximation to the sampling distribution of a Studentized statistic. For small tail probabilities of a Studentized statistic, relative errors of the bootstrap approximation are more relevant than absolute errors. In this section we apply Theorem 7.4 to study the relative error in the case of the t-statistic $T_n = \sqrt{n}(\bar{X}_n - \mu)/s_n$ and its bootstrap version $T_n^* = \sqrt{n}(\bar{X}_n^* - \bar{X}_n)/s_n^*$, where $X_1, \ldots, X_n$ are i.i.d. with mean μ and $X_1^*, \ldots, X_n^*$ are i.i.d. drawn from the empirical distribution $\hat{F}_n$ of $\{X_1, \ldots, X_n\}$.

Theorem 16.4. *If* $E|X_1|^{2+\delta} < \infty$ *for some* $0 < \delta \le 1$, *then with probability* 1,

$$\frac{P(T_n^* \ge x|\hat{F}_n)}{P(T_n \ge x)} = 1 + o(1) \quad \textit{and} \quad \frac{P(T_n^* \le -x|\hat{F}_n)}{P(T_n \le -x)} = 1 + o(1) \tag{16.20}$$

uniformly in $0 \le x \le o(n^{\delta/(4+2\delta)})$.

Proof. We only consider the first part of (16.20) and assume $\mu = 0$ without loss of generality. In view of (15.1), the distribution functions of T_n and S_n/V_n are related via

$$\{T_n \ge x\} = \left\{ S_n \ge x \left(\frac{n}{n + x^2 - 1} \right)^{1/2} V_n \right\}. \tag{16.21}$$

From Theorem 7.4, it follows that, uniformly in $0 \le x \le o(n^{\frac{\delta}{2(2+\delta)}})$,

$$\frac{P(T_n \ge x)}{1 - \Phi(x)} = 1 + o(1). \tag{16.22}$$

For the bootstrap distribution, we can apply Theorem 7.4 again (see (16.21) and Remark 7.9) to obtain

$$\left| \frac{P(T_n^* \ge x|\hat{F}_n)}{1 - \Phi(x)} - 1 \right| \le \frac{A(1+x)^{2+\delta}}{d_{n,\delta}^{*\,2+\delta}}, \tag{16.23}$$

for $0 \le x \le d_{n,\delta}^*$ where, letting $E^*(\cdot) = E(\cdot|\hat{F}_n)$,

$$d^*_{n,\delta} = n^{\delta/(4+2\delta)} \frac{E^*(|X_1^*|^2)^{1/2}}{(E^*|X_1^*|^{2+\delta})^{1/(2+\delta)}}$$
$$= n^{\delta/(4+2\delta)} \frac{(n^{-1}\sum_{i=1}^n X_i^2)^{1/2}}{(n^{-1}\sum_{i=1}^n |X_i|^{2+\delta})^{1/(2+\delta)}}.$$

By the strong law of large numbers,

$$d^*_{n,\delta}/n^{\delta/(4+2\delta)} \to (EX_1^2)^{1/2}/(E|X_1|^{2+\delta})^{1/(2+\delta)} \quad a.s. \quad \text{as } n \to \infty. \tag{16.24}$$

From (16.21)–(16.24), (16.20) follows. □

Theorem 16.4 states that the bootstrap provides an accurate approximation of moderate deviation probabilities for the t-statistics. Jing et al. (1994) have obtained a similar result under the much stronger assumption $E\exp(tX_1^2) < \infty$ for some $t > 0$. If $\delta = 1$, the region where (16.20) is valid becomes $0 \le x \le o(n^{1/6})$, which is smaller than $0 \le x \le o(n^{1/3})$ obtained by Jing et al. (1994) under the much stronger assumption that X_1^2 has a finite moment generating function.

16.5 Supplementary Results and Problems

1. Let $X_1, \ldots, X_n$ be i.i.d $\alpha \times 1$ random vectors with characteristic function ψ. Let $\hat{\psi}_n(t) = n^{-1}\sum_{k=1}^n \exp(it'X_k)$ denote the *empirical characteristic function* (i.e., characteristic function of the empirical distribution):
   (a) Make use of the Bennett–Hoeffding inequalities (Theorem 2.17) to obtain an exponent bound for $P(|\hat{\psi}_n(t) - \psi(t)| > \varepsilon)$ for every t.
   (b) Let $\alpha > 0$, $\Delta_n(t) = \hat{\psi}_n(t) - \psi(t)$. Cover $\{t : \|t\| \le n^\alpha\}$ by cubes I_j, $1 \le j \le (2n^\alpha/\varepsilon)^d$, with centers t_j and width ε. Show that
   $$\sup_{\|t\| \le n^\alpha} |\Delta_n(t)| \le \max_{1 \le j \le (2n^\alpha/\varepsilon)^d} |\Delta_n(t_j)| + \varepsilon \left\{ E\|X_1\| + n^{-1}\sum_{i=1}^n \|X_i\| \right\}.$$
   Hint: $|e^{iu} - 1| \le |u|$, and more generally, $|e^{iu} - 1 - \cdots - \frac{(iu)^{n-1}}{(n-1)!}| \le \frac{|u|^n}{n!}$ for $u \in \mathbb{R}$.
   (c) Make use of (a) and (b) to show that if $E\|X_1\| < \infty$ then
   $$\sup_{\|t\| \le n^\alpha} |\hat{\psi}_n(t) - \psi(t)| > \varepsilon \to 0 \ a.s.$$
   (d) Show that if $E\|X_1\| < \infty$ and ψ satisfies Cramér's condition $\limsup_{\|t\| \to \infty} |\psi(t)| < 1$, then $\limsup_{n\to\infty}(\sup_{\|t\| \le n^\alpha} |\hat{\psi}_n(t)|) < 1$ $a.s.$ for every $\alpha > 0$.
2. Consider the problem, mentioned at the end of Sect. 16.2, of constructing a confidence interval for the correlation coefficient ρ of a bivariate distribution F based on a sample of i.i.d. random vectors $X_1, \ldots, X_n$ drawn from F. Let $\hat{\rho}$ denote the sample correlation coefficient:

(a) As noted in Sect. 16.2.1, one can use the delta method to show that $\hat{\rho}$ is asymptotically normal if $EX_1^4 < \infty$. Carry out the details and use the method of moments to derive $\widehat{\text{se}}$ in this case.

(b) Note that whereas ρ and $\hat{\rho}$ do not exceed 1 in absolute value, the random variable $\widehat{\text{se}}$ in (a) is not bounded when F does not have bounded support. Instead of ρ, consider the transformed parameter

$$\theta = \frac{1}{2}\log\left(\frac{1+\rho}{1-\rho}\right), \tag{16.25}$$

which ranges from $-\infty$ to ∞ (instead of from -1 to 1). Use the delta method to estimate the standard error of $\hat{\theta} = \frac{1}{2}\log[(1+\hat{\rho})/(1-\hat{\rho})]$.

(c) Show that in the case where F is bivariate normal, $\sqrt{n}(\hat{\theta}-\theta)$ has a limiting standard normal distribution. Thus, the nonlinear transformation (16.25) approximately normalizes the estimate $\hat{\rho}$ and is often called a *variance-stabilizing* transformation.

One way to improve the bootstrap-t confidence interval for ρ is to construct the bootstrap-t interval for θ and then to transform it back to a confidence interval for $\rho = (e^{2\theta}-1)/(e^{2\theta}+1)$. Bootstrap-$t$ confidence intervals are not *transformation-respecting*; it makes a difference which scale is used to construct the interval. In the case of the correlation coefficient ρ, the transformation (16.25) is known to be variance-stabilizing and normalizing, if F is bivariate normal (and therefore bypasses the need to estimate the standard error), and also works well for more general F. Efron and Tibshirani (1993, pp. 164–165) describe a bootstrap method to find an appropriate variance-stabilizing transformation and construct a bootstrap-t interval for the transformed parameter so that the inverse transformation then yields the confidence interval for the original parameter.

3. Let $X_1,\dots,X_n$ be i.i.d. random variables with $E|X_1|^3 < \infty$ and a common continuous distribution function F. Let $X_{(1)} \le \dots \le X_{(n)}$ denote the order statistics and let $\psi : [0,1] \to \mathbb{R}$ be four times continuously differentiable. Consider the linear combination

$$S_n = \sum_{i=1}^{n} \psi(i/n)X_{(i)} = n\int_{-\infty}^{\infty} x\psi(F_n(x))\,dF_n(x)$$

of order statistics, where F_n is the empirical distribution function. Let $\mu = \int_{-\infty}^{\infty} x\psi(F(x))dF(x)$,

$$\sigma^2 = 2\int\int_{s<t} \psi(F(s))\,\psi(F(t))\,F(s)\,(1-F(t))\,dsdt.$$

Then $(S_n - n\mu)/\sqrt{n} \Rightarrow N(0,\sigma^2)$, and Lai and Wang (1993, pp. 525–526) have expressed $(S_n - n\mu)/\sqrt{n}$ as an asymptotic U-statistic. Give a consistent estimate $\hat{\sigma}^2$ of σ and express $(S_n - n\mu)/(\sqrt{n}\hat{\sigma})$ as an asymptotic U-statistic.

Chapter 17
Pseudo-Maximization in Likelihood and Bayesian Inference

The self-normalized statistics in Chaps. 15 and 16 are Studentized statistics of the form $(\hat{\theta}-\theta)/\widehat{\text{se}}$, which are generalizations of the t-statistic $\sqrt{n}(\bar{X}_n-\mu)/s_n$ for testing the null hypothesis that the mean of a normal distribution is μ when the variance σ^2 is unknown and estimated by the sample variance s_n^2. In Sect. 17.1 we consider another class of self-normalized statistics, called *generalized likelihood ratio* (GLR) statistics, which are extensions of likelihood ratio (LR) statistics (for testing simple hypotheses) to composite hypotheses in parametric models. Whereas LR statistics are martingales under the null hypothesis, GLR statistics are no longer martingales but can be analyzed by using LR martingales and the pseudo-maximization technique of Chap. 11. The probabilistic technique of pseudo-maximization via the method of mixtures has a fundamental statistical counterpart that links likelihood to Bayesian inference; this is treated in Sect. 17.2.

17.1 Generalized Likelihood Ratio Statistics

Let $X_1, X_2, \ldots$ be observations drawn from a probability measure P under which f_1 is the marginal density of X_1 and for $i \geq 2$, the conditional distribution of X_i given $X_1, \ldots, X_{i-1}$ has density function $f_i(\cdot|X_1, \ldots, X_{i-1})$ with respect to some measure ν_i. To test a simple null hypothesis $H_0 : f_i = p_i$ versus a simple alternative hypothesis $H_1 : f_i = q_i$, the likelihood ratio test based on a sample $X_1, \ldots, X_n$ of fixed size n rejects H_0 if

$$\text{LR}_n = \prod_{i=1}^{n} \{q_i(X_i|X_1, \ldots, X_{i-1})/p_i(X_i|X_1, \ldots, X_{i-1})\} \tag{17.1}$$

exceeds the threshold c for which the Type I error probability $P_{H_0}\{\text{LR}_n \geq c\}$ is equal to some prescribed α. The Neyman–Pearson lemma says that among all tests whose Type I error probability does not exceed α, the *likelihood ratio test* is most powerful

Probability and its Applications,
© Springer-Verlag Berlin Heidelberg 2009

in the sense that it maximizes the probability of rejecting the null hypothesis (called *power*) under the alternative hypothesis. Note that $\{\mathrm{LR}_m, m \geq 1\}$ is a martingale, with mean 1, under P_{H_0}.

One can also control the Type II error probability (or 1−power) of the likelihood ratio test by choosing the sample size n appropriately. Instead of using a fixed sample size n, an alternative approach is to continue sampling until LR_n shows enough evidence against H_0 or H_1. In the case of i.i.d. X_t, this is the idea behind Wald's *sequential probability ratio test* (SPRT), which stops sampling at stage

$$N = \inf\{n \geq 1 : \mathrm{LR}_n \geq B \text{ or } \mathrm{LR}_n \leq A\} \quad (\inf \emptyset = \infty), \tag{17.2}$$

and which rejects H_0 if $\mathrm{LR}_N \geq B$, and H_1 if $\mathrm{LR}_N \leq A$, upon stopping, where $0 < A < 1 < B$. Wald and Wolfowitz (1948) showed that $E_{H_0}(N)$ and $E_{H_1}(N)$ are both minimized among all tests, sequential or otherwise, of H_0 versus H_1 whose Type I and Type II error probabilities do not exceed those of the SPRT; see Sect. 18.1.1.

In parametric models in which f_i depends on an unknown parameter θ, the null hypothesis H_0 can be described by $\theta \in \Theta_0$, which is not simple unless Θ_0 is a singleton. Let Θ denote the parameter space, i.e., the set of possible values of θ. Since θ is unknown, a simple extension of the likelihood ratio (17.1) is to estimate θ by maximum likelihood under $H_0 : \theta \in \Theta_0$ and under $H_1 : \theta \in \Theta - \Theta_0$ and replace θ in $f_{i,\theta}$ by separate maximum likelihood estimates (MLE) under H_0 and H_1. A variant of this approach leads to the generalized likelihood ratio statistic

$$\mathrm{GLR}_n = \frac{\sup_{\theta \in \Theta} \prod_{i=1}^n f_{i,\theta}(X_i|X_1,\ldots,X_n)}{\sup_{\theta \in \Theta_0} \prod_{i=1}^n f_{i,\theta}(X_i|X_1,\ldots,X_n)}. \tag{17.3}$$

This test statistic is asymptotically pivotal under H_0. In fact, a classical result, due to Wilks (1938), states that under certain regularity conditions, $2\log(\mathrm{GLR}_n)$ has a limiting χ^2_{d-p}-distribution, where d is the dimension of θ and p is its effective dimension under H_0. More precisely, Θ is typically a d-dimensional manifold and Θ_0 is a p-dimensional submanifold of Θ. Further discussion and some recent developments concerning the asymptotic theory of GLR_n are given in the next chapter (Sect. 18.2).

17.1.1 The Wilks and Wald Statistics

The test statistic $\Lambda_n := 2\log(\mathrm{GLR}_n)$ is often called *Wilks' statistic*. Let $\hat{\theta}$ be the MLE of θ, and $\hat{\theta}_0 \in \Theta_0$ be the constrained MLE under H_0. Let

$$l_n(\theta) = \sum_{i=1}^n \log f_{i,\theta}(X_i|X_1,\ldots,X_{i-1}) \tag{17.4}$$

be the log-likelihood function. Under H_0, Λ_n is asymptotically equivalent to *Wald's statistic*

$$\mathrm{W}_n := (\hat{\theta} - \hat{\theta}_0)'\hat{V}^{-1}(\hat{\theta} - \hat{\theta}_0), \tag{17.5}$$

where $\hat{V}$ is an estimate of the asymptotic convariance matrix of $\hat{\theta}$ under H_0. In the case of simple null hypothesis $\Theta_0 = \{\theta^0\}$, $\hat{\theta}_0 = \theta^0$, $\hat{V}^{-1} = -\nabla^2 l_n(\hat{\theta})$ and W_n reduces to the square of a Studentized statistic, which has a limiting χ^2_d distribution and is therefore asymptotically pivotal; see Problem 17.2. We next consider more general $\Theta_0 = \{(\gamma^0, \lambda) : \lambda \in \mathbb{R}^{d-p}\}$, in which λ can be viewed as a nuisance parameter (similar to σ in the t-test). Let $\theta = (\gamma, \lambda)$,

$$I_{\psi\lambda} = \left(-\frac{\partial^2}{\partial\gamma_i \partial\lambda_j} l_n(\gamma, \lambda)\right)_{1\le i\le p, 1\le j\le d-p}, \qquad I_{\psi\psi} = \left(-\frac{\partial^2}{\partial\gamma_i \partial\lambda_j} l_n(\gamma, \lambda)\right)_{1\le i,j\le p}$$

and define $I_{\lambda\lambda}, I_{\lambda\gamma}$ similarly; note that $I_{\lambda\gamma} = I'_{\gamma\lambda}$. Then by the asymptotic theory of MLE,

$$\hat{V}^{-1} = I_{\gamma\gamma}(\hat{\theta}) - I_{\gamma\lambda}(\hat{\theta}) I^{-1}_{\lambda\lambda}(\hat{\theta}) I_{\lambda\gamma}(\hat{\theta}). \tag{17.6}$$

The asymptotic equivalence between Λ_n and W_n can be derived by applying a two-term Taylor expansion to $l_n(\theta)$, first around $(\hat{\gamma}, \hat{\lambda})$ and then around $(\gamma^0, \hat{\lambda}^0)$, where $\hat{\lambda}^0$ is the MLE under the constraint $\gamma = \gamma^0$. Approximating the likelihood function by making use of the asymptotic normality of $(\hat{\gamma}, \hat{\lambda})$, it can be shown that

$$\hat{\lambda}^0 = \hat{\lambda} + I^{-1}_{\lambda\lambda}(\theta^0) I_{\lambda\gamma}(\theta^0)(\hat{\gamma} - \gamma^0) + o_p(n^{1/2}), \tag{17.7}$$

where the superscript 0 denotes the true parameter value; see Problem 17.4. Moreover, from the Taylor expansions of $l_n(\hat{\theta}) - l_n(\theta^0)$ and $l_n(\gamma^0, \hat{\lambda}^0) - l_n(\theta^0)$, it follows that

$$\begin{aligned}
\Lambda_n &= 2\{l_n(\hat{\theta}) - l_n(\theta^0)\} - 2\left\{l_n(\gamma^0, \hat{\lambda}^0) - l_n(\gamma^0, \lambda^0)\right\} \\
&= \left((\hat{\gamma} - \gamma^0)', (\hat{\lambda} - \lambda^0)'\right) \nabla^2 l_n(\theta^0) \begin{pmatrix} \hat{\gamma} - \gamma^0 \\ \hat{\lambda} - \lambda^0 \end{pmatrix} \\
&\quad - \left(0, (\hat{\lambda}^0 - \lambda^0)'\right) \nabla^2 l_n(\theta^0) \begin{pmatrix} 0 \\ \hat{\lambda} - \hat{\lambda}^0 \end{pmatrix} + o_p(1).
\end{aligned}$$

Combining this with (17.7) then yields

$$\begin{aligned}
\Lambda_n &= (\hat{\lambda} - \lambda^0)' \left\{I_{\gamma\gamma}(\theta^0) - I_{\gamma\lambda}(\theta^0) I^{-1}_{\lambda\lambda}(\theta^0) I_{\lambda\gamma}(\theta^0)\right\} (\hat{\lambda} - \lambda^0) + o_p(1) \\
&= \mathrm{W}_n + o_p(1),
\end{aligned} \tag{17.8}$$

in view of (17.5) and (17.6) for the present setting with $\Theta_0 = \{(\gamma^0, \lambda) : \lambda \in \mathbb{R}^{d-p}\}$; see Problem 17.5.

Although Wilks' and Wald's statistics are asymptotically equivalent under H_0, Wilks' statistic is often preferred because it is already self-normalized and does not require estimation of the asymptotic covariance matrix under H_0. As pointed out in Sect. 16.2.2, the linear approximations in deriving asymptotic standard error formulas may be poor when $\hat{\theta}$ is not sufficiently close to θ and therefore the χ^2-approximation to W_n is often less accurate than that to Λ_n for the sample sizes encountered in practice.

17.1.2 Score Statistics and Their Martingale Properties

Since $E_{\theta^0}\{\frac{\partial}{\partial\theta}\log f_{t,\theta}(X_t \mid X_1,\dots,X_{t-1})\mid_{\theta=\theta^0} \mid X_1,\dots,X_{t-1}\} = 0$ for $i = 1,\dots,d$ (see Problem 17.1), the score statistics

$$S_n(\theta^0) = \nabla l_n(\theta^0) = \sum_{t=1}^{n} \nabla \log f_{t,\theta}(X_t \mid X_1,\dots,X_{t-1})\mid_{\theta=\theta^0} \tag{17.9}$$

form a martingale with respect to the filtration $\mathscr{F}_n := \sigma(X_1,\dots,X_n)$. Martingale central limit theorems (see Sect. 15.3.1) can therefore be used to show that under certain regularity conditions, $S_n(\theta^0)/\sqrt{n}$ has a limiting normal distribution with mean 0 and covariance matrix V. Moreover, likelihood theory shows that V can be consistently estimated by $(-\nabla^2 l_n(\theta^0))/n$ or $(-\nabla^2 l_n(\hat\theta))/n$; see Problem 17.2. Hence an alternative to the Wilks' and Wald's tests of $H_0 : \theta = \theta^0$ is the *score test* (also called Rao's test) that rejects H_0 if the Studentized score statistic $(-\nabla^2 l_n(\theta^0))^{-1/2} S_n(\theta^0)$ exceeds the normal quantile $z_{1-\alpha}$ for a one-sided level-α test of H_0, or if $(S_n(\theta^0))'(-\nabla^2 l_n(\theta_0))^{-1} S_n(\theta^0)$ exceeds the $1-\alpha$ quantile of the χ^2_d distribution for a two-sided level-α test of H_0.

Assuming $\hat\theta$ to be consistent, its asymptotic normality follows from that of the score statistic (17.9) since

$$0 = \nabla l_n(\hat\theta) \approx \nabla l_n(\theta^0) + (\nabla^2 l_n(\theta^0))(\hat\theta - \theta^0),$$

implying that

$$(-\nabla^2 l_n(\theta^0))^{\frac{1}{2}}(\hat\theta - \theta^0) \approx (-\nabla^2 l_n(\theta^0))^{-\frac{1}{2}} \nabla l_n(\theta^0). \tag{17.10}$$

Note that the left-hand side of (17.10) is the self-normalized Wald statistic while the right-hand side is the self-normalized score statistic. Although the Hessian matrix $-\nabla l_n^2(\theta^0)$ is commonly used for self-normalization in (17.10), their asymptotically equivalent variants also can be used. In the case of i.i.d. X_t, the score statistic (17.9) reduces to $\sum_{t=1}^n Y_t$, where $Y_t = \nabla \log f_\theta(X_t)\mid_{\theta=\theta^0}$, and using $(\sum_{t=1}^n Y_t Y_t')^{1/2}$ instead of $(-\nabla^2 l_n(\theta^0))^{1/2}$ to self-normalize the score statistic $\sum_{t=1}^n Y_t$ leads to the self-normalized sum which is a multivariate extension of that considered in Part I; see also Sect. 15.2.3 for the advantages of using $(\sum_{t=1}^n Y_t Y_t')^{1/2}$ to self-normalize the score statistic in this case.

17.2 Penalized Likelihood and Bayesian Inference

The GLR statistic (17.3) for testing the null hypothesis H_0 that a d-dimensional parameter vector θ belongs to be a p-dimensional space Θ_0 with $p < d$ can be regarded as a special case of the problem of choosing the dimension of a model. Instead of using a hypothesis testing approach to this dimension selection problem, an alternative approach, introduced by Schwarz (1978), is to use a Bayesian formulation that puts a prior distribution first on the set of parametric models and then on θ given the parametric model. Using Laplace's method described in Sect. 11.1, Schwarz has shown that the Bayes solution can be approximated by a *penalized likelihood criterion* that involves $l_n(\hat{\theta}^{(j)})$ plus a penalty term that involves the dimension d_j of the model, where $\hat{\theta}^{(j)}$ denotes the MLE under the constraint that it belongs to Θ_j with dimension d_j. Besides model selection, this section also uses pseudo-maximization via the method of mixtures to relate GLR and Bayes tests.

17.2.1 Schwarz's Bayesian Selection Criterion

Schwarz considers the special case of i.i.d. d-dimensional random vectors X_i whose common density function belongs to the exponential family

$$f_\theta(x) = \exp(\theta' x - \psi(\theta)) \tag{17.11}$$

with respect to some measure ν or $\mathbb{R}^d$. The natural parameter space $\{\theta \in \mathbb{R}^d : \int e^{\theta' x} d\nu(x) < \infty\}$ is a convex subset of $\mathbb{R}^d$. Since $\int f_\theta(x) d\nu(x) = 1$, $e^{\psi(\theta)} = \int e^{\theta' x} d\nu(x)$. Suppose that for $1 \le j \le J$, Θ_j consists of vectors with a known subvector γ_j of dimension $d - d_j$, and the prior distribution assigns probability α_j to Θ_j and has a density function $\pi_j(\lambda)$, with respect to Lebesgue measure on $\mathbb{R}^{d_j}$, for the remaining subvector $\lambda \in \mathbb{R}^{d_j}$ of θ. Partitioning $\bar{X}_n$ into corresponding subvectors $\bar{X}_n^{(1)}$, $\bar{X}_n^{(2)}$ with respective dimensions $d - d_j$ and d_j, the posterior probability in favor of Θ_j is proportional to

$$\begin{aligned} p_j :=& \alpha_j \int e^{n\left(\gamma_j' \bar{X}_n^{(1)} + \lambda' \bar{X}_n^{(2)}\right)} \pi_j(\lambda) d\lambda \\ &\sim \alpha_j (2\pi/n)^{d_j/2} e^{l_n(\hat{\theta}^{(j)})} \pi_j\left(\hat{\theta}^{(j)}\right) \Big/ \left\{\det\left(\nabla^2 \psi\left(\hat{\theta}^{(j)}\right)\right)\right\}^{1/2} \end{aligned} \tag{17.12}$$

where $l_n(\theta) = n(\theta' \bar{X}_n - \psi(\theta))$ is the log-likelihood function (17.4) and $\hat{\theta}^{(j)}$ is the maximizer of $l_n(\theta)$ over Θ_j. Note that the maximization is in fact over the subvector λ which belongs to a convex subset C_j of $\mathbb{R}^{d_j}$ so that $\int \exp(\gamma_j' x^{(1)} + \lambda' x^{(2)}) d\nu(x) < \infty$ for all $\lambda \in C_j$. The asymptotic approximation in (17.12) follows from Laplace's asymptotic formula (11.4), assuming that π_j is positive and continuous on C_j. Suppose the loss for choosing the wrong model is $a > 0$ and there is no loss for choosing the correct model. Since the posterior probability in favor of Θ_j is $p_j/(p_1 + \cdots + p_J)$, the Bayes rule chooses the J that has the largest p_j. Using the asymptotic approximation in (17.12) therefore leads to the *Bayes information criterion*

$$\mathrm{BIC}(j) = l_n\left(\hat{\theta}^{(j)}\right) - \frac{d_j}{2} \log n, \tag{17.13}$$

choosing the model j with the largest BIC(j). Note that BIC(j) penalizes the maximized likelihood $l_n(\hat{\theta}^{(j)})$ by using the penalty $(\log n)/2$ for each additional dimension.

Extension of the preceding argument to more general d_j-dimensional submanifolds Θ_j of Θ involves geometric integration and generalization of Laplace's method, which will be presented in Sect. 18.2. Some authors define BIC(j) as $-l_n(\hat{\theta}^{(j)}) + \frac{d_j}{2}\log n$ and therefore choose the model that minimizes such BIC(j). There is a connection between BIC(j) and the p-value of the GLR test of Θ_1, the lowest-dimensional model, versus Θ_j $(1 \le j \le J)$, as shown by Siegmund (2004) who has also modified the BIC to handle non-regular cases, such as change-point models with the number of change-points to be chosen by modified BIC.

17.2.2 Pseudo-Maximization and Frequentist Properties of Bayes Procedures

As illustrated by (17.12), Bayes procedures are asymptotically equivalent to penalized likelihood procedures and therefore have the same asymptotic frequentist properties. For parametric models involving a family of measures P_θ indexed by a parameter vector $\theta \in \mathbb{R}^d$, frequentist properties of a procedure refer to its properties under P_{θ_0}, where θ_0 is the true parameter. In the case of Schwarz's Bayesian information criterion for model selection, a well-known frequentist property is its strong consistency, i.e., with probability 1, the BIC chooses the lowest-order true model.

A well-known frequentist property of Bayes procedures is that, under certain regularity conditions, the posterior distribution of θ given $X_1,\ldots,X_n$ is asymptotically normal with mean $\hat{\theta}$ and covariance matrix $(-\nabla^2 l_n(\hat{\theta}))^{-1}$ $a.s.[P_{\theta_0}]$, where $l_n(\theta)$ is the log-likelihood function (17.4) and $\hat{\theta}$ is the MLE. In particular, if the X_i are i.i.d., then

$$P_{\theta_0}\left\{\mathscr{L}\left[\left(-\nabla^2 l_n(\hat{\theta}_n)\right)^{1/2}(\theta-\hat{\theta})\Big|(X_1,\ldots,X_n)\right] \Longrightarrow N(0,1)\right\} = 1 \qquad (17.14)$$

when f_θ satisfies certain regularity conditions. The notation $\mathscr{L}(\cdot|X_1,\ldots,X_n)$ in (17.14) denotes the posterior distribution given $(X_1,\ldots,X_n)$, which is a random measure. Thus, (17.14) says that this random measure converges weakly to $N(0,1)$ with probability 1. To illustrate the underlying ideas, consider the one-parameter $(d=1)$ exponential family of densities $f_\theta(x) = e^{\theta x-\psi(\theta)}$ with measure to some measure ν and assume that θ has prior distribution with density function π with respect to Lebesgue measure. Then the posterior density of θ given $X_1,\ldots,X_n$ is

$$\frac{e^{n(\theta\bar{X}_n-\psi(\theta))}\pi(\theta)}{\int_{-\infty}^{\infty} e^{n(\lambda\bar{X}_n-\psi(\lambda))}\pi(\lambda)d(\lambda)},$$

and applying Laplace's asymptotic formula (11.2) to the denominator shows that the posterior density is concentrated around $\hat{\theta}$ and that it is asymptotically equivalent to

$$\begin{aligned}&\{n\psi''(\hat\theta)/2\pi\}^{\frac{1}{2}}\exp\left\{n\left[(\theta-\hat\theta)\bar X-\left(\psi(\theta)-\psi(\hat\theta)\right)\right]\right\}\\&\quad\sim\{n\psi''(\hat\theta)/2\pi\}^{\frac{1}{2}}\exp\left\{-n\psi''(\hat\theta)(\theta-\hat\theta)^2/2\right\},\end{aligned}\tag{17.15}$$

since $\psi'(\hat\theta)=\bar X$. Noting that $-l_n''(\hat\theta)=n\psi'(\hat\theta)$, we obtain from (17.15) that the posterior distribution of $(-l_n''(\hat\theta))^{1/2}(\theta-\hat\theta)$ converges weakly to the standard normal distribution $a.s.[P_{\theta_0}]$. Under certain regularity conditions, the preceding argument can also be applied to more general parametric families f_θ (with multivariate θ) for which $\hat\theta\to\theta_0$ $a.s.[P_{\theta_0}]$. In fact, by assuming $\log f_\theta(x)$ to have $K+3$ continuous partial derivatives with respect to θ, Johnson (1970) has derived asymptotic expansions of the form $\Phi(\theta)+\sum_{j=1}^K n^{-j/2}\gamma_j(\theta;X_1,\ldots,X_n)$ for the posterior distribution function of $(-l_n''(\hat\theta))^{1/2}(\theta-\hat\theta)$, with an error of the order $O(n^{-(K+1)/2})$ $a.s.[P_{\theta_0}]$; he has also extended the result to the case of Markov-dependent X_t.

Under the regularity conditions assumed above, $\hat\theta\to\theta_0$ $a.s.[P_{\theta_0}]$. Since the posterior distribution is asymptotically normal with mean $\hat\theta$, this implies that the posterior distribution converges weakly to the point mass at θ_0 $a.s.[P_{\theta_0}]$. Most generally, assume that the parameter space Θ is a complete separable metric space with metric ρ and let μ be a prior distribution on Θ. Let P denote the probability measure under which θ has distribution μ and conditional on θ, the X_t are generated from the probability measure P_θ, and let E denote expectation with respect to P. By the martingale convergence theorem, for any bounded Borel-measurable function $\varphi:\Theta\to\mathbb{R}$,

$$E\left[\varphi(\theta)\mid X_1,\ldots,X_n\right]\longrightarrow E\left[\varphi(\theta)\mid X_1,X_2,\ldots\right]\ a.s.\tag{17.16}$$

Suppose there exists a measurable function $f:\mathscr{X}\to\Theta$ such that $E\rho(\theta,f(X_1,X_2,\ldots,X_n))=0$, where $\mathscr{X}$ denotes the sample space. Then for μ-almost every $\theta_0\in\Theta$, the posterior distributions of θ given $X_1,\ldots,X_n$ converges weakly to δ_{θ_0} (the point mass at θ_0); see Le Cam and Yang (1990, pp. 148–149).

The preceding gives positive consistency results on the frequentist properties of Bayes procedures. Section 7.5 of Le Cam and Yang (1990) summarizes negative results, which arise in nonparametric problems and in parametric models when the prior measure is too "thin" around the true parameter θ_0.

17.3 Supplementary Results and Problems

1. Let $X_1,\ldots,X_n$ be n observations for which the joint density function f_θ depends on an unknown d-dimensional parameter vector θ, whose true value is denoted by θ^0. Show that

   (a) $E(\nabla\log f_\theta(X_1,\ldots,X_n)|_{\theta=\theta^0})=0$,
   (b) $E(-\nabla^2 l_n(\theta^0))=\mathrm{Cov}(\nabla l_n(\theta^0))$

   under suitable regularity conditions, where l_n is the log-likelihood function. State the regularity conditions you assume.

2. Show that under suitable regularity conditions, $\{S_n(\theta^0), \mathscr{F}_n, n \geq 1\}$ is a martingale, where $S_n(\theta^0)$ is the score statistic (17.9) and $\mathscr{F}_n$ is the σ-field generated by $X_1, \ldots, X_n$, and that $S_n(\theta^0)/\sqrt{n}$ has a limiting normal distribution with mean 0 and covariance matrix V, which can be consistently estimated by $(-\nabla^2 l_n(\theta^0))/n$ or $(-\nabla^2 l_n(\hat{\theta}))/n$. Hence show that (a) the Wald statistic (17.5) for testing $H_0 : \theta = \theta^0$ has a limiting χ^2_d-distribution, and (b) the maximum likelihood estimator $\hat{\theta}$ is asymptotically normal.
3. Consider a $d \times d$ nonsingular matrix

$$A = \begin{pmatrix} A_{11} & A_{12} \\ A_{21} & A_{22} \end{pmatrix},$$

where A_{11} is $p \times p$ $(p < d)$. Assume that A_{22} and $\tilde{A}_{11} := A_{11} - A_{12}A_{22}^{-1}A_{21}$ are nonsingular:

(a) Show that A^{-1} is given by

$$\begin{pmatrix} \tilde{A}_{11}^{-1} & -\tilde{A}_{11}^{-1}A_{12}A_{22}^{-1} \\ -A_{22}^{-1}A_{21}\tilde{A}_{11}^{-1} & A_{22}^{-1} + A_{22}^{-1}A_{21}\tilde{A}_{11}^{-1}A_{12}A_{22}^{-1} \end{pmatrix}.$$

(b) Show that $\det(A) = \det(A_{22})\det(A_{11} - A_{12}A_{22}^{-1}A_{21})$.
(c) Use (a) and (b) to show that if $Y \sim N(\mu, V)$ is partitioned as

$$Y = \begin{pmatrix} Y_1 \\ Y_2 \end{pmatrix}, \quad \mu = \begin{pmatrix} \mu_1 \\ \mu_2 \end{pmatrix}, \quad V = \begin{pmatrix} V_{11} & V_{12} \\ V_{21} & V_{22} \end{pmatrix},$$

where Y_1 and μ_1 have dimension $p < d$ and V_{11} is $p \times p$, then the conditional distribution of Y_1 given $Y_2 = y_2$ is

$$N\left(\mu_1 + V_{12}V_{22}^{-1}(y_2 - \mu_2),\ V_{11} - V_{12}V_{22}^{-1}V_{21}\right).$$

4. Prove (17.7). Make use of Problem 3 and the asymptotic normality of $\hat{\theta}$ in Problem 2 to prove (17.6). State your assumptions.
5. Prove (17.8). Make use of Problems 2 and 3 to show that Λ_n has a limiting χ^2_{d-p}-distribution when $\theta^0 \in \Theta_0 = \{(\gamma^0, \lambda) : \lambda \in \mathbb{R}^{d-p}\}$. State your assumptions.

Chapter 18
Sequential Analysis and Boundary Crossing Probabilities for Self-Normalized Statistics

In Sect. 17.1 we have described likelihood ratio statistics and Wald's sequential probability ratio test (SPRT). The likelihood ratio statistics for testing simple hypotheses are then extended to generalized likelihood ratio (GLR) statistics for testing composite hypotheses. However, corresponding extensions of the SPRT have not been considered. On the other hand, Sect. 15.1.3 mentions a sequential extension of the t-test. In fact, shortly after Wald's 1945 introduction of the SPRT, there were several proposals to extend the SPRT for testing the mean of a normal distribution when the variance is unknown, but these tests are different from the repeated t-test in Sect. 15.1.3. Section 18.1 reviews these different approaches to constructing sequential t-tests and provides a general class of sequential GLR tests of composite hypotheses. It also develops certain "information bounds" whose attainment characterizes the asymptotic optimality of a sequential test. In Sect. 18.1 we show that sequential GLRs attain these information bounds and are therefore asymptotically optimal for parametric models. In the case of nonparametric or semiparametric models, we modify these ideas to construct sequential score tests (involving self-normalized test statistics) that are asymptotically optimal for testing local alternatives. Whereas Sect. 11.2 has described the method of mixtures to derive bounds for certain boundary crossing probabilities, Sect. 18.2 refines this method to derive more precise asymptotic formulas for boundary crossing probabilities in various sequential testing applications. The essence of the refinement lies in a generalization of Laplace's method that involves tubular neighborhoods of extremal manifolds. Section 18.2 also describes another approach that applies these geometric integration ideas more directly to saddlepoint approximations of density functions of random walks with i.i.d. or Markov-dependent increments. Instead of analytic approximations, one can compute the boundary crossing probabilities by Monte Carlo, and Sect. 18.3 describes efficient importance sampling methods for Monte Carlo evaluation of boundary crossing probabilities. These importance sampling methods are also shown to be related to the method of mixtures.

V.H. de la Peña et al., *Self-Normalized Processes: Limit Theory and Statistical Applications*, Probability and its Applications,

© Springer-Verlag Berlin Heidelberg 2009

18.1 Information Bounds and Asymptotic Optimality of Sequential GLR Tests

18.1.1 Likelihood Ratio Identities, the Wald–Hoeffding Lower Bounds and their Asymptotic Generalizations

The likelihood ratio statistics in Sect. 17.1 are closely related to change of measures; in fact, (17.1) is the Radon–Nikodym derivative of the measure under H_1 relative to that under H_0. The optimality of the likelihood ratio test (Neyman–Pearson lemma) is a consequence of this change of measures. Regarding a test of H_0 versus H_1 as a function φ from the sample space $\mathscr{X}$ into $[0,1]$ (i.e., $\varphi(X_1,\dots,X_n)$ is the probability of rejecting H_0), the likelihood ratio test φ^* can be characterized by $\varphi^* = 1$ if $\mathrm{LR}_n > c$ and $\varphi^* = 0$ if $\mathrm{LR}_n < c$. Since $(\varphi^* - \varphi)(\mathrm{LR}_n - c) \geq 0$, $E_0\{(\varphi^* - \varphi)\mathrm{LR}_n\} \geq cE_0(\varphi^* - \varphi)$. Changing the measures for P_1 to P_0 then yields

$$E_1(\varphi^* - \varphi) = E_0\{(\varphi^* - \varphi)\mathrm{LR}_n\} \geq cE_0(\varphi^* - \varphi), \tag{18.1}$$

in which the equality is a special case of Wald's likelihood ratio identity described below. From (18.1), it follows that if the Type I error of φ does not exceed that of φ^* (i.e., $E_0\varphi \leq E_0\varphi^*$), then $E_1\varphi^* \geq E_1\varphi$, proving the Neyman–Pearson lemma.

Wald (1945) extended the preceding argument involving change of measures to derive (1) Type I and Type II error probability bounds of the SPRT with stopping rule (17.2) and (2) lower bounds for the expected sample sizes $E_{H_0}(T)$ and $E_{H_1}(T)$ of any test (sequential or otherwise) of simple H_0 vs. simple H_1 with prescribed Type I and Type II error probabilities. More generally, let $(\Omega,\mathscr{F})$ be a measurable space and P,Q be probability measures on $(\Omega,\mathscr{F})$. Let $\{\mathscr{F}_n\}$ be an increasing sequence of sub-σ-fields of $\mathscr{F}$, and P_n and Q_n be the restrictions of P and Q, respectively, to $\mathscr{F}_n$. Assuming that P_n is absolutely continuous with respect to Q_n for every n, let $L_n = dP_n/dQ_n$ denote the Radon–Nikodym derivative. Let T be a stopping time with respect to $\{\mathscr{F}_n\}$. Then for all $F \in \mathscr{F}_T$,

$$\begin{aligned} P(F\cap\{T<\infty\}) &= E_Q\{L_T I(T<\infty,F)\}, \\ Q(F\cap\{T<\infty\}) &= E_P\{L_T^{-1} I(T<\infty,F)\}. \end{aligned} \tag{18.2}$$

When $L_n = \prod_{i=1}^n (g_1(X_i)/g_0(X_i))$ is the likelihood ratio LR_n of i.i.d. observations $X_1,\dots,X_n$ having common density function $f \in \{g_0, g_1\}$ with respect to some dominating measure m, (18.2) is known as *Wald's likelihood ratio identity*.

To derive Type I and Type II error probability bounds of the SPRT of $H_0: f = g_0$ vs. $H_1: f = g_1$ that stops sampling at stage N defined by (17.2), Wald (1945) noted that since $P_i(N<\infty) = 1$ under the natural assumption that $P_i\{g_1(X_1) \neq g_0(X_1)\} > 0$ for $i = 0, 1$, (18.2) yields

$$P_0\{L_N \geq B\} \leq B^{-1}P_1\{L_N \geq B\}, \quad P_1\{L_N \leq A\} \leq AP_0\{L_N \leq A\}, \tag{18.3}$$

and $\le$ can be replaced by $=$ in (18.3) if L_N has to fall on either boundary exactly (i.e., there is no "overshoot"). Ignoring overshoots, he made use of both approximate equalities in (18.3) to solve for the error probabilities $\alpha = P_0\{L_N \ge B\}$ and $\beta = P_1\{L_N \le A\}$:

$$\alpha \approx \frac{1-A}{B-A}, \qquad \beta \approx A\left(\frac{B-1}{B-A}\right).$$

Let T be the stopping rule of a test of H_0 vs. H_1 with error probabilities α, β, and let d_T denote its terminal decision rule ($d_T = i$ if H_i is accepted, $i = 0, 1$). Wald's likelihood ratio identity yields

$$\begin{aligned}\alpha &= P_0(d_T = 1) = E_1\left\{L_T^{-1} I(d_T = 1)\right\} \\ &= E_1\{e^{-\log L_T} | d_T = 1\} P_1(d_T = 1) \ge \exp\{-E_1(\log L_T | d_T = 1)\} P_1(d_T = 1) \\ &= \exp\left\{-E_1\left[(\log L_T) I(d_T = 1)\right]/(1-\beta)\right\}(1-\beta),\end{aligned}$$

in which $\ge$ follows from Jensen's inequality. Therefore

$$-E_1\left[(\log L_T) I(d_T = 1)\right] \le (1-\beta)\log(\alpha/(1-\beta)).$$

A similar argument also gives $-E_1[(\log L_T) I(d_T = 0)] \le \beta \log((1-\alpha)/\beta)$. Adding the two inequalities then yields

$$(1-\beta)\log\frac{\alpha}{1-\beta} + \beta\log\frac{1-\alpha}{\beta} \ge -E_1(\log L_T) = -E_1\left(\sum_{t=1}^{T}\log\frac{g_1(X_t)}{g_0(X_t)}\right) = -\mu_1 E_1 T,$$

by Wald's equation (assuming that $E_1 T < \infty$; see Problem 18.1), where $\mu_i = E_i[\log(g_1(X_1)/g_0(X_1))]$. This proves Wald's lower bound for $E_1(T)$ and a similar argument can be used to prove that for $E_0(T)$, i.e.,

$$\begin{aligned}E_1(T) &\ge \mu_1^{-1}\left\{(1-\beta)\log\left(\frac{1-\beta}{\alpha}\right) + \beta\log\left(\frac{\beta}{1-\alpha}\right)\right\}, \\ E_0(T) &\ge (-\mu_0)^{-1}\left\{(1-\alpha)\log\left(\frac{1-\alpha}{\beta}\right) + \alpha\log\left(\frac{\alpha}{1-\beta}\right)\right\},\end{aligned} \tag{18.4}$$

noting that $\mu_1 > 0 > \mu_0$ under the assumption $P_i\{g_1(X_1) \ne g_0(X_1)\} > 0$ for $i = 0, 1$. Since the right-hand sides of (18.4) are Wald's approximations, ignoring overshoots, to $E_1(N)$ and $E_0(N)$, Wald (1945) concluded that the SPRT should minimize both $E_0(T)$ and $E_1(T)$ among all tests that have Type I and Type II errors α and β, respectively, at least approximately when the overshoots are ignored. Later, Wald and Wolfowitz (1948) used dynamic programming arguments to prove that the SPRT is indeed optimal.

Hoeffding (1960) extended Wald's arguments to derive lower bounds for $E(T)$ when the sequential test of H_0 versus H_1 has error probabilities α and β, under another measure that has density function g with respect to ν. One such lower bound involves the Kullback–Leibler information numbers $I(g, g_i) =$

$E[\log(g(X_1)/g_i(X_1))]$. Let $\tau^2 = E\{\log[g_1(X_1)/g_0(X_1)] - I(g,g_0) + I(g,g_1)\}^2$, $\zeta = \max\{I(g,g_0), I(g,g_1)\}$. Then

$$E(T) \geq \left\{ \left[-\zeta \log(\alpha+\beta) + (\tau/4)^2 \right]^{1/2} - \tau/4 \right\}^2 \Big/ \zeta^2. \tag{18.5}$$

The derivation of the lower bounds (18.4) and (18.5) depends heavily on the fact that $\log L_n$ is a sum of i.i.d. random variables. Lai (1981) has provided the following asymptotic extension of Hoeffding's lower bound to the general setting in Sect. 17.1 for likelihood ratio statistics.

Theorem 18.1. *Let P be a probability measure under which $(X_1,\dots,X_n)$ has joint density function $p_n(x_1,\dots,x_n)$ with respect to ν_n, for all $n \geq 1$. Assume that $(X_1,\dots,X_n)$ has joint density function $p_{in}(x_1,\dots,x_n)$ with respect to ν_n for all $n \geq 1$ under H_i, $i = 0,1$. For $0 < \alpha, \beta < 1$, let $\mathscr{T}(\alpha,\beta)$ be the class of tests (T, d_T) of H_0 versus H_1 based on the sequence $\{X_n\}$ and satisfying the error probability constraints*

$$P_0\{d_T \text{ rejects } H_0\} \leq \alpha, \qquad P_1\{d_T \text{ rejects } H_1\} \leq \beta. \tag{18.6}$$

Define $L_n^{(i)} = p_n(X_1,\dots,X_n)/p_{in}(X_1,\dots,X_n)$. Assume that there exist finite constants η_0 and η_1 such that

$$\eta_0 \geq 0, \qquad \eta_1 \geq 0, \qquad \max\{\eta_0, \eta_1\} > 0, \tag{18.7}$$

and

$$n^{-1} \log L_n^{(i)} \to \eta_i \text{ a.s. } [P] \qquad \text{for } i = 0,1. \tag{18.8}$$

(a) For every $0 < \delta < 1$, as $\alpha + \beta \to 0$,

$$\inf_{(T,d_T) \in \mathscr{T}(\alpha,\beta)} P[T > \delta \min\{|\log\alpha|/\eta_0, |\log\beta|/\eta_1\}] \to 1, \tag{18.9}$$

where $a/0$ is defined as ∞ for $a > 0$.
(b) For $0 < \alpha, \beta < 1$, let $C_{\alpha,\beta}$ and $D_{\alpha,\beta}$ be positive constants such that

$$\log C_{\alpha,\beta} \sim |\log\alpha|, \quad \log D_{\alpha,\beta} \sim |\log\beta| \quad \text{as } \alpha+\beta \to 0. \tag{18.10}$$

Define

$$T_{\alpha,\beta} = \inf\{n \geq 1 : L_n^{(0)} \geq C_{\alpha,\beta} \text{ or } L_n^{(1)} \geq D_{\alpha,\beta}\} \quad (\inf \emptyset = \infty). \tag{18.11}$$

Let $(T_{\alpha,\beta}, d^*)$ be the test which stops sampling at stage $T_{\alpha,\beta}$ and rejects H_0 iff $L_{T_{\alpha,\beta}}^{(0)} \geq C_{\alpha,\beta}$. Then as $\alpha + \beta \to 0$,

$$\frac{T_{\alpha,\beta}}{\min\{|\log\alpha|/\eta_0, |\log\beta|/\eta_1\}} \to 1 \text{ a.s. } [P]. \tag{18.12}$$

Moreover, the error probabilities of the test $(T_{\alpha,\beta}, d^*)$ satisfy

$$
\begin{aligned}
P_0\left[(T_{\alpha,\beta}, d^*) \text{ rejects } H_0\right] &\le C_{\alpha,\beta}^{-1} P\left[(T_{\alpha,\beta}, d^*) \text{ rejects } H_0\right], \\
P_1\left[(T_{\alpha,\beta}, d^*) \text{ rejects } H_1\right] &\le D_{\alpha,\beta}^{-1} P\left[(T_{\alpha,\beta}, d^*) \text{ rejects } H_1\right].
\end{aligned} \tag{18.13}
$$

Proof. Let $l_n^{(i)} = \log L_n^{(i)}$. Let $0 < \delta < 1$ and $\bar{\delta} > 1$ such that $\delta\bar{\delta} < 1$. Let m be the greatest integer $\le \delta \min\{|\log\alpha|/\eta_0, |\log\beta|/\eta_1\}$. Then for $(T, d_T) \in \mathscr{T}(\alpha, \beta)$,

$$
\begin{aligned}
\alpha &= \int_{\{T<\infty,\, d_T \text{ rejects } H_0\}} \exp\left(-l_T^{(0)}\right) dP \\
&\ge \int_{\{T\le m,\, l_T^{(0)} \le \bar{\delta}\eta_0 m,\, d_T \text{ rejects } H_0\}} \exp\left(-l_T^{(0)}\right) dP \\
&\ge \exp(-\bar{\delta}\eta_0 m) P\left[T \le m,\, l_T^{(0)} \le \bar{\delta}\eta_0 m, d_T \text{ rejects } H_0\right].
\end{aligned} \tag{18.14}
$$

Since $\bar{\delta}\eta_0 m \le \delta\bar{\delta}|\log\alpha|$, it follows from (18.14) that

$$
\begin{aligned}
P[T \le m, d_T \text{ reject } H_0] &\le \alpha^{1-\delta\bar{\delta}} + P[T \le m,\, l_T^{(0)} > \bar{\delta}\eta_0 m] \\
&\le \alpha^{1-\delta\bar{\delta}} + P\left[\max_{j\le m} l_j^{(0)} > \bar{\delta}\eta_0 m\right].
\end{aligned} \tag{18.15}
$$

Using a similar argument, we also obtain that

$$
P[T \le m, d_T \text{ rejects } H_1] \le \beta^{1-\delta\bar{\delta}} + P\left[\max_{j\le m} l_j^{(1)} > \bar{\delta}\eta_1 m\right]. \tag{18.16}
$$

From (18.15) and (18.16), it follows that

$$
\begin{aligned}
\sup_{(T,d_T)\in\mathscr{T}(\alpha,\beta)} P[T \le m] &\le \alpha^{1-\delta\bar{\delta}} + \beta^{1-\delta\bar{\delta}} + P\left[\max_{j\le m} l_j^{(0)} > \bar{\delta}\eta_0 m\right] \\
&\quad + P\left[\max_{j\le m} l_j^{(1)} > \bar{\delta}\eta_1 m\right].
\end{aligned} \tag{18.17}
$$

Since $j^{-1} l_j^{(i)} \to \eta_i$ a.s. $[P]$ for $i = 0, 1$ and $\bar{\delta} > 1$, (18.9) follows from (18.17).

The a.s. asymptotic behavior (18.12) of $T_{\alpha,\beta}$ follows easily from (18.8) and (18.11). The bounds in (18.13) for the error probabilities of $(T_{\alpha,\beta}, d^*)$ can be proved by essentially the same argument as those in (18.3). □

18.1.2 Asymptotic Optimality of 2-SPRTs and Sequential GLR Tests

The test $(T_{\alpha,\beta}, d^*)$ in Theorem 18.1 is a general form of the 2-SPRT introduced by Lorden (1976) for the case of i.i.d. X_t, with common density function g_0 under H_0, g_1 under H_1, and g under P. Let $n_{\alpha,\beta} = \inf_{(T,d_T)\in\mathscr{T}_{\alpha,\beta}} E(T)$. Under the assumption that $E\{\log^2[g(X_1)/g_0(X_1)] + \log^2[g(X_1)/g_1(X_1)]\} < \infty$, Lorden (1976) showed that $ET_{\alpha,\beta} - n_{\alpha,\beta} \to 0$ as $\alpha + \beta \to 0$. For the special case of a normal family with mean θ, he also showed numerically that $ET_{\alpha,\beta}$ is close to Hoeffding's lower bound (18.5). This provides an asymptotic solution, with $o(1)$ error, to the Kiefer and Weiss (1957) problem of minimizing the expected sample size $E_{\theta^*}(T)$ at given θ^* subject to error probability constraints of the test (T, d_T) at θ_0 and θ_1 in a one-parameter exponential family of densities $f_\theta(x) = e^{\theta x - \psi(\theta)}$ with respect to some measure ν on $\mathbb{R}$.

Ideally, the θ^* where we want to minimize the expected sample size of the 2-SPRT

$$T^* = \inf\left\{n : \prod_{i=1}^{n} \left(f_{\theta^*}(X_i)/f_{\theta_0}(X_i)\right) \geq A_0 \quad \text{or} \quad \prod_{i=1}^{n} \left(f_{\theta^*}(X_i)/f_{\theta_1}(X_i)\right) \geq A_1\right\} \tag{18.18}$$

should be chosen to be the true parameter value θ that is unknown. For the problem of testing $H_0 : \theta \leq \theta_0$ versus $H_1 : \theta \geq \theta_1$ $(> \theta_0)$ in an exponential family, replacing θ^* in (18.18) by its maximum likelihood estimate $\hat{\theta}_n$ at stage n leads to Schwarz's (1962) test which he derived as an asymptotic solution to the Bayes problem of testing H_0 versus H_1 with 0–1 loss and cost ε per observation, as $\varepsilon \to 0$ while θ_0 and θ_1 are fixed. For the case of a normal mean θ, Chernoff (1961, 1965) derived a different and considerably more complicated approximation to the Bayes test of $H_0' : \theta < \theta_0$ versus $H_1' : \theta > \theta_0$. In fact, setting $\theta_1 = \theta_0$ in Schwarz's test does not yield Chernoff's test. This disturbing discrepancy between the asymptotic approximations of Schwarz (assuming an indifference zone) and Chernoff (without an indifference zone separating the one-sided hypotheses) was resolved by Lai (1988), who gave a unified solution (to both problems) that uses a stopping rule of the form

$$\hat{N} = \inf\left\{n : \max\left[\sum_{i=1}^{n} \log \frac{f_{\hat{\theta}_n}(X_i)}{f_{\theta_0}(X_i)}, \sum_{i=1}^{n} \log \frac{f_{\hat{\theta}_n}(X_i)}{f_{\theta_1}(X_i)}\right] \geq g(\varepsilon n)\right\} \tag{18.19}$$

for testing H_0 versus H_1, and setting $\theta_1 = \theta_0$ in (18.19) for the test of H_0' versus H_1'. The function g in (18.19) satisfies $g(t) \sim \log t^{-1}$ as $t \to 0$ and is the boundary of an associated optimal stopping problem for the Wiener process. By solving the latter problem numerically, Lai (1988) also gave a closed-form approximation to the function g.

This unified theory for composite hypotheses provides a bridge between asymptotically optimal sequential and fixed sample size tests. In the fixed sample size case discussed in Sect. 17.1, the Neyman–Pearson approach replaces the likelihood ratio by the generalized likelihood ratio (GLR), which is also used in (18.19) for the sequential test. Since the accuracy of $\hat{\theta}_n$ as an estimate of θ varies with n, (18.19) uses a time-varying boundary $g(\varepsilon n)$ instead of the constant boundary in (18.18) (with $A_0 = A_1$) where θ is completely specified. Simulation studies and asymptotic analysis have shown that $\hat{N}$ is nearly optimal over a broad range of parameter values θ, performing almost as well as (18.18) that assumes θ to be known; see Lai (1988). This broad range covers both fixed alternatives, at which the expected sample size is of the order $O(|\log \varepsilon|)$, and local alternatives θ approaching θ_0 as $\varepsilon \to 0$, at which the expected sample size divided by $|\log \varepsilon|$ tends to ∞. In other words, $\hat{N}$ can adapt to the unknown θ by learning it during the course of the experiment and incorporating the diminishing uncertainties in its value into the stopping boundary $g(\varepsilon n)$. Lai and Zhang (1994) have extended these ideas to construct nearly optimal sequential GLR tests of one-sided hypotheses concerning some smooth scalar function of the parameter vector in multiparameter exponential families, with an indifference zone separating the null and alternative hypotheses and also without an indifference zone. Lai (1997) has provided further extension to a general class of loss functions and prior distributions, thereby unifying (18.19) with another type of sequential tests involving mixture likelihood ratios which were introduced by Robbins (1970); see Sect. 11.3 for Robbins' applications of these mixture likelihood ratio statistics.

In practice, one often imposes an upper bound M and also a lower bound m on the total number of observations. With $M/m \to b > 1$ and $\log \alpha \sim \log \beta$, we can replace the time-varying boundary $g(\varepsilon n)$ in (18.19) by a constant threshold c since $g(t) \sim \log t^{-1}$ and $\log n = \log m + O(1)$ for $m \le n \le M$. The test of $H_0 : \theta = \theta_0$ with stopping rule

$$\tilde{N} = \inf\left\{ n \ge m : \left[\prod_{i=1}^{n} f_{\hat{\theta}_n}(X_i)\right] \Big/ \left[\prod_{i=1}^{n} f_{\theta_0}(X_i)\right] \ge e^c \right\} \wedge M, \qquad (18.20)$$

which corresponds to (18.19) with $\theta_1 = \theta_0$, $g(\varepsilon n)$ replaced by c, and n restricted between m and M, is called a *repeated GLR test*. The test rejects H_0 if the GLR statistic exceeds e^c upon stopping. Whereas (18.20) considers the simple null hypothesis $\theta = \theta_0$ in the univariate case, it is straightforward to extend the repeated GLR test to multivariate θ and composite null hypothesis $H_0 : \theta \in \Theta_0$, by simply replacing $\prod_{i=1}^{n} f_{\theta_0}(X_i)$ in (18.20) by $\sup_{\theta \in \Theta_0} \prod_{i=1}^{n} f_\theta(X_i)$. A particular example is the repeated t-test with stopping rule (15.15), and its multivariate extension is the repeated T^2-test; see Problem 18.2.

The relative simplicity of (15.15) and its multivariate extension that have asymptotically optimal properties is in sharp contrast to the earlier attempts in extending the SPRT to sequential t-, χ^2-, F-, T^2-statistics; see Ghosh (1970). These attempts began with Sect. 6 of Wald (1945), who suggested using weight functions to handle composite hypotheses so that one can still work with likelihood ratios. When H_0 is simple, say $\theta = 0$, but H_1 is composite, Wald proposed to integrate the likelihood over the alternative hypothesis and to consider the integrated Type II error in applying the SPRT with $L_n = \{\int \prod_{i=1}^{n} f_\theta(X_i) w(\theta) d\theta\} / \{\prod_{i=1}^{n} f_0(X_i)\}$. The likelihood ratio identity can again be used to approximate the Type I error and the integrated Type II error of the test, as in Sect. 18.1.1 for Wald's SPRT. When H_0 is also composite, he proposed to use the SPRT with

$$L_n = \left\{ \int \prod_{i=1}^{n} f_\theta(X_i) w_1(\theta) d\theta \right\} \Big/ \left\{ \prod_{i=1}^{n} f_\theta(X_i) w_0(\theta) d\theta \right\},$$

for which the likelihood ratio identity can again be used to approximate the integrated error probabilities. Recognizing that one usually would like to have $\sup_{\theta \in H_0}$

$\alpha(\theta) \le \alpha$ instead of $\int \alpha(\theta)w(\theta)d\theta \le \alpha$ (where $\alpha(\theta)I(\theta \in H_0)$ and $\beta(\theta)I(\theta \in H_1)$ denote the Type I and Type II error probabilities), he showed in the case of testing $H_0 : \mu = 0$ vs. $H_1 : |\mu/\sigma| = \delta$, for the mean μ of a normal distribution with unknown variance σ^2, that it is possible to choose weight functions w_0 and w_1 such that

$$\sup_{\theta \in H_0} \alpha(\theta) = \int \alpha(\theta) w_0(\theta) d\theta, \qquad \sup_{\theta \in H_1} \beta(\theta) = \int \beta(\theta) w_1(\theta) d\theta.$$

This is the Wald–Arnold–Goldberg–Rushton sequential t-test; see David and Kruskal (1956) who proved that the test terminates with probability 1 for all choices of μ and $\sigma > 0$.

The weight function approach, which has been used to derive the sequential t-, T^2- or F-tests, can be replaced by an alternative approach that reduces composite hypotheses to simple ones by the principle of invariance. If G is a group of transformations leaving the problem invariant, then the distribution of a maximal invariant depends on P only through its orbit. Therefore, by considering only invariant sequential tests, the hypotheses become simple; see Chap. 6 of Ghosh (1970). This is, therefore, a special case of the SPRT with stopping rule (17.2), in which $\mathrm{LR}_n = p_{1n}(U_n)/p_{0n}(U_n)$, where $U_n = U_n(X_1, \ldots, X_n)$ is a maximal invariant with respect to G based on $X_1, \ldots, X_n$ and p_{in} is the density function of this maximal invariant under H_i $(i = 0, 1)$. In the case of the sequential t-test of $H_0 : \mu = 0$ for the mean μ of a normal distribution with unknown variance σ^2, G is the group of scale changes $x \mapsto cx$ $(c > 0)$ and U_n is the t-statistic $\sqrt{n}\bar{X}_n/s_n$ in Chap. 15. Thus, even though the X_i are i.i.d., the U_n are no longer i.i.d. and classical random walk results like Wald's equation are no longer applicable. This makes questions such as whether the SPRT based on the maximal invariants terminates with probability 1 and its expected sample sizes at the null and alternative hypotheses much harder than in the i.i.d. case. On the other hand, the simple bounds (18.3) and Wald-type approximations for the error probabilities still hold for the SPRT in the dependent case.

For the repeated GLR test (18.20) that has at most M observations, the issue of termination with probability 1 becomes trivial. Although the simple bounds (18.3) and related approximations are no longer applicable to the repeated GLR test, we can still use the likelihood ratio identity involving mixture of densities together with the pseudo-maximization method to analyze the error probabilities of the test. This and another technique that uses saddlepoint approximation and geometric integration will be described in the next section. To obtain asymptotic approximations for the expected sample sizes of SPRTs for the general setting of dependent random variables, Lai (1981, p. 326) make use of (18.12) and uniform integrability after strengthening the a.s. convergence in (18.8) into r-quick convergence. A sequence of random variables Y_n is said to converge to 0 *r-quickly* if $EL_\varepsilon^r < \infty$ for every $\varepsilon > 0$, where $L_\varepsilon = \sup\{n \ge 1 : |Y_n| \ge \varepsilon\}$; note that $Y_n \to 0$ a.s. can be restated as $P\{L_\varepsilon < \infty\} = 1$ for every $\varepsilon > 0$ (see Lai, 1976b). For the special case of invariant SPRTs (that use invariance to reduce composite hypotheses to simple ones) or repeated GLR tests based on i.i.d. observations, Lai and Siegmund (1979) have derived asymptotic expansions for $E(T_{\alpha,\beta})$ or $E(\tilde{N})$, up to the $o(1)$ term, by making use of nonlinear renewal theory; see Problem 18.4 for a sketch of the basic ideas.

18.2 Asymptotic Approximations via Method of Mixtures and Geometric Integration

18.2.1 Boundary Crossing Probabilities for GLR Statistics via Method of Mixtures

By using the method of mixtures, Siegmund (1977) has derived an asymptotic approximation to the Type I error probability of the repeated t-test of $H_0 : \mu = 0$ for the mean μ of a normal distribution with unknown variance σ^2. Under the group of scale changes $x \mapsto cx$ $(c > 0)$, a maximal invariant is $(Y_2, \ldots, Y_n)$, where $Y_i = X_i/X_1$. By conditioning on X_1/σ, the density of $(Y_2, \ldots, Y_n)$ under $\mu/\sigma = \theta$ can be easily shown to be

$$\int_{-\infty}^{\infty} \frac{|x|^{n-1}}{(2\pi)^{n/2}} \exp\left[-\frac{x^2}{2}\sum_{i=1}^{n}(y_i-\theta)^2\right] dx, \qquad \text{with } y_1 = 1.$$

Let P_θ be the probability measure induced by the sequence $(Y_2, Y_3, \ldots)$ and let $Q = \int_{-\infty}^{\infty} P_\theta d\theta/\sqrt{2\pi}$. The likelihood ratio of $(Y_2, \ldots, Y_n)$ under Q relative to P_0 is therefore

$$\begin{aligned}
&\frac{1}{\sqrt{2\pi}}\int_{-\infty}^{\infty} \frac{\int_{-\infty}^{\infty}|x|^{n-1}\exp\left[-\frac{x^2}{2}\sum_{i=1}^{n}(Y_i-\theta)^2\right]dx}{\int_{-\infty}^{\infty}|x|^{n-1}\exp\left[-\frac{x^2}{2}\sum_{i=1}^{n}Y_i^2\right]dx}d\theta \\
&= n^{-1/2}\left\{\left(\sum_{i=1}^{n}Y_i^2\right)\Big/\sum_{i=1}^{n}(Y_i-\bar{Y}_n)^2\right\}^{n/2} \\
&= \frac{1}{\sqrt{n}}\exp\left\{\frac{n}{2}\log\left(1+\frac{\bar{X}_n^2}{s_n^2}\right)\right\} = \frac{1}{\sqrt{n}}e^{\ell_n},
\end{aligned}$$

where $\ell_n = \frac{1}{2}n\log(1+\bar{X}_n^2/s_n^2)$.

Consider the stopping rule (15.15) of the repeated t-test (τ, d_τ), in which the terminal decision rule rejects $H_0 : \mu = 0$ if $\ell_\tau \geq c$. Letting $m = [\delta c]$ and $M = [ac]$, the Type I error of the test is

$$P_0(\ell_\tau \geq c) = P_0(\ell_m \geq c) + \int_{-\infty}^{\infty} E_\theta\left\{\sqrt{\tau}e^{-\ell_\tau}I(\ell_\tau \geq c,\ \tau > m)\right\}d\theta \Big/ \sqrt{2\pi}, \quad (18.21)$$

by applying the likelihood ratio identity (18.2). The first summand in (18.21), $P_0(\ell_m \geq c)$, can be represented as the tail probability of the t_{m-1}-distribution since $\sqrt{m}\bar{x}_m/s_m$ has the t_{m-1}-distribution under H_0. To analyze the second summand, note that $e^{-\ell_\tau} = e^{-c}e^{-(\ell_\tau - c)}$ and Siegmund (1977) uses the law of large numbers and nonlinear renewal theory to show that, as $c \to \infty$,

$$E_\theta\left\{\sqrt{\frac{\tau}{c}}e^{-(\ell_\tau - c)}I(\ell_\tau \geq c, [ac] \geq \tau > [\delta c])\right\}$$
$$\to \begin{cases} 0 & \text{if } \log(1+\theta^2) \notin [2a^{-1}, 2\delta^{-1}] \\ \left[\frac{2}{\log(1+\theta^2)}\right]^{1/2} \lim_{c\to\infty} E_\theta\left[\exp\{-(\ell_\tau - c)\}\right] & \text{if } \theta \neq 0 \text{ and } \frac{2}{a} < \log(1+\theta^2) < \frac{2}{\delta}. \end{cases} \tag{18.22}$$

Problem 18.3 provides a sketch of the basic ideas; in particular, the existence and characterization of the limiting distribution of the overshoot $\ell_\tau - c$ follows from nonlinear renewal theory. The monographs by Woodroofe (1982) and Siegmund (1985) give a systematic introduction to asymptotic approximations to Type I error probabilities of sequential GLR tests obtained by using this approach that involves change of measures from P_0 to $\int P_\theta dG(\theta)$, the likelihood ratio identity, and the renewal-theoretic formula for the overshoot term.

A basic feature of this approach is that the approximations depend crucially on the fact that stopping occurs at the first time τ when the likelihood ratio or GLR statistic ℓ_τ exceeds some threshold c. Thus ℓ_τ is equal to c plus an excess over the boundary whose limiting distribution can be obtained using renewal theory. When the test statistic used is not ℓ_τ, the arguments break down. Since they are based on the fact that $\ell_\tau = c + O_p(1)$, where the $O_p(1)$ term is the overshoot, these arguments are also not applicable when τ is replaced by a fixed sample size n. Moreover, whereas the role of ℓ_τ in change-of-measure arguments is quite easy to see when the null hypothesis is simple, it becomes increasingly difficult to work with ℓ_τ when the region defining a composite null hypothesis becomes increasingly complex. In the next section we describe another approach, developed by Chan and Lai (2000), which can be applied not only to likelihood ratio or GLR statistics but also to other functions of the sufficient statistics in a multiparameter exponential family, and which is applicable to both sequential and fixed sample size tests.

18.2.2 A More General Approach Using Saddlepoint Approximations and Geometric Integration

Chan and Lai (2000) consider the following three classes of large deviation probabilities, which they tackle by integrating saddlepoint approximations to the density functions of sums of i.i.d. random vectors over tubular neighborhoods of certain extremal manifolds that are related to Laplace's method. Let $X_1, X_2, \ldots$ be i.i.d. d-dimensional non-lattice random vectors whose common moment generating function is finite in some neighborhood of the origin. Let $S_n = X_1 + \cdots + X_n$, $\mu_0 = EX_1$ and $\Theta = \{\theta : Ee^{\theta' X} < \infty\}$. Assume that the covariance matrix of X_1 is positive definite. Let $\psi(\theta) = \log(Ee^{\theta' X})$ denote the cumulant generating function of X_1. Let Λ be the closure of $\nabla\psi(\Theta)$, Λ^o be its interior, and denote the boundary of Λ by $\partial\Lambda$ $(=\Lambda-\Lambda^o)$. Then $\nabla\psi$ is a diffeomorphism from Θ^o onto Λ^o. Let $\theta_\mu=(\nabla\psi)^{-1}(\mu)$. For $\mu\in\Lambda^o$, define

$$\phi(\mu)=\sup_{\theta\in\Theta}\left\{\theta'\mu-\psi(\theta)\right\}=\theta_\mu'\mu-\psi(\theta_\mu). \tag{18.23}$$

The function ϕ is the convex dual of ψ and is also known as the *rate function* in large deviations theory. Let $g:\Lambda\to\mathbb{R}$ and define the stopping time

$$T_c=\inf\left\{k\geq n_0: kg(S_k/k)>c\right\}, \tag{18.24}$$

where n_0 corresponds to a prescribed minimal sample size. Chan and Lai (2000) develop asymptotic approximations to the large deviation probabilities

$$P\{T_c\leq n\},\quad P\left\{ng(S_n/n)>c\right\},\quad P\left\{\min_{k\leq n}\left[(n-k)\beta+kg(S_k/k)\right]>c\right\}, \tag{18.25}$$

with $n\sim ac$ and $n_0\sim\delta c$ as $c\to\infty$, for some $a>\delta>0$ such that $g(\mu_0)<1/a$ for the first two probabilities, and $\beta>1/a$ and $g(\mu_0)=0$ for the third probability.

For the first probability in (18.25), the large deviation principle suggests that $\log P\{ng(S_n/n)>c\}$ is asymptotically equivalent to $-n\inf\{\phi(\mu):g(\mu)>c/n\}$ as $c\to\infty$. This in turn suggests that $\log(\sum_{\delta c\leq n\leq ac}P\{ng(S_n/n)>c\})$ is asymptotically equivalent to $-\min_{\delta c\leq n\leq ac}\inf_{g(\mu)>c/n}n\phi(\mu)$, which, upon interchanging the min and inf signs, is asymptotically equivalent to

$$-\inf_{g(\mu)>1/a}\frac{c\phi(\mu)}{\min(1/\delta,g(\mu))}=-\frac{c}{r},\quad\text{where}\quad r=\sup_{g(\mu)>1/a}\frac{\min(\delta^{-1},g(\mu))}{\phi(\mu)}.$$

Hence $P\{T_c\leq ac\}=e^{-c/r+o(c)}$ as $c\to\infty$. Chan and Lai (2000, Theorem 1) assume the following regularity conditions to obtain a more precise asymptotic approximation:

(A1) g is continuous on Λ^o and there exists $\varepsilon_0>0$ such that

$$\sup_{a^{-1}<g(\mu)<\delta^{-1}+\varepsilon_0}g(\mu)/\phi(\mu)=r<\infty.$$

(A2) $M_\varepsilon:=\{\mu:a^{-1}<g(\mu)<\delta^{-1}+\varepsilon$ and $g(\mu)/\phi(\mu)=r\}$ is a q-dimensional oriented manifold for all $0\leq\varepsilon\leq\varepsilon_0$, where $q\leq d$.

(A3) $\liminf_{\mu\to\partial\Lambda}\phi(\mu)>(\delta r)^{-1}$ and there exists $\varepsilon_1>0$ such that $\phi(\mu)>(\delta r)^{-1}+\varepsilon_1$ if $g(\mu)>\delta^{-1}+\varepsilon_0$.

(A4) g is twice continuously differentiable in some neighborhood of M_{ε_0} and $\sigma(\{\mu:g(\mu)=\delta^{-1}$ and $g(\mu)/\phi(\mu)=r\})=0$, where σ is the volume element measure of M_{ε_0}.

Spivak (1965) provides a concise introduction to q-dimensional manifolds in $\mathbb{R}^d$ and integration on these manifolds. Assumptions (A1)–(A3) imply that $\sup_{g(\mu)>a^{-1}}$

$\min(\delta^{-1}, g(\mu))/\phi(\mu)$ can be attained on the q-dimensional manifold M_0. The first part of (A3) implies that there exists $\varepsilon^* > 0$ such that

$$M^* := \{\mu : a^{-1} \le g(\mu) \le \delta^{-1} + \varepsilon^*, g(\mu)/\phi(\mu) = r\} \tag{18.26}$$

is a compact subset of Λ; it clearly holds if $\phi(\mu) \to \infty$ as $\mu \to \partial\Lambda$, which is usually the case. For $\mu \in M_0$, let $TM_0(\mu)$ denote the tangent space of M_0 at μ and let $TM_0^{\perp}(\mu)$ denote its orthogonal complement (i.e., $TM_0^{\perp}(\mu)$ is the normal space of M_0 at μ). Let $\rho(\mu) = \phi(\mu) - g(\mu)/r$. By (A1) and (A3), ρ attains on M_{ε_0} its minimum value 0 over $\{\mu : \alpha^{-1} < g(\mu) < \delta^{-1} + \varepsilon_0\}$, and therefore

$$\nabla\rho(\mu) = 0 \quad \text{and} \quad \nabla^2\rho(\mu) \quad \text{is nonnegative definite for } \mu \in M_0. \tag{18.27}$$

Let $\Pi_{\mu}^{\perp}$ denote the $d \times (d-q)$ matrix whose column vectors form an orthonormal basis of $TM_0^{\perp}(\mu)$. Then the matrix $\nabla_{\perp}^2\rho(\mu) := (\Pi_{\mu}^{\perp})'\nabla^2\rho(\mu)\Pi_{\mu}^{\perp}$ is nonnegative definite for $\mu \in M_0$. Letting $|\cdot|$ denote the determinant of a nonnegative definite matrix, we shall also assume that:

(A5) $\inf_{\mu \in M_0} |\nabla_{\perp}^2\rho(\mu)| > 0$, with $\rho = \phi - g/r$, where we set $|\nabla_{\perp}^2\rho(\mu)| = 1$ in the case $d - q = 0$.

Under (A1)–(A5), Chan and Lai (2000) first consider the case where X_1 has a bounded continuous density function (with respect to Lebesgue measure) so that S_n/n has the saddlepoint approximation

$$P\{S_n/n \in d\mu\} = (1 + o(1))\,(n/2\pi)^{d/2}\left|\sum(\mu)\right|^{-1/2} e^{-n\phi(\mu)}\,d\mu, \tag{18.28}$$

where $\sum(\mu) = \nabla^2\psi(\theta)|_{\theta=\theta_\mu}$ and the $o(1)$ term is uniform over compact subsets of Λ^o; see Borovkov and Rogozin (1965), Barndorff-Nielsen and Cox (1979) and Jensen (1995) for the proofs and applications of these saddlepoint approximations. Note that Sect. 15.1.2 already gives a concrete example of such saddlepoint approximations, with $X_i = (Y_i, Y_i^2)$ associated with Student's t-statistic based on $Y_i, \ldots, Y_n$. Let

$$\begin{aligned} f(\mu)d\mu &= P\{T_c \le ac, S_{T_c}/T_c \in d\mu\} \\ &= \sum_{\delta c \le n \le ac} P\{S_n/n \in d\mu\} I_{\{ng(\mu) > c\}} \\ &\quad \times P\{kg(S_k/k) < c \qquad \text{for all } \delta c \le k < n | S_n/n \in d\mu\}. \end{aligned} \tag{18.29}$$

Making use of (18.28) and (18.29), they first show that

$$P\{T_c \le ac\} = \int_{\mathbb{R}^d} f(\mu)\,d\mu \sim \int_{U_{c^{-1/2}\log c}} f(\mu)\,d\mu, \tag{18.30}$$

where U_η is a tubular neighborhood of M_0 with radius η, and then perform the integration in (18.30) over $U_{c^{-1/2}\log c}$. This is basically an extension of Laplace's asymptotic method in Sect. 11.1 to manifolds. Specifically, we say that

$$U_\eta = \{y+z : y \in M_0,\ z \in TM_0^\perp(y) \quad \text{and} \quad \|z\| \le \eta\} \tag{18.31}$$

is a *tubular neighborhood* of M_0 with radius η if the representation of the elements of U_η in (18.31) is unique. The integral in (18.30) uses the *infinitesimal change of volume* in differential geometry; see Gray (1990) for a comprehensive treatment. From Lemmas 3.13, 3.14 and Theorem 3.15 of Gray (1990), it follows that as $\eta := c^{-1/2}\log c \to 0$,

$$\int_{U_\eta} f(\mu)\,d\mu \sim \int_{M_0}\left\{\int_{z\in TM_0^\perp(y),\|z\|\le\eta} f(y+z)\,dz\right\} d\sigma(y). \tag{18.32}$$

The inner integral in (18.32) can be evaluated asymptotically by making use of (18.28) and (18.29), and combining the result with (18.30) yields the asymptotic formula for $P\{T_c \le ac\}$ in the following theorem. While the preceding analysis has assumed that X_1 has a bounded continuous density function, Chan and Lai (2000, pp. 1651–1652) replace this assumption by the much weaker assumption that X_1 be non-lattice. By partitioning Λ into suitably small cubes, they use change of measures (see (18.33) below) and a local limit theorem (see Sect. 2.3.2) to modify the preceding analysis, replacing "$\in d\mu$" above by "$\in I_\mu$", where I_μ denotes a cube centered at μ.

Theorem 18.2. *Let $X_i^{(\mu)}$ be i.i.d. such that*

$$P\{X_i^{(\mu)} \in dx\} = e^{\theta_\mu' x - \psi(\theta_\mu)}\,dF(x), \tag{18.33}$$

where F is the distribution of X_1, and let $S_n(\mu) = \sum_{i=1}^n\{\theta_\mu' X_i^{(\mu)} - \psi(\theta_\mu)\}$. Let $\Sigma(\mu) = \nabla^2\psi(\theta)|_{\theta=\theta_\mu}$. Suppose X_1 is non-lattice and $g : \Lambda \to \mathbb{R}$ satisfies (A1)–(A5) with $a > \delta$, $g(\mu_0) < a^{-1}$ and $n_0 \sim \delta c$. Let $\gamma(\mu) = \int_0^\infty e^{-y} P\{\min_{n\ge1} S_n(\mu) > y\}dy$. Then as $c \to \infty$, $P\{T_c \le ac\}$ is asymptotically equivalent to

$$\left(\frac{c}{2\pi r}\right)^{q/2} e^{-c/r}\int_{M_0}\gamma(\mu)(\phi(\mu))^{-(q/2+1)}\left|\Sigma(\mu)\right|^{-1/2}|\nabla_\perp^2\rho(\mu)|^{-1/2}\,d\sigma(\mu),$$

where $\nabla_\perp^2\rho$ is introduced in (A5).

For the second and third probabilities in (18.25) with $g(\mu_0) < b$, Chan and Lai (2000) impose the following conditions in lieu of (A1)–(A5):

(B1) g is continuous on Λ^o and $\inf\{\phi(\mu) : g(\mu) \ge b\} = b/r$.
(B2) g is twice continuously differentiable on $\{\mu \in \Lambda^o : b - \varepsilon_0 < g(\mu) < b + \varepsilon_0\}$ for some $\varepsilon_0 > 0$.
(B3) $\nabla g(\mu) \ne 0$ on $N := \{\mu \in \Lambda^o : g(\mu) = b\}$, and $M := \{\mu \in \Lambda^o : g(\mu) = b,\ \phi(\mu) = b/r\}$ is a smooth p-dimensional manifold (possibly with boundary) for some $0 \le p \le d-1$.
(B4) $\liminf_{\mu\to\partial\Lambda}\phi(\mu) > br^{-1}$ and $\inf_{g(\mu)>b+\delta}\phi(\mu) > br^{-1}$ for every $\delta > 0$.

For the notion of smooth submanifolds (with or without boundaries), see Spivak (1965). Under (B2) and (B3), N is a $(d-1)$-dimensional manifold and $TN^\perp(\mu)$ is a one-dimensional linear space with basis vector $\nabla g(\mu)$. Making use of (B1)–(B4), Chan and Lai (2000, p. 1665) show that

$$\inf_{\mu\in M}\|\nabla\phi(\mu)\|>0, \quad (\nabla g(\mu))'\nabla\phi(\mu)>0 \quad \text{and} \quad \nabla\phi(\mu)\in TN^{\perp}(\mu) \quad \text{for all } \mu\in M. \tag{18.34}$$

Hence $\nabla\phi(\mu) = s\nabla g(\mu)$ with $s = \|\nabla\phi(\mu)\|/\|\nabla g(\mu)\|$. Let $e_1(\mu) = \nabla\phi(\mu)/\|\nabla\phi(\mu)\|$ and let $\{e_1(\mu), e_2(\mu), \ldots, e_{d-p}(\mu)\}$ be an orthonormal basis of $TM^{\perp}(\mu)$. Define a $d\times(d-p-1)$ matrix Π_μ (in the case $d>p+1$) by $\Pi_\mu = (e_2(\mu)\ldots e_{d-p}(\mu))$ and a positive number $\xi(\mu)$ by

$$\xi(\mu)=\begin{cases} 1/\|\nabla\phi(\mu)\| & \text{if } d=p+1 \\ \left|\Pi_\mu'\left\{\Sigma^{-1}(\mu)-s\nabla^2 g(\mu)\right\}\Pi_\mu\right|^{-\frac{1}{2}}/\|\nabla\phi(\mu)\| & \text{if } d>p+1. \end{cases}$$

The following assumption is analogous to (A5):

(B5) $\inf_{\mu\in M}|\Pi_\mu'\{\Sigma^{-1}(\mu)-s\nabla^2 g(\mu)\}\Pi_\mu|>0$ if $d>p+1$.

Theorem 18.3. *Suppose X_1 is non-lattice and $g:\Lambda\to\mathbb{R}$ satisfies (B1)–(B5). Let $b>g(\mu_0)$. Then as $n\to\infty$,*

$$\begin{aligned} P\{g(S_n/n)>b\} &\sim P\{g(S_n/n)\geq b\} \\ &\sim (2\pi)^{-(p+1)/2}n^{(p-1)/2}e^{-bn/r}\int_M \xi(\mu)|\Sigma(\mu)|^{-1/2}d\sigma(\mu). \end{aligned}$$

Theorem 18.4. *Suppose X_1 is non-lattice, $g:\Lambda\to\mathbb{R}$ satisfies (B1)–(B5) and $g(\mu_0)=0$. Let $\beta>b>0$. Define $X_i^{(\mu)}$ as in Theorem 18.2 and let $W_n(\mu)=\sum_{i=1}^n\{\theta_\mu'(X_i^{(\mu)}-\mu)+s(b-\beta)\}$. Let $w(\mu)=\int_0^\infty e^{-y}P\{\max_{n\geq1}W_n(\mu)<y\}dy$. Then as $n\to\infty$,*

$$\begin{aligned} P\left\{\min_{k\leq n}[(n-k)\beta+kg(S_k/k)]>bn\right\} &\sim P\left\{\min_{k\leq n}[(n-k)\beta+kg(S_k/k)]\geq bn\right\} \\ &\sim (2\pi)^{-(p+1)/2}n^{-(p-1)/2}e^{-bn/r}\int_M \xi(\mu)w(\mu)|\Sigma(\mu)|^{-1/2}d\sigma(\mu). \end{aligned}$$

The proofs of Theorems 18.3 and 18.4 are given in Chan and Lai (2000, pp. 1653–1654). We summarize here the main ideas in the proof of Theorem 18.3, as the proof of Theorem 18.4 is similar. Assume that $r=1$ and X_1 has a bounded continuous density. Recall that $e_1(y),\ldots,e_{d-p}(y)$ form an orthonormal basis of $TM^{\perp}(y)$ and that $\nabla g(y)$ is a scalar multiple of $e_1(y)$, for every $y\in M$. For $y\in M$ and $\max_{1\leq i\leq d-p}|v_i|\leq(\log n)^{-1}$, since $g(y)=b$ and $(\nabla g(y))'\sum_{i=1}^{d-p}v_ie_i(y)=v_1\|\nabla\phi(y)\|/s$, Taylor's expansion yields

$$\begin{aligned} g\left(y+\sum_{i=1}^{d-p}v_ie_i(y)\right) &= b+v_1\|\nabla\phi(y)\|/s+O(v_1^2)+v'\Pi_y'\nabla^2 g(y)\Pi_y v/2+o(\|v\|^2) \\ &> b \quad \text{if } v_1\|\nabla\phi(y)\|/s>c(v)+o(\|v\|^2)+O(v_1^2), \end{aligned} \tag{18.35}$$

where $v = (v_2, \dots, v_{d-p})'$ and $c(v) = -v'\Pi_y'\nabla^2 g(y)\Pi_y v/2$. Let

$$V_n = \left\{ y + \sum_{i=1}^{d-p} v_i e_i(y) : y \in M, \max_{1 \le i \le d-p} |v_i| \le (\log n)^{-1}, \; v_1 \|\nabla \phi(y)\|/s > c(v) \right\}.$$

By (18.28), $P\{S_n/n \in V_n\}$ is equal to $(1+o(1))(n/2\pi)^{d/2} \int_{V_n} |\Sigma(\mu)|^{-1/2} e^{-n\phi(\mu)} d\mu$. We can use the infinitesimal change of volume function over tubular neighborhoods as in (18.32) to evaluate the integral. Making use of (18.28) together with (B1) and (B4), it can be shown that $P\{g(S_n/n) > b\} = P\{S_n/n \in V_n\} + o(n^{-q}e^{-bn})$ for every $q > 0$.

Chan and Lai (2000, Sect. 4) have also extended the preceding ideas to derive approximations to moderate deviation probabilities. One may argue that, for the usual significance levels of hypothesis tests, the probabilities of large deviations in (18.25) seem to be too small to be of practical relevance. The moderate deviation theory in Chan and Lai (2000) basically shows that the large deviation approximations can be extended to probabilities of moderate deviations. More importantly, large deviation approximations are important for multiple testing situations, as shown by Chan and Lai (2002, 2003) in applications to change-point problems and limiting distribution of scan statistics, for which the i.i.d. setting above is also extended to Markov chains on general state spaces.

18.2.3 Applications and Examples

Consider the multiparameter exponential family with density function $\exp(\theta' x - \psi(\theta))$ with respect to some probability measure F. The natural parameter space is Θ. Let Θ_1 be a q_1-dimensional smooth submanifold of Θ and Θ_0 be a q_0-dimensional smooth submanifold of Θ_1 with $0 \le q_0 < q_1 \le d$. The GLR statistics for testing the null hypothesis $H_0 : \theta \in \Theta_0$ versus the alternative hypothesis $H_1 : \theta \in \Theta_1 - \Theta_0$ are of the form $ng(S_n/n)$, where

$$g(x) = \phi_1(x) - \phi_0(x), \qquad \text{with } \phi_i(x) = \sup_{\theta \in \Theta_i} \left(\theta' x - \psi(\theta)\right). \tag{18.36}$$

Then $g(x) \le \phi(x)$ and equality is attained if and only if $\phi_1(x) = \phi(x)$ and $\phi_0(x) = 0$. Since $\nabla\psi$ is a diffeomorphism, $\Lambda_i = \nabla\psi(\Theta_i)$ is a q_i-dimensional submanifold of Λ^o. Note that $\phi(x) = \phi_1(x)$ iff $x_1 \in \Lambda_1$. Consider the repeated GLR test with stopping rule $T_c \wedge [ac]$ where T_c is defined in (18.24) with g given by (18.36) and $n_0 \sim \delta c$. To evaluate the Type I error probability at θ_0, we can assume, by choosing the underlying probability measure F as that associated with θ_0 and by replacing X_i by $X_i - \nabla\psi(\theta_0)$, that $\theta_0 = 0$, $\psi(0) = 0$ and $\nabla\psi(0) = 0$. Then (A1)–(A5) hold with $r = 1$ and $q = q_1 - q_0$ under certain regularity conditions and therefore we can apply Theorem 18.2 to approximate the Type I error probability $P_0\{T_c \le ac\}$.

Example 18.5. Consider the repeated t-test of the null hypothesis H_0 that the common mean of i.i.d. normal observations $Y_1, Y_2, \dots$ is 0 when the variance is unknown. Here $X_i = (Y_i, Y_i^2)$, $S_n/n = (n^{-1}\sum_1^n Y_i, n^{-1}\sum_1^n Y_i^2)$, $\Lambda^o = \{(y,v) : v > y^2\}$ and $\Lambda_0 = \{(0,v) : v > 0\}$. The GLR statistics are of the form $ng(S_n/n)$, where $g(y,v) = \frac{1}{2}\log(v/(v-y^2))$, and the repeated t-test rejects H_0 if $ng(S_n/n) > c$ for some $\delta c \le n \le ac$. The test is invariant under scale changes, so we can consider the Type I error probability when $\mathrm{Var}(Y_i) = 1$. Since $\phi(y,v) = [v - 1 - \log(v-y^2)]/2$, (A1)–(A4) are satisfied with $r = 1$ and

$$M_\varepsilon = \left\{(y,1) : 1 - \exp\left(-2(\delta^{-1}+\varepsilon)\right) > y^2 > 1 - \exp(-2a^{-1})\right\}.$$

Moreover, (A5) holds since $\nabla_\perp^2 \rho(y,1) = 1/2$ for $(y,1) \in M_0$.

Suppose next that instead of the stopping rule $T_c \wedge [ac]$, the GLR test of H_0 is based on a sample of fixed size n. The test rejects H_0 if $g(S_n/n) > b$, where g is defined by (18.36). To evaluate the Type I error probability at θ_0, there is no loss of generality in assuming that $\theta_0 = 0$, $\psi(0) = 0$ and $\nabla\psi(0) = 0$. Then (B1)–(B5) hold with $r = 1$ and $p = q_1 - q_0 - 1$ under certain regularity conditions, so Theorem 18.3 can be used to approximate the Type I error probability $P_0\{g(S_n/n) > b\}$. A different choice of g in Theorem 18.3 also gives an approximation to the Type II error probability $P_\theta\{g(S_n/n) \le b\}$ with $g(\nabla\psi(\theta)) > b$. Specifically, let $\tilde{g}(\mu) = g(\nabla\psi(\theta)) - g(\mu)$, $\tilde{b} = g(\nabla\psi(\theta)) - b$, and apply Theorem 18.3 with g, b replaced by $\tilde{g}$, $\tilde{b}$.

Theorems 18.2 and 18.3 can also be applied to analyze error probabilities of tests that are not based on likelihood ratio statistics. For example, consider the repeated t-test of Example 18.5 when the underlying distribution is actually non-normal. Here $g(y,v) = -\frac{1}{2}\log(1-y^2/v)$ is an increasing function of $|y|/\sqrt{v}$, which increases as v decreases. Thus change of measures for the probabilities in (18.25) can be restricted to $\{(\theta_1,\theta_2) : \theta_2 < 0\}$, on which $Ee^{\theta_1 Y + \theta_2 Y^2} < \infty$ without any moment conditions on Y. In this general setting,

$$\phi(y,v) = \sup_{\gamma\in\mathbb{R},\lambda>0}\left\{\gamma y - \lambda v - \log Ee^{\gamma Y - \lambda Y^2}\right\} \qquad \text{for } v \ge y^2;$$

see (10.1). Write $g(y,v) = G(|y|/\sqrt{v})$. For $0 \le t \le 1$, define $F(t) = \inf_{v>0}\phi(t\sqrt{v},v) = \phi(t\sqrt{v_t},v_t)$. Then

$$\sup_{v\ge y^2} g(y,v)/\phi(y,v) = \sup_{0\le t\le 1}\left[G(t)/\min\{F(t),F(-t)\}\right].$$

In the normal case considered in Example 18.5, $G = F$ since $v - 1 - \log v$ has minimum value 0. For non-normal Y, suppose $r = \sup_{0\le t\le 1}[G(t)/\min\{F(t),F(-t)\}]$ is attained at $t^* \in (0,1)$ and $a^{-1} < G(t^*) < \delta^{-1}$. Then (A1)–(A5) hold with $q = 0$ and $M_\varepsilon = \{(t^*\sqrt{v_{t^*}}, v_{t^*})\}$, or $\{(-t^*\sqrt{v_{t^*}}, v_{t^*})\}$, or $\{(t^*\sqrt{v_{t^*}}, v_{t^*}), (-t^*\sqrt{v_{t^*}}, v_{t^*})\}$ according as $F(t^*) < F(-t^*)$, or $F(-t^*) < F(t^*)$, or $F(t^*) = F(-t^*)$. Hence application of Theorem 18.2 yields a large deviation approximation to $P\{T_c \le ac\}$ even when the underlying distribution to which the repeated t-test is applied does not have finite pth absolute moment for any $p > 0$, which is similar to the large deviation theory in Chap. 3 for fixed sample size n.

Remark 18.6 (Asymptotic efficiencies of fixed sample size tests). Chernoff (1952), Bahadur (1960, 1967, 1971) and Hoeffding (1965) have used large deviation approximations for Type I and Type II error probabilities to evaluate asymptotic efficiencies of fixed sample size tests. Theorem 18.3 provides a much more precise approximation for these error probabilities. In the case of linear hypotheses about a multivariate normal mean, such refined large deviation approximations in the literature have been derived from well-developed exact distribution theory in the normal case; see Groeneboom (1980, pp. 71–90). Chernoff and Hoeffding consider the Type I error probability α_n, and the Type II error probability β_n at a fixed alternative θ_1, of a typical test of $H_0 : \theta = \theta_0$, as the sample size $n \to \infty$ so that both α_n and β_n approach 0 exponentially fast. By introducing the *Chernoff index*

$$\lambda = \lim_{n \to \infty} n^{-1} \log \max(\alpha_n, \beta_n), \tag{18.37}$$

Chernoff (1952) defines the asymptotic efficiency of a test δ_1 relative to a test δ_2 by

$$e(\delta_1, \delta_2) = \lambda_1 / \lambda_2, \tag{18.38}$$

and gives examples of the index λ for standard normal, chi-square and binomial tests. Hoeffding (1965) considers tests of multinomial probabilities and shows that GLR tests have the minimal index and are therefore asymptotically efficient.

Bahadur (1960) makes use of the attained significance levels (or p-values) for stochastic comparison of test statistics. Let T_n be a test statistic based on i.i.d. observations $X_1, \ldots, X_n$ such that large values of T_n show significant departures from the null hypothesis $H_0 : \theta \in \Theta_0$. Letting $G_n(t) = \sup_{\theta \in \Theta_0} P_\theta(T_n > t)$, the attained significance level of T_n is $\pi_n := G_n(T_n)$. In typical cases, π_n converges weakly as $n \to \infty$ under P_θ for $\theta \in \Theta_0$, and there exists $c(\theta) > 0$ such that

$$-2n^{-1} \log \pi_n \longrightarrow c(\theta) \quad a.s.[P_\theta] \tag{18.39}$$

at $\theta \notin \Theta_0$. If (18.39) holds, $c(\theta)$ is called the *Bahadur slope* of T_n at θ. If the a.s. convergence in (18.39) is replaced by $\xrightarrow{P_\theta}$ (i.e., convergence in probability), then $c(\theta)$ is called the "weak Bahadur slope" at θ. The larger the value of $c(\theta)$, the faster T_n tends to reject H_0. For two sequences of test statistics $T_n^{(1)}$ and $T_n^{(2)}$, the *Bahadur efficiency* of $T_n^{(1)}$ relative to $T_n^{(2)}$ at alternative θ is given by the ratio $c_1(\theta)/c_2(\theta)$, where $c_i(\theta)$ is the Bahadur slope of $T_n^{(i)}$ at θ. See Bahadur (1960, 1967, 1971), Akritas and Kourouklis (1988) and He and Shao (1996) for results on Bahadur slopes and efficiencies and their derivations from large and moderate deviation theories.

Theorem 18.4 can be used to evaluate the Type II error probability of the sequential test that rejects H_0 if $kg(S_k/k) > c$ for some $k \le ac$. Suppose $g(\mu_0) > a^{-1}$.

Then the Type II error probability of the test at the alternative with $EX_1 = \mu_0$ can be expressed in the following form to which Theorem 18.4 is applicable:

$$P_{\mu_0}\left\{\max_{k\le n} kg(S_k/k) \le c\right\} = P_{\mu_0}\left\{\min_{k\le n}\left[ng(\mu_0) - kg(S_k/k)\right] \ge ng(\mu_0) - c\right\}$$
$$= P_{\mu_0}\left\{\min_{k\le n}\left[(n-k)\beta + k\tilde{g}(S_k/k)\right] \ge bn\right\},$$

where $\beta = g(\mu_0)$, $\tilde{g} = g(\mu_0) - g$ and $b = g(\mu_0) - a^{-1}$.

Example 18.7. With the same notation and assumptions as Example 18.5, consider the Type II error probability of the repeated t-test at the alternative where $E(Y_i) = \gamma \ne 0$ and $\mathrm{Var}(Y_i) = 1$. Thus $E(Y_i^2) = 1 + \gamma^2$. Suppose $\gamma > 0$ and $g(\gamma, 1+\gamma^2) > a^{-1}$. Let $b = g(\gamma, 1+\gamma^2) - a^{-1}$,

$$\tilde{g}(y,v) = g(\gamma, 1+\gamma^2) - g(y,v) = \left\{\log(1+\gamma^2) - \log\left(v/(v-y^2)\right)\right\}/2.$$

Since the logarithm of the underlying density function is $-(y-\gamma)^2/2 - \log(\sqrt{2\pi})$, the rate function now takes the form

$$\phi(y,v) = \left[v - 1 - \log(v - y^2) - 2\gamma y + \gamma^2\right]/2.$$

Since ϕ is strictly convex with its global minimum at $(\gamma, 1+\gamma^2)$ and since $g(\gamma, 1+\gamma^2) > a^{-1}$, the minimum of ϕ over the region $\{(y,v) : g(y,v) \le a^{-1}\}$ occurs at $v = \alpha y^2$ with α satisfying $g(1,\alpha) = a^{-1}$, or equivalently, $\alpha/(\alpha-1) = e^{2/a}$. Since $\phi(y, \alpha y^2) = \{\alpha y^2 - 1 - \log(\alpha - 1) - \log y^2 - 2\gamma y + \gamma^2\}/2$ is minimized at $y_a := (\gamma + \sqrt{\gamma^2 + 4\alpha})/2\alpha$, (B1) holds with $\tilde{g}$ in place of g and $b/r = \phi(\mu_a)$ and (B3) holds with M consisting of the single point $\mu_a := (y_a, \alpha y_a^2)$. Moreover, (B2), (B4) and (B5) also hold (with $\tilde{g}$ in place of g). Hence Theorem 18.4 can be applied to give the Type II error probability of the repeated t-test: As $c \to \infty$,

$$P_\gamma\left\{\max_{2\le k\le ac} kg(S_k/k) \le c\right\} \sim (ac/2\pi)^{1/2}\xi(\mu_a)w(\mu_a)\left|\sum(\mu_a)\right|^{-1/2} e^{-ac\phi(\mu_a)}.$$

18.3 Efficient Monte Carlo Evaluation of Boundary Crossing Probabilities

The likelihood ratio identity and the method of mixtures in Sect. 18.2.1 can be used to compute boundary crossing probabilities directly, by Monte Carlo simulations using importance sampling, instead of relying on asymptotic approximations (which may be inaccurate for the given sample size and may involve difficult numerical integration) developed from the method. When an event A occurs with a small probability (e.g., 10^{-4}), generating 100 events would require a very large number of simulations (e.g., 1 million) for direct Monte Carlo computation of $P(A)$. To circumvent this difficulty, one can use importance sampling instead of direct Monte

Carlo, changing the measure P to Q under which A is no longer a rare event and evaluating $P(A) = E_Q LI(A)$ by $m^{-1}\sum_{i=1}^m L_i I(A_i)$, where $(L_1, I(A_1)), \ldots, (L_m, I(A_m))$ are m independent samples drawn from the distribution Q, with L_i being a realization of the likelihood ratio statistic $L := dP/dQ$, which is the importance weight. We next discuss how Q should be chosen to produce an efficient Monte Carlo estimate.

Let p_n denote the probability in Theorem 18.3, i.e., $p_n = P\{g(S_n/n) > b\}$ with $b > g(\mu_0)$. Glasserman and Wang (1997) have pointed out that importance sampling which uses the same change of measures as that used in deriving large deviations approximations may perform much worse than direct Monte Carlo for nonlinear functions g. They consider the case $d = 1$ and $g(x) = x^2$, for which

$$p_n = P\{|S_n|/n > \sqrt{b}\} = P\{|S_n| > an\}, \tag{18.40}$$

where $a = \sqrt{b} > |\mu_0|$ and $a \in \Lambda^o$. Suppose $\phi(a) < \phi(-a)$. Then $p_n \sim P\{S_n > an\}$ and

$$n^{-1}\log L_n \xrightarrow{P} -\phi(a) = \lim_{n\to\infty} n^{-1}\log P\{|S_n| > an\}, \tag{18.41}$$

where $L_n = dP_n/dP_{a,n}$ and $P_{\mu,n}$ denotes the probability measure under which $X_1, \ldots,$ X_n are i.i.d. from the exponential family (18.33) with natural parameter θ_μ. Since $n^{-1}\log P\{S_n > an\} \to -\phi(a)$ and $n^{-1}\log P\{S_n < -an\} \to -\phi(-a)$,

$$\mathrm{Var}_P I(|S_n| > an) \sim P\{|S_n| > an\} = e^{-\{\phi(a)+o(1)\}n}. \tag{18.42}$$

Consider importance sampling of $\{|S_n| > an\}$ by using $Q_n = P_{a,n}$, a choice that is "consistent with large deviations" in the terminology of Glasserman and Wang (1997, p. 734), who have also shown that for the case $\theta_a + \theta_{-a} > 0$,

$$\lim_{n\to\infty} E_{Q_n} L_n^2 I(|S_n| > an) = \infty. \tag{18.43}$$

Comparison of (18.43) with (18.42) shows that Monte Carlo computation of p_n by using importance sampling from $P_{a,n}$ in this case is much worse than direct Monte Carlo.

To simulate the tail probability $p_n = P\{g(S_n/n) > b\}$ under (B1)–(B5), Chan and Lai (2007) propose to use importance densities of the form

$$\tilde{w}_n(\mu) = \tilde{\beta}_n e^{-n\phi(\mu)} I(g(\mu) > b), \qquad \mu \in \Lambda, \tag{18.44}$$

where $\tilde{\beta}_n$ is a normalizing constant such that $\int_\Lambda \tilde{w}_n(\mu)d\mu = 1$. Specifically, they propose to generate i.i.d. $(X_1^{(i)}, \ldots, X_n^{(i)})$, $i = 1, \ldots, m$, from the importance sampling measure

$$Q_n^* = \int_\Lambda P_{\mu,n}\tilde{w}_n(\mu)\,d\mu \tag{18.45}$$

and estimate p_n by

$$\hat{p}_n = m^{-1}\sum_{i=1}^m L_n^{(i)} I\left(g(S_n^{(i)}/n) > b\right), \quad \text{where } S_n^{(i)} = X_1^{(i)} + \cdots + X_n^{(i)}, \tag{18.46}$$

and $L_n^{(i)} = dP_n/dQ_n^* = (\int_\Lambda e^{\theta_\mu' S_n^{(i)} - n\psi(\theta_\mu)} \tilde{w}_n(\mu)d\mu)^{-1}$. Note that $\hat{p}_n$ is unbiased for p_n. Let G be a distribution function on $\mathbb{R}^d$ satisfying $F(A) > 0 \Rightarrow G(A) > 0$ for any Borel set $A \subset \mathbb{R}^d$. Assume that $\lambda(\theta) := \log[\int e^{\theta' x} G(dx)] < \infty$ for all $\|\theta\| \le \theta_1$ and let $\Gamma = \{\theta : \lambda(\theta) < \infty\}$. For $\theta \in \Gamma$, define a probability distribution G_θ on $\mathbb{R}^d$ by

$$dG_\theta(x) = \exp\{\theta' x - \lambda(\theta)\}\, dG(x), \tag{18.47}$$

and let $Q_{\mu,n}$ denote the measure under which $X_1, \ldots, X_n$ are i.i.d. with distribution function $G_{(\partial\lambda)^{-1}(\mu)}$. Let W_n be a probability measure and define the mixture

$$Q_n = \int_{\nabla\lambda(\Gamma)} Q_{\mu,n}\, dW_n(\mu). \tag{18.48}$$

Chan and Lai (2007) have proved the following result on the asymptotic optimality of the importance sampling measure (18.45).

Theorem 18.8. *Assume that g satisfies (B1)–(B5). Then for any distribution function G on $\mathbb{R}^d$ such that $\int e^{\theta' x} dG(x) < \infty$ for θ in some neighborhood of the origin,*

$$\liminf_{n\to\infty} E_{Q_n}\left[\left(\frac{dP_n}{dQ_n}\right)^2 I(g(S_n/n) > b)\right] \Big/ (\sqrt{n} p_n^2) > 0,$$

where Q_n is defined from G via (18.47) *and* (18.48)*. Moreover, defining Q_n^* by* (18.44) *and* (18.45)*, we have*

$$E_{Q_n^*}\left[\left(\frac{dP_n}{dQ_n^*}\right)^2 I(g(S_n/n) > b)\right] = O(\sqrt{n} p_n^2).$$

Hence Q_n^* is asymptotically efficient.

Chan and Lai (2007) have modified (18.47) and (18.48) to give a similar importance sampling measure that is asymptotically efficient for Monte Carlo computation of the probability $p_c = P\{T_c \le ac\}$ in Theorem 18.2. They have also described how these importance sampling methods can be implemented in practice and have provided numerical results on their performance. Moreover, extensions to Markov-dependent X_i are also given in Chan and Lai (2007).

18.4 Supplementary Results and Problems

1. (a) Let $X_1, X_2, \ldots$ be i.i.d. random variables and let T be a stopping time such that $ET < \infty$. Show that if $EX_1 = EX_1^+ - EX_1^-$ is well-defined (i.e., EX_1^+ and EX_1^- are not both infinite), then $E(\sum_{i=1}^T X_i) = (ET)(EX_1)$. This is often called *Wald's equation*. Moreover, by using a truncation argument, show that if $EX_1 = 0$, then $E\{(\sum_{i=1}^T X_i)^2\} = (ET)EX_1^2$, which is often called "Wald's equation for the second moment."

(b) Using the notation of Sect. 18.1.1, show that if $P_i\{g_1(X_1) \neq g_0(X_1)\} > 0$ for $i = 0, 1$, then $E_i N < \infty$ and that ignoring overshoots, the stopping time N of the SPRT with Type I and Type II error probabilities α and β attain the lower bounds in (18.4).

(c) Suppose that under the true probability measure P, the X_i are i.i.d. such that $E\{\log(g_1(X_1)/g_0(X_1))\} = 0$ while $P\{g_1(X_1) \neq g_0(X_1)\} > 0$. Show that $EN < \infty$ and use Wald's equations in (a) to derive approximations (ignoring overshoots) for $P\{L_N \geq B\}$ and EN.

2. (a) Show that the repeated t-test with stopping rule (15.15) is a repeated GLR test for testing the null hypothesis that the mean of a normal distribution is 0 when its variance is unknown.

(b) Generalize (a) to the case of a multivariate normal distribution and thereby derive the repeated T^2-test.

3. *Renewal theory, nonlinear extension and applications.* Let $X, X_1, X_2, \ldots$ be i.i.d. random variables with $EX_1 = \mu > 0$, and let $S_n = X_1 + \cdots + X_n$, $S_0 = 0$. Define

$$\tau(b) = \inf\{n \geq 1 : S_n \geq b\}, \quad \tau_+ = \inf\{n \geq 1 : S_n > 0\}, \quad U(x) = \sum_{n=0}^{\infty} P\{S_n \leq x\}, \tag{18.49}$$

and call X *arithmetic* if its support is of the form $\{0, \pm d, \pm 2d, \ldots\}$, where the largest such d is called its *span*. The function U is called the *renewal function*, and *Blackwell's renewal theorem* says that if X is a.s. positive, then

$$U(x+h) - U(x) \to h/\mu \qquad \text{as } x \to \infty, \tag{18.50}$$

for any $h > 0$ in the case of non-arithmetic X, and for $h = d$ and x being an integral multiple of d when X is arithmetic with span d. The renewal theorem provides a key tool to prove the following results on $(\tau(b), S_{\tau(b)} - b)$:

- *As $b \to \infty$ (through multiples of d in the lattice case), $S_{\tau(b)} - b$ converges in distribution; in fact*

$$P\{S_{\tau(b)} - b > y\} \to 1 - H(y) := (ES_{\tau_+})^{-1} \int_y^{\infty} P(S_{\tau_+} > x)dx. \tag{18.51}$$

Moreover, if $\operatorname{Var} X = \sigma^2 < \infty$, *then*

$$\lim_{b \to \infty} E(S_{\tau(b)} - b) = \begin{cases} ES_{\tau_+}^2/(2ES_{\tau_+}) & \text{if } X \text{ is non-arithmetic} \\ ES_{\tau_+}^2/(2ES_{\tau_+}) + d/2 & \text{if } X \text{ is arithmetic with span } d. \end{cases} \tag{18.52}$$

- *As $b \to \infty$ (through multiples of d in the lattice case),*

$$P\left\{\tau(b) \leq b/\mu + x(b\sigma^2/\mu^3)^{1/2}, S_{\tau(b)} - b \leq y\right\} \to \Phi(x)H(y) \tag{18.53}$$

for all $x \in \mathbb{R}$ and $y > 0$; i.e., $(\tau(b) - b/\mu)/(b\sigma^2/\mu^3)^{1/2}$ is asymptotically standard normal and asymptotically independent of the overshoot $S_{\tau(b)} - b$.

Lai and Siegmund (1977) have extended (18.51) and (18.53) to the case where the random walk S_n above is replaced by $Z_n = \tilde{S}_n + \zeta_n$, where $\tilde{S}_n$ is a random walk whose increments are i.i.d. with a positive mean and ζ_n is *slowly changing* in the sense that $\max_{1 \le i \le n} |\zeta_i|/\sqrt{n} \xrightarrow{P} 0$ and for every $\varepsilon > 0$ there exist n^* and $\delta > 0$ such that

$$P\left\{ \max_{1 \le k \le n\delta} |\zeta_{n+k} - \zeta_n| > \varepsilon \right\} < \varepsilon \qquad \text{for all } n \le n^*. \tag{18.54}$$

They have shown that (18.51) and (18.53) still hold with $\tau(b)$ replaced by $T(b) = \inf\{n \ge 1 : Z_n \ge b\}$, $S_{\tau(b)}$ replaced by $Z_{T(b)}$ and S_{τ_+} replaced by $\tilde{S}_{\tau_+}$. This extension of (18.51) and (18.53) covers a wide variety of applications in which the statistics are nonlinear functions of sample mean vectors and can be represented as $\tilde{S}_n + \zeta_n$, where ζ_n is the remainder in a Taylor series expansion:

(a) Let $Y, Y_1, Y_2, \dots$ be i.i.d. random variables with $EY^2 < \infty$ and let $Z_n = ng(\sum_{i=1}^n Y_i/n)$, where g is positive and twice continuously differentiable in a neighborhood of EY. By making use of the strong law of large numbers and Taylor's expansion around EY, show that Z_n can be expressed in the form $\tilde{S}_n + \zeta_n$, where

$$\tilde{S}_n = ng(EY) + g'(EY) \sum_{i=1}^n (Y_i - EY)$$

and ζ_n is slowly changing.

(b) Let $0 \le \gamma < 1$ and assume that $EY > 0$. Show that $T_c := \inf\{n \ge 1 : \sum_{i=1}^n Y_i \ge cn^\gamma\}$ can be re-expressed as $T(b) = \inf\{n \ge 1 : Z_n \ge b\}$, with $b = c^{1/(1-\gamma)}$ and Z_n of the form in (a).

(c) Show that ℓ_n in Sect. 18.2.1 can be written in the form $\tilde{S}_n + \zeta_n$ and hence prove (18.22).

4. Let $X_1, X_2, \dots$ be i.i.d. with $EX_1 = \mu > 0$, and define $\tau(b)$ by (18.49). Show that $E\tau(b) < \infty$ and $\lim_{b\to\infty} E\tau(b)/b = 1/\mu$, first in the case $P\{X_1 \le c\} = 1$ for some $c > 0$, and then in general by a truncation argument. Assuming furthermore that $\mathrm{Var} X_1 = \sigma^2 < \infty$ and that X_1 is non-arithmetic, make use of Wald's equation and (18.52) to show that

$$E\tau(b) = \frac{1}{\mu}\left\{ b + \frac{ES_{\tau_+}^2}{2ES_{\tau_+}} + o(1) \right\} \qquad \text{as } b \to \infty. \tag{18.55}$$

Lai and Siegmund (1979) have extended Blackwell's renewal theorem (18.50) to $U(x)$ in which S_n is replaced by $Z_n = \tilde{S}_n + \zeta_n$ in (18.49), where ζ_n is slowly changing and satisfies some additional assumptions, including that ζ_n converges in distribution to ζ. Letting $\tilde{\mu} = E\tilde{S}_1$, they have used this result to show that (18.55) can be extended to

$$ET(b) = \frac{1}{\tilde{\mu}}\left\{b - E\zeta + \frac{E\tilde{S}_{\tau_+}^2}{2E\tilde{S}_{\tau_+}} + o(1)\right\} \qquad \text{as } b \to \infty, \tag{18.56}$$

where $T(b) = \inf\{n \geq 1 : Z_n \geq b\}$.

References

Abramovitch, L. and Singh, K. (1985). Edgeworth corrected pivotal statistics and the bootstrap. *Ann. Statist.* **13**, 116–132.

Akritas, M. G. and Kourouklis, S. (1988). Local Bahadur efficiency of score tests. *J. Statist. Plan. Infer.* **19**, 187–199.

Alberink, I. B. and Bentkus, V. (2001). Berry–Esseen bounds for von-Mises and U-statistics. *Lith. Math. J.* **41**, 1–16.

Alberink, I. B. and Bentkus, V. (2002). Lyapounov type bounds for U-statistics. *Theory Probab. Appl.* **46**, 571–588.

Azuma, K. (1967). Weighted sums of certain dependent random variables. *Tôhoku Math. J. (2)* **19**, 357–367.

Bahadur, R. R. (1960). Stochastic comparison of tests. *Ann. Math. Statist.* **31**, 276–295.

Bahadur, R. R. (1967). Rates of convergence of estimates and test statistics. *Ann. Math. Statist.* **38**, 303–324.

Bahadur, R. R. (1971). Some limit theorems in statistics. In *Regional Conference Series in Appl. Math., vol. 4.* SIAM, Philadelphia.

Bahadur, R. R. and Ranga Rao, R. (1960). On deviations of the sample mean. *Ann. Math. Statist.* **31**, 1015–1027.

Barlow, M. T., Jacka, S. D. and Yor, M. (1986). Inequalities for a pair of processes stopped at a random time. *Proc. London Math. Soc. Ser.* III **53**, 152–172.

Barndorff-Nielsen, O. and Cox, D. R. (1979). Edgeworth and saddlepoint approximations with statistical applications (with discussion). *J. R. Statist. Soc. Ser. B* **41**, 279–312.

Bentkus, V., Bloznelis, M. and Götze, F. (1996). A Berry–Esseen bound for Student's statistic in the non-i.i.d. case. *J. Theoret. Probab.* **9**, 765–796.

Bentkus, V. and Götze, F. (1996). The Berry–Esseen bound for Student's statistic. *Ann. Probab.* **24**, 491–503.

Bercu, B., Gassiat, E. and Rio, E. (2002). Concentration inequalities, large and moderate deviations for self-normalized empirical processes. *Ann. Probab.* **30**, 1576–1604.

Bercu, B. and Touati (2008). Exponential inequalities for self-normalized martingales with applications. *Ann. Appl. Probab.* **18**, 1848–1869.

Bertoin, J. (1998). Darling–Erdős theorems for normalized sums of i.i.d. variables close to a stable law. *Ann. Probab.* **26**, 832–852.

Bhattacharya, R. N. and Ghosh, J. K. (1978). On the validity of the formal Edgeworth expansion. *Ann. Statist* **6**, 494–451.

Bhattacharya, R. N. and Ranga Rao, R. (1976). *Normal Approximation and Asymptotic Expansion.* Wiley, New York.

Bickel, P. J., Götze, F. and van Zwet, W. R. (1986). The Edgeworth expansion for U-statistics of degree 2. *Ann. Statist.* **14**, 1463–1484.

Bingham, N. H., Goldie, C. M. and Teugels, J. L. (1987). *Regular Variation*. Cambridge University Press, Cambridge.

Borovkov, A. A. and Rogozin, B. A. (1965). On the multidimensional central limit theorem. *Theory Probab. Appl.* **10**, 55–62.

Borovskikh, Y. V. and Weber, N. C. (2003a). Large deviations of U-statistics I. *Lietuvos Matematikos Rinkinys* **43**, 13–37.

Borovskikh, Y. V. and Weber, N. C. (2003b). Large deviations of U-statistics II. *Lietuvos Matematikos Rinkinys* **43**, 294–316.

Breiman, L. (1968). *Probability*. Addison-Wesley, Reading, MA.

Caballero, M. E., Fernández, B. and Nualart, D. (1998). Estimation of densities and applications. *Theoret. Probab.* **11**, 831–851.

Cao, H. Y. (2007). Moderate deviations for two sample t-statistics. *ESAIM P&S* **11**, 264–271.

Chan, H. P. and Lai, T. L. (2000). Asymptotic approximations for error probabilities of sequential or fixed sample size tests in exponential families. *Ann. Statist.* **28**, 1638–1669.

Chan, H. P. and Lai, T. L. (2002). Boundary crossing probabilities for scan statistics and their applications to change-point detection. *Method. Comput. Appl. Probab.* **4**, 317–336.

Chan, H. P. and Lai, T. L. (2003). Saddlepoint approximations and nonlinear boundary crossing probabilities of Markov random walks. *Ann. Appl. Probab.* **13**, 395–429.

Chan, H. P. and Lai, T. L. (2007). Efficient importance sampling for Monte Carlo evaluation of exceedance probabilities. *Ann. Appl. Probab.* **17**, 440–473.

Chen, L. H. Y. and Shao, Q. M. (2001). A non-uniform Berry–Esseen bound via Stein's method. *Probab. Th. Related Fields* **120**, 236–254.

Chen, L. H. Y. and Shao, Q. M. (2007). Normal approximation for nonlinear statistics using a concentration inequality approach. *Bernoulli* **13**, 581–599. doi:10.3150/07-BEJ5164

Chernoff, H. (1952). A measure of asymptotic efficiency for tests of a hypothesis based on the sum of observations. *Ann. Math. Statist.* **23**, 493–507.

Chernoff, H. (1961). Sequential tests for the mean of a normal distribution. In *Proc. 4th Berkeley Sympos. Math. Statist. and Prob. I*, 79–91. University of California Press, Berkeley.

Chernoff, H. (1965). Sequential tests for the mean of a normal distribution. III. (Small t). *Ann. Math. Statist.* **36**, 28–54.

Chistyakov, G. P. and Götze, F. (2004a). Limit distributions of Studentized means. *Ann. Probab.* **32**, 28–77.

Chistyakov, G. P. and Götze, F. (2004b). On bounds for moderate deviations for Student's statistic. *Theory Probab. Appl.* **48**, 528–535.

Chow, Y. S. and Teicher, H. (1988). *Probability Theory*, 2nd edition. Springer, New York.

Csörgő, M. and Révész, P. (1981). *Strong Approximations in Probability and Statistics*. Academic, New York.

Csörgő, M., Szyszkowicz, B. and Wang, Q. (2003a). Donsker's theorem for self-normalized partial sums processes. *Ann. Probab.* **31**, 1228–1240.

Csörgő, M., Szyszkowicz, B. and Wang, Q. (2003b). Darling-Erdős theorems for self-normalized sums. *Ann. Probab.* **31**, 676–692.

Csörgő, M., Szyszkowicz, B. and Wang, Q. (2004). On weighted approximations and strong limit theorems for self-normalized partial sums processes. In *Asymptotic Methods in Stochastics*, Fields Inst. Commun., **44**, 489–521. AMS, Providence, RI.

Csörgő, M., Szyszkowicz, B. and Wang, Q. (2008). On weighted approximations in $D[o, 1]$ with applications to self-normalized partial sum process. *Acta. Math. Hungarica* **121**, 307–332.

Daniels, H. E. and Young, G. A. (1991). Saddlepoint approximation for the Studentized mean, with an application to the bootstrap. *Biometrika* **78**, 169–179.

Darling, D. A. and Erdős, P. (1956). A limit theorem for the maximum of normalized sums of independent random variables. *Duke Math. J.* **23**, 143–155.

David and Kruskal (1956). The WAGR sequential t-test reaches a decision with probability one. *Ann. Math. Statist.* **27**, 797–805.

de la Peña, V. H. (1999). A general class of exponential inequalities for martingales and ratios. *Ann. Probab.* **27**, 537–564.

de la Peña, V. H. and Giné, E. (1999). *Decoupling: From Dependence to Independence*. Springer, New York.

de la Peña, V. H., Klass, M. J. and Lai, T. L. (2000). Moment bounds for self-normalized martingales. In *High Dimensional Probability II* (E. Giné, D. M. Mason and J. A. Wellner, eds.) 1–11. Birkhauser, Boston.

de la Peña, V. H., Klass, M. J. and Lai, T. L. (2004). Self-normalized processes: Exponential inequalities, moment bounds and iterated logarithm laws. *Ann. Probab.* **32**, 1902–1933.

de la Peña, V. H., Klass, M. J. and Lai, T. L. (2007). Pseudo-maximization and self-normalized processes. *Probability Surveys* **4**, 172–192.

de la Peña, V. H., Klass, M. J. and Lai, T. L. (2008). Theory and applications of multivariate self-normalized processes. Department of Statistics Technical Report, Stanford University.

Dembo, A. and Shao, Q. M. (2006). Large and moderate deviations for Hotelling's T^2-statistic. *Elect. Comm. Probab.* **11**, 149–159.

Dembo, A. and Zeitouni, O. (1998). *Large Deviations Techniques and Applications*, 2nd edition. Springer, New York.

Durrett, R. (2005). *Probability: Theory and Examples*, 3rd edition. Thomson-Brooks/Cole, Belmont, CA.

Efron, B. (1969). Student's t-test under symmetry conditions. *J. Am. Statist. Assoc.* **64**, 1278–1302.

Efron, B. (1981). Censored data and the bootstrap. *J. Am. Statist. Assoc.* **89**, 452–462.

Efron, B. and Stein, C. (1981). The jackknife estimate of variance. *Ann. Statist.* **9**, 586–596.

Efron, B. and Tibshirani, R. J. (1993). *An Introduction to the Bootstrap (Monographs on Statistics and Applied Probability)*. Chapman and Hall, New York.

Einmahl, U. (1989). The Darling–Erdős theorem for sums of i.i.d. random variables. *Probab. Th. Rel. Fields* **82**, 241–257.

Elliott, R. J. (1982). *Stochastic Calculus and Applications*, Springer, New York.

Erdös, P. (1942). On the law of the iterated logarithm. *Ann. Math.* **43**, 419–436.

Feller, W. (1946). A Limit theorem for random variables with infinite moments. *Am. J. Math.* **68**, 257–262.

Freedman, D. (1973). Another note on the Borel–Cantelli lemma and the strong law, with the Poisson approximation as a by-product. *Ann. Probab.* **6**, 910–925.

Ghosh, B. K. (1970). *Sequential Tests of Statistical Hypotheses*. Addison-Wesley, Reading, MA.

Giné, E. and Götze, F. (2004). On standard normal convergence of the multivariate Student t-statistic for symmetric random vectors. *Electron. Comm. Probab.* **9**, 162–171.

Giné, E., Götze, F. and Mason, D. M. (1997). When is the Student t-statistic asymptotically standard normal? *Ann. Probab.* **25**, 1514–1531.

Glasserman, P. and Wang, Y. (1997). Counterexamples in importance sampling for large deviation probabilities. *Ann. Appl. Probab.* **7**, 731–746.

Gosset (Student), W. S. (1908). On the probable error of a mean. *Biometrika* **6**, 1–25.

Graversen, S. E. and Peskir, G. (2000). Maximal inequalities for the Ornstein–Uhlenbeck process. *Proc. Am. Math. Soc.* **128**, 3035–3041.

Gray, A. (1990). *Tubes*. Addison Wesley, New York.

Griffin, P. S. (2002). Tightness of the Student t-statistic. *Elect. Comm. Probab.* **7**, 181–190.

Griffin, P. S. and Kuelbs, J. D. (1989). Self-normalized laws of the iterated logarithm. *Ann. Probab.* **17**, 1571–1601.

Griffin, P. S. and Kuelbs, J. D. (1991). Some extensions of the laws of the iterated logarithm via self-normalized. *Ann. Probab.* **19**, 380–395.

Groeneboom, P. (1980). *Large Deviations and Asymptotic Efficiencies*. Mathematical Centre Tracts, **118**. Mathematisch Centrum, Amsterdam.

Gross, S. T. and Lai, T. L. (1996). Bootstrap methods for truncated and censored data. *Stat. Sinica* **6**, 509–530.

Hahn, M. G. and Klass, M. J. (1980). Matrix normalization of sums of random vectors in the domain of attraction of the multivariate normal. *Ann. Probab.* **8**, 262–280.

Hall, P. (1988). Theoretical comparison of bootstrap confidence intervals. *Ann. Statist.* **16**, 927–953.

Hall, P. (1992). *The Bootstrap and Edgeworth Expansion*. Springer, New York.

Hall, P. and Wang, Q. (2004). Exact convergence rate and leading term in central limit theorem for student's t statistic. *Ann. Probab.* **32**, 1419–1437.

He, X. and Shao, Q. M. (1996). Bahadur efficiency and robustness of Studentized score tests. *Ann. Inst. Statist. Math.* **48**, 295–314.

Hoeffding, W. (1948). A class of statistics with asymptotically normal distribution. *Ann. Math. Statist.* **19**, 293–325.

Hoeffding, W. (1960). Lower bounds for the expected sample size and the average risk of a sequential procedure. *Ann. Math. Statist.* **31**, 352–368.

Hoeffding, W. (1965). Asymptotically optimal tests for multinomial distributions. *Ann. Math. Statist.* **36**, 369–405.

Horváth, L. and Shao, Q. M. (1996). Large deviations and law of the iterated logarithm for partial sums normalized by the largest absolute observation. *Ann. Probab.* **24**, 1368–1387.

Hu, Z., Shao, Q. M. and Wang, Q. (2008). Cramér type large deviations for the maximum of self-normalized sums. Preprint.

Ibragimove, I. A. and Linnik, Y. V. (1971). *Independent and Stationary Sequences of Random Variables.* Wolters-Noordhoff, Groningen.

Jensen, J. L. (1995). *Saddlepoint Approximations.* Oxford University Press, New York.

Jing, B. Y., Feuerverger, A. and Robinson, J. (1994). On the bootstrap saddlepoint approximations. *Biometrika* **81**, 211–215.

Jing, B. Y., Shao, Q. M. and Wang, Q. Y. (2003). Self-normalized Cramér type large deviations for independent random variables. *Ann. Probab.* **31**, 2167–2215.

Jing, B. Y., Shao, Q. M. and Zhou, W. (2004). Saddlepoint approximation for Student's t-statistic with no moment conditions. *Ann. Statist.* **32**, 2679–2711.

Jing, B. Y., Shao, Q. M. and Zhou, W. (2008). Towards a universal self-normalized moderate deviation. *Trans. Am. Math. Soc.* **360**, 4263–4285.

Johnson, R. A. (1970). Asymptotic expansions associated with posterior distributions. *Ann. Math. Statist.* **41**, 851–864.

Karatzas, I. and Shreve, S. E. (1991). *Brownian Motion and Stochastic Calculus*, 2nd edition. Springer, New York.

Kiefer, J. and Weiss, L. (1957). Some properties of generalized sequential probability ratio tests. *Ann. Math. Statist.* **28**, 57–74.

Kikuchi, M. (1991). Improved ratio inequalities for martingales. *Studia Math.* **99**, 109–113.

Koroljuk, V. S. and Borovskich, Y. V. (1994). *Theory of U-Statistics*. Kluwer, Dordrecht.

Lai, T. L. (1976). Boundary crossing probabilities for sample sums and confidence sequences. *Ann. Probab.* **4**, 299–312.

Lai, T. L. (1976a). Boundary crossing probabilities for sample sums and confidence sequences. *Ann. Probab.* **4**, 299–312.

Lai, T. L. (1976b). On r-quick convergence and a conjecture of Strassen. *Ann. Probab.* **4**, 612–627.

Lai, T. L. (1981). Asymptotic optimality of invariant sequential probability ratio tests. *Ann. Statist.* **9**, 318–333.

Lai, T. L. (1988). Nearly optimal sequential tests of composite hypotheses. *Ann. Statist.* **16**, 856–886.

Lai, T. L. (1997). On optimal stopping problems in sequential hypothesis testing. *Stat. Sinica* **7**, 33–51.

Lai, T. L. (2003). Stochastic approximation. *Ann. Statist.* **31**, 391–406.

Lai, T. L. and Robbins, H. (1981). Consistency and asymptotic efficiency of slope estimates in stochastic approximation schemes. *Z. Wahrsch. Verw. Gebiete* **56**, 329–360.

Lai, T. L., Robbins, H. and Wei, C. Z. (1978). Strong consistency of least squares estimates in multiple regression. *Proc. Natl. Acad. Sci. USA* **75**, 3034–3036.

Lai, T. L., Robbins, H. and Wei, C. Z. (1979). Strong consistency of least squares estimates in multiple regression II. *J. Multivar. Anal.* **9**, 343–361.

Lai, T. L., Shao, Q. M. and Wang, Q. Y. (2008). Cramér-type large deviations for Studentized U-statistics. Department of Statistics Technical Report, Stanford University.

Lai, T. L. and Siegmund, D. (1977). A non-linear renewal theory with applications to sequential analysis I. *Ann. Statist.* **5**, 946–954.
Lai, T. L. and Siegmund, D. (1979). A non-linear renewal theory with applications to sequential analysis II. *Ann. Statist.* **7**, 60–76.
Lai, T. L. and Wang, J. Q. (1993). Edgeworth expansions for symmetric statistics with applications to bootstrap methods. *Stat. Sinica* **3**, 517–542.
Lai, T. L. and Wei, C. Z. (1982). Least squares estimates in stochastic regression models with applications to identification and control of dynamic systems. *Ann. Statist.* **10**, 154–166.
Lai, T. L. and Wei, C. Z. (1986). Extended least squares and their applications to adaptive control and prediction in linear systems. *IEEE Trans. Automat. Control* **31**, 898–906.
Lai, T. L. and Xing, H. (2008). *Statistical Models and Methods for Financial Markets*. Springer, New York.
Lai, T. L. and Zhang, L. M. (1994). A modification of Schwarz's sequential likelihood ratio tests in multivariate sequential analysis. *Sequential Anal.* **13**, 79–96.
Le Cam, L. and Yang, G. (1990). *Asymptotics in Statistics. Some Basic Concepts*. Springer Series in Statistics. Springer, New York.
Lenglart, E. (1977). Relation de domination entre deux processus. *Ann. Inst. Henri Poincaré* **13**, 171–179.
LePage, R., Woodroofe, M. and Zinn, J. (1981). Convergence to a stable distribution via order statistics. *Ann. Probab.* **9**, 624–632.
Logan, B. F., Mallows, C. L., Rice, S. O. and Shepp, L. A. (1973). Limit distributions of self-normalized sums. *Ann. Probab.* **1**, 788–809.
Logan, B. F. and Shepp, L. A. (1968). Real zeros of random polynomials II. *Proc. London Math. Soc.* **18**, 308–314.
Lorden, G. (1976). 2-SPRT's and the modified Kiefer–Weiss problem of minimizing an expected sample size. *Ann. Statist.* **4**, 281–291.
Maller, R. (1981). On the law of the iterated logarithm in the infinite variance case. *J. Austral. Math. Soc. Ser. A* **30**, 5–14.
Martsynyuk, Y. V. (2007a). Central limit theorems in linear structural errors-in-variables models with explanatory variables in the domain of attraction of the normal law. *Electron. J. Statist.* **1**, 195–222.
Martsynyuk, Y. V. (2007b). New multivariate central limit theorems in linear structural and functional errors-in-variables models. *Electron. J. Statist.* **1**, 347–380.
Mason, D. M. (2005). The asymptotic distribution of self-normalized triangular arrays. *J. Theoret. Probab.* **18**, 853–870.
Mason, D. M. and Zinn, J. (2005). When does a randomly weighted self-normalized sum converge in distribution? *Electron. Comm. Probab.* **10**, 70–81.
Montgomery-Smith, S. J. (1993). Comparison of sums of independent identically distributed random vectors. *Probab. Math. Statist.* **14**, 281–285.
Petrov, V. V. (1965). On the probabilities of large deviations for sums of independent random variables. *Theory Probab. Appl.* **10,** 287–298.
Petrov, V. V. (1975). *Sums of Independent Random Variables.* Springer, New York.
Rebolledo, R. (1980). Central limit theorems for local martingales. *Z. Wahrsch. verw. Geb.* **51**, 269–286.
Revuz, D. and Yor, M. (1999). *Continuous Martingales and Brownian Motion*, 3rd edition. Springer, New York.
Robbins, H. (1970). Statistical methods related to the law of the iterated logarithm. *Ann. Math. Statist.* **41**, 1397–1409.
Robbins, H. and Siegmund, D. (1970). Boundary crossing probabilities for the Wiener process and sample sums. *Ann. Math. Statist.* **41**, 1410–1429.
Schwarz, G. (1962). Asymptotic shapes of Bayes sequential testing regions. *Ann. Math. Statist.* **33**, 224–236.
Schwarz, G. (1978). Estimating the dimension of a model. *Ann. Statist.* **6**, 461–464.

Sepanski, S. J. (1994). Probabilistic characterizations of the generalized domain of attraction of the multivariate normal law. *J. Theor. Probab.* **7**, 857–866.

Sepanski, S. J. (1996). Asymptotics for multivariate t-statistics for random vectors in the generalized domain of attraction of the multivariate normal law. *Statist. Probab. Lett.* **30**, 179–188.

Shao, Q. M. (1989). On a problem of Csörgő and Révész. *Ann. Probab.* **17**, 809–812.

Shao, Q. M. (1995). Strong approximation theorems for independent random variables and their applications. *J. Multivar. Anal.* **52**, 107–130.

Shao, Q. M. (1997). Self-normalized large deviations. *Ann. Probab.* **25**, 285–328.

Shao, Q. M. (1998). Recent developments in self-normalized limit theorems. In *Asymptotic Methods in Probability and Statistics* (B. Szyskowicz, ed.), 467–480. Elsevier, Amsterdam.

Shao, Q. M. (1999). Cramér type large deviation for Student's t-statistic. *J. Theoret. Probab.* **12**, 387–398.

Shao, Q. M. (2005). An explicit Berry–Esseen bound for Student's t-statistic via Stein's method. In *Stein's Method and Applications*, Lect. Notes Ser. Inst. Math. Sci. Natl. Univ. Singap., vol. 5, 143–155. Singapore University Press, Singapore.

Shorack, G. R. and Wellner, J. A. (1986). *Empirical Processes with Applications to Statistics*. Wiley, New York.

Siegmund, D. (1977). Repeated significance tests for a normal mean. *Biometrika* **64**, 177–189.

Siegmund, D. (1985). *Sequential Analysis*. Springer, New York.

Siegmund, D. (2004). Model selection in irregular problems: Applications to mapping quantitative trait loci. *Biometrika* **91**, 785–800.

Spivak, M. (1965). *Calculus on Manifolds*. W. A. Benjamin, New York.

Stein, C. (1972). A bound for the error in the normal approximation to the distribution of a sum of dependent random variables. In *Proc. Sixth Berkeley Symp. Math. Stat. Prob., 2*, 583–602. University of California Press, Berkeley.

Stone, C. (1965). A local limit theorem for nonlattice multi-dimensional distribution functions. *Ann. Math. Statist.* **36**, 546–551.

Stout, W. F. (1970). A martingale analogue of Kolmogorov's law of the iterated logarithm. *Z. Wahrsch. verw. Gebiete* **15**, 279–290.

Stout, W. F. (1973). Maximal inequalities and the law of the iterated logarithm. *Ann. Probab.* **1**, 322–328.

Strassen, V. (1966). A converse to the law of the iterated logarithm. *Z. Wahrsch. verw. Gebiete* **4**, 265–268.

van der Vaart, A. and Wellner, J. A. (1996). *Weak Convergence and Empirical Processes: With Applications to Statistics*. Springer, New York.

Vu, H. T. V., Maller, R. A. and Klass, M. J. (1996). On the Studentisation of random vectors. *J. Multivariate Anal.* **57**, 142–155.

Wald, A. (1945). Sequential tests of statistical hypotheses. *Ann. Math. Statist.* **16**, 117–186.

Wald, A. and Wolfowitz, J. (1948). Optimum character of the sequential probability ratio test. *Ann Math. Statist.* **19**, 326–339.

Wang, Q. (2005). Limit theorems for self-normalized large deviation. *Electron. J. Probab.* **10**, 1260–1285.

Wang, Q. and Jing, B. Y. (1999). An exponential non-uniform Berry–Esseen bound for self-normalized sums. *Ann. Probab.* **27**, 2068–2088.

Wang, Q., Jing, B.-Y. and Zhao, L. (2000). The Berry–Esséen bound for Studentized statistics. *Ann. Probab.* **28**, 511–535.

Wang, Q. and Weber, N. C. (2006). Exact convergence rate and leading term in the central limit theorem for U-statistics. *Stat. Sinica* **16**, 1409–1422.

Wilks, S. S. (1938). The large-sample distribution of the likelihood ratio for testing composite hypotheses. *Ann. Math. Statist.* **9**, 60–62.

Woodroofe, M. (1982). *Nonlinear Renewal Theory in Sequential Analysis*. CBMS-NSF Regional Conference Series in Applied Mathematics, vol. 39. Society for Industrial and Applied Mathematics (SIAM), Philadelphia.

Index

Made in the
USA
Middletown, DE